MW00513042

THE LIFE AND TIMES OF
ERNEST BEVIN

VOLUME TWO

BOOKS BY
ALAN BULLOCK

Hitler: A Study in Tyranny
The Liberal Tradition (with Maurice Shock)
The Life and Times of Ernest Bevin Vol. I
The Life and Times of Ernest Bevin Vol. II

Ernest Bevin in March 1942, aged 61.

ALAN BULLOCK

THE LIFE AND TIMES OF
ERNEST BEVIN

VOLUME TWO

Minister of Labour
1940 – 1945

HEINEMANN : LONDON

William Heinemann Ltd
LONDON MELBOURNE TORONTO
CAPE TOWN AUCKLAND

First published 1967

Printed and bound in Great Britain by
Bookprint Limited, London and Crawley

Contents

List of Illustrations

Acknowledgements

THE AUTHOR AND PUBLISHERS desire to thank the following for permission to quote copyright material: Lord Attlee; Lord Citrine; the late Lord Morrison; the late Lord Chuter-Ede; the Beaverbrook Foundation; Mr. Randolph Churchill; the Rt. Hon. Julian Amery, M.P.; Mrs. Frida Laski; the Passfield Trustees and the literary executors of the late R. H. Tawney.

The Controller of Her Majesty's Stationery Office (History of the Second World War: U.K. Civil Series); Constable & Co., Ltd. (*Winston Churchill, The Struggle for Survival* by Lord Moran); Victor Gollancz Ltd. (*Harold Laski* by Kingsley Martin); Hutchinson Ltd. (*Distinguished for Talent* by Woodrow Wyatt); Frederick Muller Ltd. (*The Fateful Years* by Lord Dalton); Cassell & Co. Ltd. (*The Second World War* by Sir Winston Churchill); William Collins Sons, Ltd. (*The Turn of the Tide* by Lord Alanbrooke); George Allen & Unwin Ltd. (*The Miners in Crisis and War* by R. Page Arnot, *Bevin* by Trevor Evans); Macmillan & Co. Ltd. (*King George VI* by Sir John Wheeler-Bennett, *The British General Election of* 1945 by R. B. McCallum and Alison Readman); MacGibbon & Kee (*Aneurin Bevan*, Vol. I by Michael Foot); Odhams Books Ltd. (*Conflict without Malice* by Emanuel Shinwell); Faber & Faber Ltd. (*Modern British Politics* by Samuel Beer); the Executors of the late John Winant and Hodder & Stoughton Ltd. (*A Letter from Grosvenor Square* by John Winant); the Executors of the late Lord Beveridge and Hodder & Stoughton Ltd. (*Power and Influence* by Lord Beveridge); Cassell & Co. Ltd. (*The War Speeches of Winston S. Churchill*, Vol. II, *The Reckoning* by Lord Avon); Lord Chandos and The Bodley Head Ltd. (*The Memoirs of Lord Chandos*); Eyre & Spottiswoode, the Beaverbrook Foundation & the Controller, H.M.S.O. (*Churchill and Beaverbrook* by Kenneth Young); William Heinemann Ltd. (*A Prime Minister Remembers* by Francis Williams).

Preface

Most books are best left to speak for themselves. I have only three points to make by way of preface.

It was my original intention to finish this work in two volumes. After I had worked on the materials, however, I came to the conclusion that it would be a mistake to compress the part Ernest Bevin played in the wartime Government into the opening chapters of a volume mainly concerned with foreign policy. The present volume therefore presents a self-contained study of Bevin's five years as Minister of Labour and a member of the War Cabinet. I now propose to complete the whole work with a third volume dealing with his years at the Foreign Office, 1945–1951.

I was fortunate to have made available to me by his executors the papers Mr. Bevin had collected from this period with a view to writing his own memoirs. I was equally fortunate in receiving much help from those who worked closely with him during the war. These two pieces of good fortune have made it possible to overcome at least some of the disadvantages of working within a period in which access to Cabinet records is not available.

Finally, I must remind the reader that this is a political biography. I hope I have never lost sight of the fact that Ernest Bevin was a man of formidable personality and I have tried to describe this and his relations with other people in Chapter 4. My purpose, however, has been to give an account of his public career, the task I was asked to undertake by his executor Arthur Deakin. To this it is worth adding that no man ever lived more fully in and for his job than Ernest Bevin —it is there that the real man is to be found—and that his private life in the last ten years of his career was not only uneventful but very much curtailed by the demands of office.

I have had help from many people who have spared time to talk about the events of these years. In view of the number involved, I hope they will not think me ungrateful if I express my thanks to them without setting out a list of their names. To this I must make two exceptions. I wish to thank Lord Attlee for talking to me several times about Mr. Bevin and for allowing me access to the papers he has deposited in the library of University College, Oxford. I also wish to thank Miss Saunders who has again been of great assistance in helping to collect much of the original material I have used.

In addition to inviting me to spend a month as their guest at the Villa Serbelloni, the Rockefeller Foundation generously provided me with a grant with which I was fortunate to secure the services of Mrs. Elizabeth Morgan as a research assistant. I have still tried to do as much of my own research as possible, but without Mrs. Morgan's help I could not have hoped to get through the mass of parliamentary and newspaper material which I have been able to use. I have also had much help from the staff of the Ministry of Labour and, like other historians before me, have found the volumes of the official History of the Second World War (Civil Series) of the greatest possible assistance.

I write an uncommonly illegible hand and there would never have been a book at all if it had not been for the skill of, first, Miss Buttar and then of my present secretary, Mrs. Janet Spincer, in discovering what it must have been that I meant to say. I am very grateful to both of them for their patience with my calligraphic shortcomings and to Mr. Arthur Turner for reading the proofs.

When the draft of the book was finished, Lord Normanbrook, Mrs. Margaret Gowing, Mrs. Morgan and Mr. D. J. Wenden read it for me. I should like to thank them most warmly for the trouble they took and for their comments which spurred me to make extensive revisions.

Other demands on my time and energies since I published the first volume had led me to fall badly behind with my writing and I should hardly have found the heart to take it up again if it had not been for the unfailing encouragement of my wife. She has borne more than her fair share of the troubles which fall upon the head of any author's wife; and she has throughout shown herself the most penetrating and understanding of critics It is with a profound sense of gratitude that I renew the dedication of this second volume to her.

ALAN BULLOCK

St. Catherine's College, Oxford

TO MY WIFE

CHAPTER I

The Coalition and the Crisis of 1940

I

IN MAY 1940 when he became Minister of Labour and National Service in the Churchill Government, Ernest Bevin had just passed the age of fifty-nine. Until his sixtieth year he had never held a ministerial post, never sat in Parliament nor even been a member of the National Executive of the Labour Party. From May 1940 until his death in April 1951 he was to remain in office, with only six weeks' interruption, throughout one of the most eventful decades in British history, and to play a part in these events second only to that of the two Prime Ministers, Churchill and Attlee. It was as unexpected a climax as anyone could have devised to the career of a man who a short time before the war had been talking of retirement.

Bevin, however, had little idea of beginning a political career when he accepted office. The situation which faced the members of the new Government left them no time to think about the future: they needed all their resolution to believe there was going to be a future at all.

Few wars have seen as sudden or complete a reversal of fortune as that which took place between the beginning of April and the end of June 1940. After the conquest of Poland in the previous autumn Hitler made no overt move for six months. This was the phoney war in which the French communiqué reported, day after day, "*Rien à signaler*". The Belgians and Dutch were still insisting on their neutrality and neither side in the West had invaded the other's territory. Then, in two campaigns neither of which lasted more than a few weeks, the Germans completely reversed the balance of advantage. Between April and June, they overran five countries, knocked France out of the war and twice drove the British into the sea, first in Norway, then at Dunkirk. All the efforts of the First World

War had failed to win the German High Command control of more than the 50 miles of the Belgian coastline: now, at a blow, Hitler commanded the whole western shore of Europe from the North Cape to the Spanish frontier and within a matter of months was able to make the Mediterranean as well impassable to British ships. From a hundred airfields the German Air Force was preparing to launch a continuous attack, not only on the sea communications on which the British depended for food, oil and raw materials, but on their ports and industrial cities, every one of which had been brought within easy bombing range. Without an ally left outside the Commonwealth, without any foothold left on the European mainland and at a marked disadvantage, economically as well as militarily, the only future the British faced, in the judgment of the rest of the world, was one of violent bombardment from the air followed by invasion and almost certain surrender or defeat. No British Government in modern times had ever found itself—within almost hours of taking office—so close to disaster.

In its first year of office it took all the new Government's energies to meet and survive the successive emergencies thrust upon them: the battles in Belgium and France, Dunkirk, the defeat of France and the loss of the French fleet, the entry of Italy into the war, the Battle of Britain, the threat of invasion, the Blitz, the German conquest of the Balkans, the loss of Greece and Crete. Yet well before the summer of 1940 was over and the Battle of Britain won, it was clear that to survive was not enough: even if Britain could get through the dangers of the autumn and winter that lay ahead undefeated, its leaders had still to answer the question, how to win. It was not a question that could wait for an answer. Even while they were trying desperately to find enough planes for the R.A.F. to hold off the Luftwaffe, enough guns to equip the Army against invasion, the Government and its advisers had to start making preparations for the time when they might hope to capture the initiative from the enemy.

In 1940 and for a long time to come, this was an economic far more than a military problem. British rearmament had only started in earnest in 1939 and after the losses of the summer, including the whole of the equipment of the British Expeditionary Force, it had still a long way to go. In August 1940 the Chiefs of Staff reported to the Prime Minister that it would not be possible to build up sufficient armed strength to go over to the offensive before the autumn of 1942.

Considering the continental resources which Germany now commanded, this was an optimistic estimate and could only be realised if during the next two years the British were able to restrict their military commitments and put their main effort into rearmament.

Rearmament on such a scale meant nothing less than reorganising the whole economy of the country, and doing this not only as fast as possible but under the handicap of black-outs and air raids and of the steady withdrawal of the younger men from industry to serve in the Armed Forces. Moreover, a change of this magnitude could not be carried through without consequences which went far beyond the organisation of war production and raised, one after another, economic or social issues—from state control of industry and industrial relations to inflation, fair shares when supplies were short and equality of sacrifice—issues which, unwisely handled, could threaten the nation's unity in waging the war and cripple it as surely as defeat in the field. These issues, far from being irrelevant or extraneous to a modern "total" war, were as much a part of the Second World War as the deployment of armies and navies.

Bevin never lost sight of the fact that the war could only be won by the defeat of the enemy in the field. From first to last—and in the second half of the war, when people were beginning to talk eagerly of the post-war world, at the cost of considerable unpopularity with a section of his own party—he never wavered in his insistence that victory in the fighting war had got to take precedence over everything else. None the less, in practice, most of his time (and as a result most of this volume) had to be devoted to the "other war", the mobilisation of the country's economic resources and the problems to which this gave rise.

The role which he was to play in this was not at all obvious in May 1940. The administration which Churchill formed to meet the crisis was a government of national union, representative of all three political parties and including virtually every political leader of consequence in the country. No such coalition would have been complete without a trade unionist among its members. But this meant little. If the record of trade-union leaders in government was anything to go by, the chances were that he would remain a passenger in the Cabinet excluded from anything more than a nominal share in power, or limited to departmental duties as George Barnes, also a wartime Minister of Labour, had been in Lloyd George's coalition.

Bevin had no experience of either war or government; he had never even sat in Parliament as many other trade unionists had, and he was not at first considered for membership of the inner War Cabinet.[1] Like other ministers, he only attended when called in for a discussion of the matters which concerned his department, and his department was one which up to that time had ranked as a second-class ministry.

None the less, within five months of joining the Government, Bevin was brought into the War Cabinet over the heads of other ministers, Labour as well as Conservative, who had far greater experience of office. And, once in the War Cabinet, he remained there through all the subsequent changes until the coalition broke up with the end of the war in Europe. Only four other men, Churchill, Attlee, Eden and Anderson, equalled this record of uninterrupted membership of the War Cabinet from the end of 1940 to May 1945 and together with these four Bevin made up the small group of men who provided the country's leadership in the greatest episode of its history.

How did this come about? There are three answers to this question, each corresponding to one of the main themes in the history that follows.

The first is the impression which Bevin's personal qualities made on Churchill. "Another minister I consorted with at this time [the summer of 1940]," Churchill wrote later, "was Ernest Bevin. . . . I was much in harmony with both Beaverbrook and Bevin in the white-hot weeks."[2] Beaverbrook was an old ally of Churchill's, but Bevin was a new discovery in whom he recognised at once a toughness of mind, a self-confidence and strength of will which matched his own. According to Lord Beaverbrook,[3] Churchill talked in the autumn of 1940, if the Germans succeeded in landing in England, of setting up a Committee of Public Safety, composed of himself, Beaverbrook and Bevin to lead the British resistance. Whether this was a serious suggestion or not, the fact that Churchill should have singled out Bevin in this way shows the opinion he had already formed of him. Here was a man who, whatever he lacked in parliamentary or

[1] This consisted of the two leaders of each of the two principal parties in the coalition, Chamberlain and Halifax, Attlee and Greenwood, with Churchill in the chair.

[2] Winston S. Churchill: *The Second World War*, Vol. II, *Their Finest Hour* (1949), p. 287.

[3] In conversation with the author, June 1961.

ministerial experience, possessed something much more important in the critical days of 1940, the temperament of a born fighter, a man whose nerve would not crack or power of decision falter in face of the storms that lay ahead.

The second reason was Bevin's position in the Labour Movement and the use he made of it. Between the wars he had established himself as the outstanding leader in the trade-union world, partly because of his power as general secretary of the biggest union, the T.&G.W.U. which he had created, partly because the same qualities which impressed Churchill impressed the General Council and the annual congress of the T.U.C. It was because of this that Churchill invited him to join the Government in the first place. But to be a powerful trade-union leader in opposition was one thing, to retain his influence with the unions when he became a minister was quite another. Instead of trying to play safe, Bevin boldly asserted his claim to be, in a special sense, the representative of the trade unions and the working class in the Cabinet and the spokesman of the Government to organised labour. It was a claim which would have proved disastrous to a lesser man but to a surprising extent he vindicated it, and in this double capacity acquired a unique authority which he retained to the end of the war.

The final reason was Bevin's success in securing for his Ministry control over manpower, thereby converting it into a key economic ministry. If manpower was bound to replace finance in wartime as the determining factor in the allocation of resources, it was by no means certain that its control would be concentrated in the Ministry of Labour or that the Minister of Labour would stay in the War Cabinet when the Chancellor of the Exchequer was left out. If there was any doubt about Bevin's position it was settled when the Ministry of Production was set up in 1942: manpower and labour were specifically excluded from its responsibilities and left, as Bevin was determined they should be from the day he accepted Churchill's invitation, in the hands of the Minister of Labour.

2

In the course of 1939 the Chamberlain Government had at last authorised a programme of rearmament which, within three years, would put Britain on something like equal terms with her enemies.

The size of the programme was adequately conceived and was substantially taken over by the Churchill Government in 1940. Its weakness lay in the late date at which it was started and in the rate at which it would begin to show results. The existing armaments industry of the country was quite inadequate to produce the volume of weapons and supplies required: new factories had to be built and equipped, management and the labour force expanded on a scale which had been hardly conceived of before. This was bound to take time. The Government assumed it had the time: it had not. The strategic timetable to which production was geared was overtaken by events. By the summer of 1940 the foundations had been laid for an output of munitions greater than at any point in the First World War, but the output itself would not appear until another year or even two had passed. Production was rising, but after nine months of war, at best, was only preventing the Germans from drawing further ahead; it had not yet begun to close the gap. When the storm broke in a clash of arms which might well decide the issue of the war then and there, the British forces had to face an enemy who enjoyed an absolute superiority in armaments.

How much more could have been done to speed up British rearmament, given the late date at which it was started, was a matter of dispute at the time, and remains so still. The Chamberlain Government had stubbornly resisted the demand for a Ministry of Supply, on the lines of Lloyd George's Ministry of Munitions, with responsibility for the rearmament of all three services. When a Ministry of Supply was finally set up in the summer of 1939—the "mule" ministry, as it was called by the critics—it took over responsibility for the Army's needs and certain common supplies, but the Admiralty and Air Ministry continued to act as the supply departments for their own Services. Each department organised its programme independently, making its own arrangements with the firms it chose to carry out the work under government contract.

Whatever the organisation at the top of war production before May 1940, however, it is unlikely that it would have overcome the lack of urgency and driving power which pervaded the Chamberlain Government's preparations for war. It was still at heart a peacetime Government, which had accepted the necessity and the cost of rearming but lacked the imagination, or the will, to recognise that this implied—as Keynes saw and urged—that the only limits to the

6

national effort in a "total" war would have to be, not financial, but physical—not whether the nation could afford to do more but whether there was anything left at all, in capital assets, natural and human resources, which could be thrown into the struggle.

Quite apart from the demands of the fighting war, the Government's rearmament programme could not be carried through without drawing all but a bare minimum of the country's manufacturing capacity and labour force into war production and as a result cutting civilian standards of living drastically. Yet the Government up to May 1940 showed great reluctance to impose the controls which would be necessary to secure this and a marked lack of confidence in the willingness of people to accept them. Even those who recognised the sort of measures which would be needed doubted whether there would ever be the will to enforce them.

This failure to mobilise the country's resources adequately can be illustrated from every sector of the economy: it shows up most clearly in the handling of manpower and the mobilisation of labour.

For twenty years Britain had been suffering from a glut of manpower expressed in unemployment. As late as April 1940, eight months after war had begun, the number of registered unemployed was still above a million, despite the call-up of one and a half million men for military service. The idea that the time would come when the lack, not of money, not even of raw materials or shipping, but of any more men (or women) to be drafted into the Services and industry would set the final limits to the British war effort—this was as unfamiliar as the idea that Britain would ever be short of coal. No attempt had been made to match the future demand for men against the resources or to foresee how these could be increased by the employment of women: no one had yet thought of a manpower budget taking the place of the financial budget as the method by which the Government, from 1941 on, controlled the allocation of the nation's resources, so many men for the Army, so many for the aircraft industry, so many for the mines. If anyone had wanted to draw up such a manpower budget, they would have found the greatest difficulty in obtaining statistics with which to do it.

The one serious attempt made to investigate the manpower situation, the Humbert Wolfe Committee,[1] showed the limitations of

[1] Humbert Wolfe, its chairman, was Deputy Secretary of the Ministry of Labour. Educated (like the author) at Bradford Grammar School and Wadham

official knowledge (the estimates were later found to be highly impressionistic) and, much more important, how little had yet been done to provide for needs already approved. The Committee took the rearmament programme fixed in 1939 and translated it into terms of the *additional* labour which would be required in the engineering industries (including shipyards and aircraft factories) to carry it out. By September 1940 the number of men and women employed in these industries would have to increase by 70 per cent above the pre-war figures, by the summer of 1941 they would have to be up by 117 per cent. This was an expansion three times as great as that achieved in the four years 1914–18, with only two years in which to do it. A second inquiry (the Stamp Survey, May 1940) estimated that the actual increase in engineering manpower in the first twelve months of the war was more likely to be 20 than 70 per cent.

What was wrong?

Some of the lessons of the 1914 war had been well learnt. This time conscription had been introduced from the beginning of the war and had increased the strength of the armed forces more than four-fold by June 1940.[1] In an attempt to prevent industry losing its skilled labour, the Government had introduced a schedule of reserved occupations in which men would be held back from military service. But protection was not enough. Little had yet been done to expand the labour force by training more men—and women—in the skills that were needed and reorganising production so that, by a process of "dilution", two skilled men could do the work of three, and more use be made of unskilled workers or women. Training centres were only half filled and the craft unions opposed any move to relax their rules and so permit dilution. Moreover, although skilled men were reserved from the call-up, there was no guarantee that they were engaged on war work and no procedure for transferring them when they were not. Far too many were still employed in industries and services producing for civilian consumption, or even unemployed, while the aircraft factories, the shipyards and munition plants were crying out for men.

The obvious course, as it seemed to some of those most concerned

---

College, Oxford, he was better known as a poet and essayist than for his mastery of labour statistics.

[1] The total strength of the U.K. Armed Forces in June 1939 was 480,000; in June 1940, 2,218,000.

with the problem, was to introduce industrial as well as military conscription and direct men to work where they were most needed. This course was advocated by Lord Stamp, the Economic Adviser to the Cabinet, and from outside Whitehall by Sir William Beveridge, both of whom urged that the Government, in order to prevent inflation, should also exercise direct control over wages. The objections to such a policy, strongly represented by the Minister of Labour, Ernest Brown, were the opposition which it was certain to encounter from the trade-union leaders (including Ernest Bevin) and the impossibility of carrying it out, in face of such opposition, for fear of industrial trouble.

The alternative to which the Ministry of Labour clung was the natural operation of economic forces, the Old Adam of individual interest. As war industry expanded and the less essential industries contracted, labour would follow the flow of work, the more so as the war industries, especially the engineering firms in the Midlands engaged on government work, were offering higher wages and bonuses. The other government departments, the Ministry of Labour argued, could help the process on if they would only spread their contracts more widely and bring in firms and areas which had manufacturing capacity and labour to spare.

This course of action—or rather of inaction—ignored the urgency of the situation. There was too little time to wait until individual interest persuaded men to change their jobs, little certainty that when they did they would take the ones where they were most needed and no reason at all to believe that the total of such individual transfers would ever come near the scale of the industrial migration required. Between June 1939 and June 1940 the percentage of the nation's labour force employed in the engineering and chemical industries had only risen from 20 to 24 per cent, and in the other essential industries and services (shipping, transport, coal, agriculture) from 31 to 32 per cent. Moreover, the higher wages and bonuses with which many firms were trying to attract workers ran right against the Government's policy of damping down wage increases in order to prevent inflation, the fear of more money being paid out in wages when the supply of goods was being restricted, and so driving up prices.

By the time Ernest Bevin became Minister of Labour no answer had been found to this dilemma, in part at least because of the attitude of Ernest Bevin, the most powerful trade-union leader in the

9

country. There was no policy for industrial manpower, not even agreement on which ministry should take responsibility for providing one, a responsibility which the Ministry of Labour was determined to avoid if it possibly could.

On the other hand, dissatisfaction with this state of affairs was growing rapidly. On 3 May, Lord Stamp submitted a memorandum to the Cabinet in which he sharply criticised the attitude of the Minister of Labour (Ernest Brown) and his department. The Government, he argued, could no longer leave the distribution of manpower to the unregulated operation of a free market. The opposition of the trade unions could not be accepted as decisive. There would have to be some control of labour and wages if the country's economy was to be effectively organised for war. The following day, Winston Churchill, then First Lord of the Admiralty, circulated a paper to the Cabinet in which he analysed a recent report on manpower in the engineering, motor and aircraft industries. The figures which he quoted showed an increase in the labour force of only 11 per cent in the last ten months. "In this fundamental group at any rate," the First Lord concluded, "we have hardly begun to organise manpower for the production of munitions."[1]

These criticisms could neither be answered nor ignored. At a meeting of senior civil servants on 8 May, the supply departments (Admiralty, Air Ministry and Ministry of Supply) united to press the representatives of the Ministry of Labour to accept full responsibility for the control of manpower, and agreed to support a recommendation to the Cabinet that additional powers should be given to the Minister for this purpose.

It was at this point that Neville Chamberlain resigned and Churchill formed his coalition, with Ernest Bevin as Minister of Labour.

Bevin was no economist, but for years he had seen what so many economists failed to see, that in refusing to take more positive steps to remedy unemployment, the State was neglecting the most precious of all its resources, the nation's human resources. Bevin was a Keynesian by the light of nature,[2] and it was a natural step from his

---

[1] H. M. D. Parker: *Manpower* (1957), p. 78.

[2] As his alliance with Keynes at the time of the Macmillan Committee showed. See Volume I, pp. 425–34.

views on unemployment in peacetime to grasping the central place of manpower in war. This was the foundation of his plans for the Ministry of Labour and one of the principal reasons why, to the surprise of his colleagues on the General Council,[1] he had agreed to accept an office which most of them regarded as a sideshow.

With Bevin in the Government there would be no lack of urgency in dealing with manpower questions, nor was there any need, with the impact of war at last coming home to people, to convince the nation that something needed to be done at once to raise the production of arms and equipment. The greatest change of the summer of 1940, and the new Government's greatest asset, was the sudden awakening of the British people from the sour and sluggish mood of the previous months, their recovery of a native resolution which many had believed lost for good. The danger now lay in the other direction. With everything suddenly beginning to expand at once, every contractor working for a government department was soon demanding more men and every government department expressing its newly found zeal by badgering the Ministry of Labour to provide them. Next year, next month even, was not enough: the war might be over and lost by then. The men had to be found so that production could be pushed up and the weapons put in the hands of the R.A.F. and the Army next week. The hour and the national mood demanded sweeping, drastic action. The danger for the new Minister of Labour was that he might be so overwhelmed by these emergency demands as to lose sight of the equal need for long-term plans, and that in his own and the nation's anxiety to see quick results, he might adopt methods which would prejudice his chances of carrying out those long-term plans when the emergency was past.

3

Bevin had been strong in his criticism of the Chamberlain Government. Now he was given the chance to show whether he could do better. The prospect did not intimidate him. He came back to

[1] Ibid., p. 652.

London from the T.U.C. meeting at Bournemouth on Whit Tuesday, 14 May and went straight from the station to Montagu House, the former mansion of the Duke of Richmond which housed the Ministry of Labour.

His predecessor had already left and the introductions of the senior staff to the new minister were made by the Parliamentary Secretary Ralph Assheton, who was later to become Chairman of the Conservative Party. When the civilities were completed and the room empty again, Bevin, flinging himself back in his chair and looking at Assheton, asked, "Well, Ralph, what do I do next?"

Taken aback by the question, Assheton gave the conventional reply that he could rely on his officials to put any proposals into effect provided he knew what he wanted to do. This suited Bevin: he might be ignorant of procedure but he had arrived with plenty of ideas about what to do. "Then if I were you," Assheton advised him, "I'd go off home and put them down on paper."

The four sheets which Bevin put in the hands of his Permanent Secretary, Tom Phillips, the next morning were not a comprehensive plan for the organisation of manpower. It is doubtful whether anyone could have drawn up such a plan at this stage of the war, still less have foreseen the changes that would have to be made in it in the next five years. What Bevin produced was a programme of action, and in getting it adopted he transformed both the attitude and the role of the Ministry of Labour.

The key to everything that followed lay in Bevin's claim that the responsibility for all manpower and labour questions must not be broken up but concentrated in the hands of a single minister, himself. Not content with reversing the previous attitude of his Ministry and demanding a responsibility which had been reluctantly accepted under pressure at the meeting on 8 May, he pressed for it to be extended to include besides the supply of labour the right to examine the use made of it, and where necessary to withdraw it, a demand which was certain to lead him into conflict both with employers and with the other government departments for which they were working. These were bold claims for a ministry which had never been more than a second-class department. They were accepted by the Cabinet however, with relief after so much indecision, and for the first time in its history the Ministry of Labour had a minister capable of making them good.

With three autonomous supply departments[1] each engaged in great efforts to accelerate the production programmes for which they were responsible and competing fiercely for everything in short supply, especially skilled labour, Bevin's own job would become ten times more difficult, unless he could get some body set up with the authority to settle which claims were to have priority. There was nothing comparable on the production side to Churchill's combination of the office of Minister of Defence with that of Prime Minister on the operational side, and the lack of this led to much trouble in the next eighteen months and to a number of proposals, none of which proved satisfactory, to fill the gap.

Bevin's first suggestion, in his paper of May 1940, was a Production Council, to consist of the three supply ministers, who between them were responsible for the munitions programme, plus the President of the Board of Trade, responsible for civilian supplies, and himself.

Bevin's Council was to be given a much wider responsibility than manpower. He wanted it to assess all the factors which played a part in war production, to see that commitments were kept in line with what was available and to take responsibility for turning strategic decisions into the necessary production programmes. His proposal was accepted, and the Council met for the first time on 22 May with a chairman (Arthur Greenwood) who was a member of the War Cabinet and with the expectation that it would play a central role in the organisation of the war effort.

The Production Council was designed to centralise the making of decisions; a second proposal of Bevin's was designed to decentralise their execution, by creating better machinery to settle as many as possible of the claims for manpower and other resources at a regional level, without having to refer everything to Whitehall. This was another instance in which Bevin saw something that needed doing, but could not get it done until much later in the war. His proposal in 1940 was to take over the twelve Area Boards established by the Ministry of Supply, each with regional representatives of the same five ministries which were to co-operate at the top in the Production Council. By putting them under his own Ministry's divisional controllers, he hoped to infuse new life into them and see that they were given the chance to do a real job.

---

[1] The Admiralty, the Ministry of Supply and the Ministry of Aircraft Production which Beaverbrook was setting up.

At the same time Bevin suggested bringing more firms under direct government control and fixing prices in order to hold down profits. Whatever merit these proposals had, they probably smacked too much of socialism to stand much chance of being taken up by a coalition Government and little came of them. He had more success, after a stiff battle with other government departments, in carrying through another proposal included in his original paper. This was to set up a Ministry of Works in order to bring some order into the chaos which the Government's rapid increase in demands was creating in the building industry.

The remainder of Bevin's paper was taken up with his plans to strengthen his own department. At the top he proposed to set up a Labour Supply Board of two industrialists and two trade unionists meeting in daily session under the Minister's chairmanship. He matched this with plans for a series of local labour supply committees drawn from management and the unions in each important centre of industry in the hope of stimulating local initiative in the settlement of labour problems. At the shop-floor level he asked for power to recruit four hundred labour supply inspectors technically qualified to investigate labour shortages and obstacles to the transfer of workers, as well as to push the two long-term policies on which he laid most stress, dilution and training.

Finally, in order to remove trade-union objections to suspending demarcation and other restrictive practices in wartime, he proposed a formal undertaking by Parliament to restore pre-war trade practices when hostilities were over.

The whole of this programme of action, drawn out in a detailed draft, was circulated to other departments on Friday, 17 May, three days after Bevin had first walked into the Ministry, and was ready for him to present to the Cabinet on Monday the 20th.

There was one striking omission from Bevin's paper: it contained no proposal for industrial conscription or the direction of labour. This was deliberate. In the dispute between the Ministry of Labour and those who wanted to apply compulsory powers to the direction of labour, the new Minister was firmly on the side of his officials, with this important difference: that what they had hitherto defended as a precaution aimed at avoiding trouble with the unions, Bevin now advocated as the necessary pre-requisite to securing the unions' co-operation.

The new Minister of Labour might be expected to tell the unions bluntly what was needed, but he had not accepted office to impose upon them measures for which he could not win their support. The Labour Party would not have agreed to join the coalition on any other basis, nor would Churchill have appointed the leading trade unionist in the country as Minister of Labour if he had intended to follow a different policy.

On 20 May, however, another paper also came up for discussion in the Cabinet. A committee of senior officials had been reviewing the additional powers the Government would need to take if the country were threatened with invasion. Among their proposals, presented by Neville Chamberlain as Lord President of the Council, was one giving the Minister of Labour and National Service authority to direct anyone in the kingdom over the age of sixteen to perform any work or service for which he was required. In view of the news from France, which hourly grew worse, the Cabinet decided to ask Parliament for the immediate grant of the powers Chamberlain suggested. Accordingly, on 22 May, while Churchill flew to France to try and rally the French resistance, Attlee introduced an Emergency Powers Bill which was passed through all its stages in a single day. In the course of the same sitting the first Defence Regulation under the new Act armed the Minister of Labour with sweeping powers to order anyone to do anything that he might require.

This was the most drastic Act ever passed by a British Parliament and in the next eighteen months it was to be the cause of much misunderstanding and recrimination between Bevin and his critics. The Emergency Powers Act was acclaimed as proof of the nation's determination to go to any lengths in prosecuting the war, and many middle-class voices were raised to urge the Government—and its trade-union Minister of Labour in particular—to show that they possessed a proper sense of urgency by applying their powers to direct labour where it was wanted. In February 1941 *The Economist* wrote in exasperation:

"There is no political check on an Executive which acts *infra vires*. But Parliament can hold a Minister guilty of a breach of trust not only for abusing or misusing his powers, but also for not using them at all."[1]

Bevin, however, stuck to his original view. He had not asked for the

1 *The Economist*, 8 February 1941.

powers he was given and he continued to regard them as a sanction to be held in reserve and used sparingly. Up to July 1941, he issued no more than 2,800 orders to individuals under his powers of direction. Since this attitude incurred a good deal of criticism, and since he clearly felt as convinced that he was right as his critics did that he was wrong, it is important to discover what lay behind it.

4

Many people ignored the fact, on which the Ministry of Labour had always laid stress, that at the beginning the big expansion of war industry would be secured less by the transfer of workers than by the change-over of firms and factories (with their existing labour force) to war production. As the expansion really got under way, many other things besides labour were in short supply—raw materials and components, machine tools and equipment, factory-floor space and new buildings—and these often accounted more for delays in production than did shortages of manpower. Managements, however, made the most of their labour difficulties, partly in order to push the blame for delays on to someone else, partly in order to put pressure on the Ministry of Labour to provide them with more men or at least to leave their labour force intact. There was a tendency, in which political prejudice no doubt played a part, to make the Ministry of Labour and its trade-union minister the whipping boy for more than its fair share of the difficulties inevitable in any forced programme of expansion.

In fact, no *general* shortage of labour appeared until the second half of 1941, and Bevin argued that there was no point in ordering more people into war industry on any large scale until the new factories and extensions were completed, the plant installed, the materials available, and the jobs there for them to do. There was no sense in making workers transfer only to keep them hanging about for weeks doing nothing or on half time because the work was not yet ready for them.

These were strong arguments against any general direction of labour at this stage of the war. But they did not apply with anything like the same effect to using direction to meet the specific shortages of *skilled* labour which were a serious handicap in 1940–41. This was the real case against Bevin and many people concluded that, in refusing to use his powers of compulsion, he paid too much attention to the opposition of the trade unions and too little to the national interest.

To this Bevin had two replies. The first was that the critics failed to understand the nature of the problem. They were still thinking in terms of the pre-war surplus of labour, assuming that the skilled men were there, if only the Minister of Labour would direct them to their particular factory. The real problem, however, with the sudden expansion of employment, was an overall shortage of skilled man-power which could only be met by a long-term programme of training, by more intensive use of machines and much less wasteful use of the labour firms already possessed.

This was one reply to which we shall have to return in the next chapter. The other was of a quite different kind. Much play was made by Bevin's opponents with the analogy between the obligation to serve in the Armed Forces and the obligation to serve where required in industry. Why, he was asked, was he prepared to apply compulsion in the first case as Minister of National Service and not in the second as Minister of Labour? The short answer was that working-class opinion had always refused to equate the two. Con-scription for military service and the enforced discipline that went ·with it once a man was in uniform, that was one thing, accepted at least in wartime as inevitable and fair: but conscription for industry and any attempt to enforce military discipline in ordinary work places was another thing altogether, to which not only the unions but the working-class opinion they represented were solidly opposed.

That this was so could hardly be disputed: even so convinced an advocate of compulsion as Beveridge was forced to admit it.[1] The question to which the critics wanted an answer was whether the

[1] In his autobiography *Power and Influence* (1953), Beveridge wrote: "On the face of things, it is unjust in a total war, that for those who work in the factories there should not be discipline, as there is discipline for those who serve in the fighting forces. But Britons have never, in practice, agreed to this view. As soldiers, sailors and airmen they accept discipline. As industrial workers they reject it, and claim personal liberty, even at the risk of weakening the war effort. It may be that, in the last resort they are right about this, however unreasonable or dangerous it seems. . . . The fighting line is for war only. The factories are for peace as well. If freedom goes there even for a while, are we sure of recovering it later? Unreasoning rejection of industrial discipline even in war, however dangerous in itself, is perhaps the last ditch against totalitarian rule for all time, in war and in peace thereafter." (p. 162)

Government, which had already announced its intention of overriding equally strong prejudices in favour of the rights of private property, was going to accept such limitations on its power of action or show itself resolute in allowing nothing to stand in the way of the national interest. Bevin thought it a politician's question. The only test, in his eyes, was a practical one, how you got more production. Workers with a sense of grievance not only made bad workers but could easily start trouble in a factory: the result of applying compulsion when it produced such a sense of grievance might be to interrupt, not increase, production.

Thirty years' experience of persuading working men to combine convinced Bevin that to start brandishing compulsory powers and ordering people about as soon as the new Government got into power would do harm to the new-found national unity out of all proportion to the amount of extra work it would produce. 1940 was a national emergency, and most people showed their recognition of the fact by the hours they were prepared to work and the effort they were ready to make. But no emergency could obliterate overnight the memory of the Depression years, of what had happened after the First World War and of the long bitter history of industrial relations. Ingrained attitudes on both sides, management's as well as the men's, were not so quickly changed. Even if the jobs they were called on to do were now officially labelled war work, the men were still working for their old employers, and the old suspicions were still there below the surface. Given time, working-class opinion would reach its own conclusions about compulsion: until then Bevin held to the belief that you would get far more work out of people if they worked willingly than if they were made resentful by compulsion.

"Given time": the question no one could answer in 1940 was whether they *would* be given time. Everything pointed to the opposite conclusion, that there was no time to spare before introducing compulsion. To refuse, as Bevin did, was to take the risk of the war being lost before the country had put forth its full strength. But, if the gamble came off and the war lasted several years, as Bevin already believed it would, then questions of morale might prove decisive and the time taken to let working-class opinion come round to compulsion of its own accord might well be justified, ten times over, by the result. This was the dilemma of 1940, as it appears now, but not for Bevin: he never hesitated to throw all his weight in favour of a long-term policy, a choice which showed uncommon confidence that summer.

5

In these circumstances, Bevin's claim to give his Ministry the sole responsibility for manpower, without employing his powers of direction over labour, might well have proved fatal to him in his first six months of office. Demands for more men were thrust at him from all directions: skilled men for the armaments and aircraft works; skilled men for the Army and the R.A.F.; men to set up coastal defences and airfields, men to repair bomb damage, men to build new factories, men and women to start production in them even before they were finished. It needed all Bevin's resilience to stand up to these demands, to turn his Ministry round and infuse not just his headquarters but his local staff up and down the country with the confidence to tackle problems with which they had never been faced before.

John Winant, soon to become American Ambassador in London, has left a description of him at work in the Ministry of Labour at the time of Dunkirk:

"He was sitting in a chair that was not built for him, a great hulk of a man. He called out to me as I passed through the door, and before I had a chance to speak, began explaining the pressures and problems that were piling up because of the fall of France. I broke in and gave him Butler's message of Italy's declaration of war. He went back in his chair for a moment and then seemed to gather all his energies, and came forward across the desk. He had already begun to plan to meet this new emergency. Both he and Churchill had the same fighting stamina for meeting reverses head on."[1]

Yet, however hard he drove himself to master the emergency, he never lost sight of the need to lay the foundations of his policy even while the battles of the summer were at their height. Within less than a month of taking office he had made three moves, each of which provided a key to the policy he meant to follow and each of which turned out to have consequences going well beyond the end of the war.

[1] John G. Winant: *A Letter from Grosvenor Square* (London, 1947), p. 155.

B

The first was to secure the co-operation of the unions. During the week in which he was preparing his proposals for the Cabinet, he had kept in close touch with Walter Citrine, the General Secretary of the T.U.C., and on 25 May he went to Central Hall, Westminster, to set out his plans in detail to a delegate conference of two thousand executive members from over a hundred and fifty unions. No similar meeting had taken place since the General Strike of 1926 (when Bevin had also addressed the assembled Executives, as a member of the strike committee[1]).

"I have to ask you," he told his audience, "virtually to place yourselves at the disposal of the State. We are Socialists and this is the test of our Socialism. It is the test whether we have meant the resolutions which we have so often passed. I do not want you to get too worried about every individual that may be in the Government. We could not stop to have an election; we could not stop to decide the issue. But this I am convinced of: if our Movement and our class rise with all their energy now and save the people of this country from disaster, the country will always turn with confidence to the people who saved them. They will pay more attention to an act of that kind than to theoretical arguments or any particular philosophy. And the people are conscious at this moment that they are in danger."[2]

He went through each of the proposals he had put to the Cabinet and explained the reasoning behind them. In return he secured for the first time the trade unions' promise of unreserved support for the Government in whatever steps might be necessary to win the war.

But this, Bevin told the delegates, was not enough; they must go back to the mines, the factories and the docks and tell their members the full story so that they too should feel themselves partners in a common enterprise. No minister had talked to the trade unions like this before: it is doubtful if the unions would have listened, or believed any other minister, if he had. But the trade-union movement had been Bevin's life, and no one in the hall doubted that he meant it when he said that he looked on the unions as part of the freedom for which the war was being fought, a view for which he found support in the fact that they were among the first institutions to be destroyed by Hitler and Mussolini on coming to power. The more serious the situation, the more strongly he believed that the only way to meet it

[1] See Vol. I, pp. 301–2.
[2] T.U.C. Report of the Special Conference of Trade Union Executives held at the Central Hall, Westminster, on Saturday, 25 May 1940, p. 18.

was to call out the latent strength of a free society, not to sap it by imitating Hitler and exercising dictatorial powers.

Bevin's plan was to go further than simply consulting the unions on matters which directly concerned them. His second move was to set about establishing, at every level, the practice of joint consultation, on equal terms, with both sides of industry, employers as well as unions, a practice which clearly reflected his efforts before the war to build up joint negotiations into something which could become a system of industrial self-government and extend further than the settlement of wages and hours. His idea was to use his powers as Minister of Labour not to issue orders to industry but to bring both sides together—always on equal terms—face them with what was required and get them to work out an agreed solution which he could put into force in statutory orders and regulations where necessary.

His predecessor at the Ministry of Labour had already set up a National Joint Advisory Council, but had made little use of it. On 22 May Bevin called together the sixty industrialists and trade-union leaders who composed the Council and, in order to impress them with the seriousness of the situation, read out the statement which Attlee was at the same time making to the House of Commons as an introduction to the Government's emergency legislation.

Bevin then went on to explain his plans, laying stress on the need for dilution and training, outlining his ideas on welfare and raising the thorny questions of wages and arbitration. If necessary, he told the Council, he would not hesitate to use the compulsory powers which had been conferred on him, but he preferred to work by consultation rather than dictation and to appeal for the voluntary co-operation of industry. He summed up the Government's attitude in a sentence which clearly reflected his own view:

"We came to the conclusion that with the good will of the T.U.C. and the Unions, and the Employers' Federation, a little less democracy and a little more trust in these difficulties, we could maintain to a very large extent intact the peacetime arrangements, merely adjusting them to suit these extraordinary circumstances."[1]

To show that he was serious in his talk of consultation Bevin asked the Advisory Council to appoint at once a smaller and more effective

[1] Quoted in Parker, pp. 95–6.

body, a Joint Consultative Committee, to advise him on the orders to be issued under his authority as Minister of Labour. This was a body which was to play a key role in his tenure of the Ministry of Labour. Composed of seven representatives from the British Employers' Confederation and seven from the T.U.C., it met thirty times in the next eighteen months and its discussions, conducted with plain speaking on both sides, ranged over the whole field of the Ministry's activities. The Minister himself almost invariably took the chair and at the Committee's first meeting on 28 May showed what he meant by consultation when he asked the Committee to consider the crucial question of wages and strikes in wartime.

Bevin himself threw out two suggestions at this first meeting: that it might be possible to set up a wages tribunal to adjust wages on a uniform basis and that independent members might be introduced into the normal procedure of wage negotiations. He plainly preferred, however, the recommendation with which the Joint Consultative Committee itself came up at its next meeting on 4 June. The Committee argued, and Bevin agreed, that the best course would be to keep the existing machinery for joint negotiation of wages between employers and unions but to ban strikes and lock-outs for the duration of the war. Instead, both sides should be required to agree, where a settlement could not be reached, to send the dispute to arbitration and bind themselves to accept the decision of the court.

Nothing could have suited Bevin better. The substance of the proposal, to continue the practice of joint negotiation reinforced by compulsory arbitration, and the procedure by which it had been arrived at, by joint consultation between the two sides of industry, were both characteristic of the policy he meant to follow. Worked out in detail and issued under his authority as the famous Order No. 1305, it provided the basis on which industrial relations in Britain were conducted for the rest of the war. So long as he remained Minister, he followed the same procedure, not only encouraging the Joint Consultative Committee to raise any matter of concern to either side but submitting for discussion and frequently for revision every important order or regulation which he proposed to issue. He did not abdicate his responsibilities as a minister: his powers of compulsion were there to be used if necessary, but they were twice as effective, in his view, because, when they were used, it was on a basis of consent not dictation.

The last of the initial moves Bevin made which illustrate the character of his policy sprang from his conviction that the right way to handle working people was to treat them, not as so many units of skilled or unskilled manpower, but as individual human beings whose efficiency and willingness to work were affected by the conditions in which they worked and lived. He brought to his office a new conception of the part which the welfare of workers, both inside and outside the factories, could play in solving labour problems, and there was no side of the Ministry's work on which he was to leave more clearly the mark of his personal interest.

As a first step, he secured the transfer from the Home Office to the Ministry of Labour of the administration of the Factories Acts and established the authority of a single ministry, his own, to handle all welfare arrangements. Inside the department he set up a new Factory and Welfare Division, offered it first to Sir William Beveridge and then, when Beveridge hesitated, appointed one of the ablest of his permanent officials, Godfrey Ince, to run it. To draw on the experience of the voluntary organisations and the trade unions he established a Factory and Welfare Advisory Board, under his own chairmanship, and started to recruit divisional and local welfare officers to get schemes started on the ground. He showed himself interested not only in the safety and health of workers in the factories, but in their living conditions as well, billets, hostels, transport, feeding arrangements and, not least, their leisure. He insisted that welfare must be treated not as a trimming but as a central part of the Government's labour-supply policy. Thanks to this early initiative on his part, when more widespread use had to be made of compulsory powers, the welfare services which then became essential were already in existence and could be rapidly developed.

Taken together, these three moves—his appeal for the co-operation of the unions, his establishment of a joint consultative committee, and his immediate creation of a Factory and Welfare Division—defined clearly the character of his policy from the beginning.

6

The worst of Bevin's problems in 1940–41 was the shortage of skilled workers in the engineering and shipbuilding industries. The phrase "a shortage of skilled workers" conceals the difficulties involved. For in practice it covered men with a score of different skills—turners, fitters, setters, sheet-metal workers, welders, shipwrights, boiler-makers—few of whom were any more interchangeable in a world of jealously guarded demarcations than a wheelwright and a radio mechanic. And the needs which the Ministry had to supply were not just so many electricians, so many turners or toolmakers, in total, but so many turners with experience of aircraft work for a factory in Birmingham, so many boilermakers with experience of marine engineering at a works in Leeds, so many electricians with experience of shipbuilding for a yard on Tyneside.

One obvious thing to do was to get the supply departments to site new factories and spread their contracts to firms in areas where there were less acute shortages of skilled labour than in the Midlands and the South East. As part of this policy Bevin worked to establish some co-ordination between the needs of the supply departments and the action of the Board of Trade in cutting down civilian production, and so releasing labour, in such trades as textiles and pottery. After much effort he achieved some success, particularly in the siting of Royal Ordnance Factories, but such a policy encountered all the familiar objections to planning the location of industry. It was not easy to persuade the supply departments at such a time to take on new and untried contractors, and nothing could alter the heavy concentration of the engineering and metal industries around Birmingham and Coventry or of shipbuilding on the Clyde and Tyne.

If the work could not be taken where there were workers to spare, then the men must be brought where they were needed. This sounded simple enough, until the practical obstacles to transfer were examined. Bevin removed one immediately by announcing that his Ministry would pay travelling and lodging allowances. Lodgings, however, were hard to come by in the main centres of war production, particularly after the bombing attacks: this pointed to the need for the sort of welfare service—including the provision of hostels—which he had already set about creating.[1] Even when lodgings were available, there were sharp differences in earnings. Shipbuilding for instance paid much lower wages than other engineering industries. Men from the tin-plate mills in South Wales where work was contracting found that they were offered much lower earnings at the drop-

[1] See below, pp. 78–84.

forging plants in the Midlands to which the Ministry wanted them to transfer. A man earning £5–10–9 a week at de Havillands aircraft factory in the Home Counties dropped to £4–2–6 for the same work if he moved to Napiers in the North West.[1] To get round this Bevin started negotiations with the engineering industry to introduce uniform rates throughout the country. He was successful in the case of the Royal Ordnance Factories and other Government establishments, but it was not until June 1941 that the Engineering Employers' Federation could be persuaded to agree with the A.E.U. that the private employers should make up the difference when men were transferred to an area where lower rates were paid.

Finally, any proposal to transfer workers from one part of the country to another, as the experience of both wars amply confirmed, was certain to run up against the deeply ingrained attachment of the British working man to his own part of the world and his dislike not only of other parts but of "foreigners" from them. London electricians, for instance, out of a job at home who agreed to go to Clydeside or the North East were back within a very short time, declaring that these were impossible places to work or live in, full of strange food, barbarous customs and people who received them with hostility. Employers were often as conservative as their workpeople and little was done to make it easy for the men who had transferred to settle down.

It was at this point, many people felt, that the Minister should have exercised his powers of direction and told men that, if they were in the Army, they would have no choice where they were to serve. Except in a small number of cases Bevin refused to do so.[2] On 5 June 1940 he published an order[3] requiring all jobs in the engineering and building industries to be filled through a labour exchange, a move to stop firms poaching from each other by offering more money, and to give the Ministry's local officers more control over the movement of labour. In August he ordered all men to register who had been employed in the key engineering trades during the past ten years: the fact that up to then no one had any idea how many men there were in the country

---

[1] Examples quoted in M.M. Postan, *British War Production* (1952), pp. 150–51.

[2] Up to July 1941 the total of individual directions he issued was no more than 2,800.

[3] This was the Undertakings (Restriction on Engagement) Order, 1940 (S.R. & O. 1940 No. 877). Besides the engineering industries it covered building and civil engineering, mining and agriculture, the problems of which are described in the next section.

with qualifications for skilled work in engineering shows how little had yet been done to develop a technique of manpower control. The register gave the labour exchanges more information on which to act, but the Minister's instructions were still to use all possible means of persuasion to fill vacancies and to issue directions only in the last resort. Even then he must be satisfied that the job to which the man was to be directed was in an industry in which wages had been properly negotiated and in an area where suitable lodgings could be found. Once a man was called up for military service, his food, clothing, accommodation, welfare, allowances were all taken care of by the Army. Industry assumed no comparable responsibilities, and this was one of the reasons why Bevin believed it impossible to treat the direction of labour as analogous to conscription for the Forces.

There is not much doubt that, if he had acted differently at this stage, he would have run into strong trade-union opposition. This was a decisive argument for him, not out of loyalty or sentiment, but for the simple reason that he had to have the union leaders' support if he was to make any headway with the more radical measures which were necessary if the shortage of skilled workers was to be overcome.

7

It was with these more radical measures in mind that Bevin had asked the Cabinet to give him the responsibility for the use as well as the supply of labour. By August he had recruited half the four hundred labour inspectors he wanted and sent them into the factories to see where the trouble lay. As he expected, they found that, in many cases, firms could manage with fewer skilled men than they were asking for if they were prepared to make better use of them by dilution.

"Dilution" was an ugly but convenient term which covered a multitude of processes, all of which were intended to economise in the use of skilled manpower, by increased mechanisation, by breaking up an operation so that part of it at least could be carried out by less skilled workers, or simply by upgrading semi-skilled or women workers and allowing them to undertake jobs hitherto jealously reserved for skilled men.

Until May 1940 neither the supply ministries nor the Ministry of Labour had been willing to take the responsibility for promoting dilution, and very little had been done. Bevin at once ended this dispute, made his own Ministry responsible and flung himself into the role of a missionary preaching self-help to industry.

It was an unpopular role and he encountered every sort of resistance.[1] Dilution roused all the conservatism and suspicion of the craft unions. To take a notorious example: in the sheet-metal shops in the Midlands—an old but dying craft threatened by the introduction of the power press and automatic tools—the men steadily refused to accept the introduction of women throughout the war. This was an exceptional case, but the attitude of mind it represented was widespread, especially in unions like the Boilermakers' which had been badly hit by unemployment between the wars. As soon as he became Minister, Bevin began negotiations with the unions to secure agreements covering dilution and the employment of women. These agreements, however, did not prevent constant disputes arising about their interpretation in individual factories, particularly over piece rates.

It was not only the men who resisted dilution. Working under pressure to fulfil current contracts, hard-pressed managements were in no mood to listen to schemes for reorganisation which in the short run would interrupt production, and stir up trouble with their workpeople. The Minister of Labour complained that the efforts of his inspectors to push dilution met with almost continuous opposition, particularly when the result might be the transfer to another firm of the skilled men freed by such methods of reorganisation.

His difficulties were increased by the fact that managements frequently had the support of the supply ministries to which they were under contract. Beaverbrook, as Minister of Aircraft Production, openly encouraged the aircraft firms to grab every skilled man they could and hold on to them at all costs. Under no circumstances, he insisted, would he agree to the transfer of skilled men outside the aircraft industry and justified his attitude by his success in producing more planes even if it meant an uneconomical use of manpower. When Bevin declared that this made it impossible for him to plan the supply of labour, Beaverbrook retorted that if Bevin did his job properly there would be no shortage.

It took a lot of persistence on Bevin's part to overcome the

[1] For a detailed discussion of dilution and the problems it involved, see Mrs. P. Inman, *Labour in the Munitions Industries* (1957), cc. 2–3.

scepticism of the supply ministries and persuade them that it was in their own interests to encourage dilution. When he started inter-departmental meetings in September 1940 to work out a scheme for co-operation between his own and the other ministries concerned, Beaverbrook refused to allow his M.A.P. officials to attend.[1] Agreement was first reached with the Ministry of Supply in October 1940 and was followed in January 1941 by similar agreements with the Ministry of Aircraft Production and the Admiralty. Many more months' effort was needed, however, to translate these agreements at the top into effective action in the factories.

With the unemployment between the wars, as Bevin frequently pointed out, managements had come to value labour cheaply. If more skilled men had been needed, a firm had only to ring up the employment exchange: now they had to accustom themselves to a situation in which labour, and especially skilled labour, was going to be more difficult to secure than any other commodity.

A similar reversal of attitude had to take place about training. Before the war it had hardly seemed worth the trouble to train more skilled workers; now that industry was clamouring for them, Bevin found it hard to persuade managements to spare skilled men to act as instructors, less-skilled men to learn, or machines on which to train them. He started another campaign of pressure and propaganda, supplementing industry's own efforts by expanding Government Training Centres (38 were open by 1941) and sponsoring courses in technical colleges. As semi-skilled labour too grew scarce, the supply of recruits for the training centres fell away and such spare labour as there was found its way direct to the factories. Since most firms of any size had by then been converted to the need for training, Bevin was ready to cut down the Government's schemes and by May 1942 the number of training centres had been reduced to twenty-five. None the less, between 1940 and 1945, 300,000 men and women completed engineering courses at the centres or in technical colleges, a sizeable addition to the country's resources of skill.

Bevin did not succeed in overcoming the engineering industries' shortage of skilled men before it was swallowed up, during 1941, in the general shortage of all kinds of manpower. Indeed the air raids in the closing months of 1940 threatened to make the problem worse by increasing the number of men leaving factories in the worst-hit

[1] Inman, p. 54.

areas and looking for jobs elsewhere. Beaverbrook, in another
exchange of letters, demanded that the Minister of Labour should
revive the Leaving Certificates[1] which had roused stubborn labour
opposition in the First World War. Bevin refused, but it was clear that
some answer would have to be found to the problem of rapid turnover in skilled labour and early in the New Year he produced his own
solution in the Essential Work Orders.[2] The subsequent history of
war production, however, was fully to vindicate his belief that the
only effective way of providing more skilled labour was by the long-
term measures of training, dilution and the employment of women,
all of which he had started to press as early as the summer of 1940.

8

Other industries, scarcely less important than engineering, required
emergency action almost at once. Agriculture and coal-mining, for
example, were already suffering a loss of workers to better-paid jobs.
To stop this drift Bevin inserted a clause in his Restriction on Engagement Order (5 June 1940) forbidding other industries to engage men
previously employed on the land or in the pits. But he was not content
to let the matter rest there. The wages in both agriculture and mining
were too low. In the former case he only consented to act at all after
the Agricultural Wages Board agreed to raise the minimum wage for
farm labourers (which, in some counties, was as low as 32 shillings a
week) to 48 shillings. Mining was a different story. The defeat of
France meant the loss of the continental market for British coal and
widespread unemployment in the mining districts. The miners'
leaders and M.P.s put strong pressure on the Government to drop
any restriction on the men seeking work elsewhere and Bevin was
persuaded, against his own better judgment and the advice of the
Secretary of Mines, Dai Grenfell, to allow men to leave the pits either
to join the Forces or find better-paid work in munitions or the construction industry. This was to prove a bad mistake and one which

[1] A section of the Munitions of War Act, 1915, prevented a munitions worker
who had left or lost his employment from obtaining another job within six weeks
unless his previous employer furnished a Leaving Certificate that he had left with
the employer's consent. Under a compulsory call-up, it was argued that this gave
an employer the right to send a man into the Army.

[2] See below, p. 57.

Bevin was bitterly to regret later when shortage of mine-workers became the most intractable of all the problems of labour supply.

If there was an industry whose idiosyncrasies he knew like the back of his hand, it was the docks. In June Bevin signed an order requiring all dockers to register,[1] and prepared a scheme for the transfer of men in case it should become necessary to divert shipping from the East and South coasts to Western ports. In July similar provisions were made for registering and, if necessary, transferring workers in the ship-building and repairing industries.

When the switch to the west coast ports had to be carried out the confusion was increased by bombing attacks and produced dangerous delays in clearing cargoes. The Prime Minister minuted that two-fifths of "the decline in the fertility of our shipping" was due to the loss of time in turning round ships and demanded that something should be done. In mid-December a Cabinet committee was appointed to improve the clearing of ports and the turn-round of ships by emergency measures. Bevin was the only member who knew the docks at first hand and had a clear idea of what to do. On his recommendation, Regional Port Directors with overriding powers were appointed for the Clyde and Mersey, and (over the protests of the Minister of Transport) were made responsible for the employment of dock labour in their ports. Bevin was certain that the dockers would never accept control if it was put in the hands of the employers. At the same time he proposed to end the casual system of employment and constitute the registered dockers in each port into a permanent labour force with a guaranteed minimum wage of £4-2-6 for a 44-hour week. This was his answer to the problem of keeping the dockers standing by when ships failed to arrive and no work was available. In return the dockers would have to be prepared to work when and where required to get the ports in the area cleared and would receive payment by result above the minimum wage.

Bevin called in Arthur Deakin and Dan Hillman[2] from his own union, the T.G.W.U., just before Christmas 1940. He told them that if the dockers would not accept the scheme, the only alternative was to call them up and clear the ports with mobile battalions. Whatever

[1] Dock Labour (Compulsory Registration) Order (S.R. & O. 1940, No. 1013), 18 June 1940.

[2] Hillman had helped Bevin organise the Bristol dockers and carters thirty years before at the beginning of his trade-union career.

had to be done, the Government meant to cut by half the time it took
to turn round ships. The three men made a number of improvements
in the scheme; then Bevin sent for the other parties involved, including
the leaders of the Glasgow dockers who had broken away from his
leadership in 1931, and obtained their agreement too. In this way he
improved the turn-round of ships and at the same time pushed
through the first stage of a reform to end casual labour in the docks
for which he had fought, often against the dockers as well as the
employers, for thirty years.

The scheme, which came into operation in March 1941, was anything but perfect: no scheme dealing with dockers, as Bevin well
knew, could ever hope to avoid trouble, and in later versions for other
ports he agreed to transfer the responsibility for employing the men
to a National Dock Labour Corporation. But it worked. After the
spring of 1941 congestion in the ports was never again a major
problem, and the principle of decasualisation remained as the basis
of the post-war reorganisation of the industry.

The other major industry with a tradition of casual labour was
building and it was no better organised than the docks to meet the
demands which war made on it. Everything that had to be done in the
summer of 1940 required buildings: airfields, army camps, coastal
defences, factories, not to mention the repair of bomb damage once
the Blitz started. The total of government orders far exceeded the
capacity of the industry to carry them out.

The first need was to get rid of obstacles to the mobility of building
labour created by different local rules and conditions of employment
as well as different wage rates. In June 1940 Bevin secured a
Uniformity Agreement between the unions and contractors providing
uniform wages and hours of work on government jobs wherever they
were situated. This was only a beginning.

It seemed obvious to Bevin that the industry would never be able
to cope with the demands made on it as long as there were twenty
different government departments each trying to run its own
building programme and clamouring for priority. What was needed
was a single programme with centralised control in the hands of a
single ministry, a proposal he had made in his first memorandum to
the Cabinet and for which he fought hard in the face of departmental
particularism. The autumn air raids settled the matter: licensing was
introduced in September and a Ministry of Works created in October.

The new ministry, however, would have made little headway in asserting its authority against such powerful departments as the War Office or the Admiralty if it had not been for the Minister of Labour, who fought its battles for it with a vigour which impressed even Whitehall.

9

The trouble from which the building industry suffered—too many demands on too few resources with no one in a position to settle which was to have priority—afflicted the whole of British war production. Improvisation was all very well for a few weeks, but, as Bevin insisted against Beaverbrook, its great advocate, industry could not be run on an emergency basis for more than a short period. Somebody had got to get hold of the production effort at the top and introduce some order into the confusion created by each of the supply departments trying to expand its programme without any regard to the limits on the resources available. No minister had a stronger motive for pressing this than the Minister of Labour, for until he could get some idea of the *total* demand for labour, he had no idea of what he would have to provide and could not begin to plan the allocation of manpower between the Services, the different branches of war industry and the needs of the civilian population.

With this in mind, he had proposed the creation of a Production Council and secured terms of reference which would enable it to act as "the pivot of the production machine". To assist it in its work, the Production Council took over four existing Cabinet committees (for manpower; materials; works and buildings; and transport) and added two more, to deal with priorities and industrial capacity.

On paper this was impressive, and the Production Council managed to iron out a number of difficulties during its brief existence. Its most important job was to devise and administer a system of priorities for war production. Pooling of stocks was discussed and a system of licensing for civil building adopted. The Industrial Capacity Committee investigated factories and plant which could be brought into war production, the Joint Materials and Production Priority Committee established a workable system for the allocation of raw materials.

All this was useful enough but fell far short of what Bevin had looked for. The Production Council was no more than an inter-departmental committee: the division of responsibility for the country's rearmament between three autonomous ministries, each preoccupied with its own programme, remained unchanged. Beaverbrook showed what he thought of the Council by never attending after its first meeting and defying any attempt to limit his freedom to take and keep everything he could lay hands on for the production of aircraft. Morrison as Minister of Supply was more co-operative, but with a weak chairman in Greenwood, it was largely left to Bevin to put some life into its proceedings.

The Production Council, it soon became clear, was not the answer. But before its replacement at the end of the year it produced two decisions which opened the way to more effective planning.

The first was the abandonment of the system of priorities in favour of the allocation of scarce resources. The War Cabinet in the last week of May 1940 had issued a direction that priority should be given to weapons which could be used against the enemy within three months, with a special priority for fighters, bombers and anti-aircraft equipment. Even before the directive was published the Production Council had to revise it in order to fit in a second category of weapons without which the Army had little chance of repelling an invasion[1] and within two months the Ministry of Supply and the Admiralty wanted to see the original directive scrapped. Beaverbrook alone, as Minister of Aircraft Production, stood out for it, seeing in the special priority for aircraft a charter of exemption from all controls. The protagonist on the other side was Bevin who demanded to know how it was possible for him to build up a labour force of skilled workers and distribute them to the best effect so long as the aircraft firms— encouraged by Beaverbrook—could flourish the priority directive as a licence to obtain and hang on to all the men they wanted whether they could keep them fully employed or not—and whatever the shortages they created elsewhere.

It was a long drawn-out battle, but the odds were on Bevin's side. More and more items, not only weapons but machine tools, electrical and radio equipment, had to be added to the priority list, and the longer it grew, the less sense it made. In September 1940 the Pro-

[1] These were anti-tank weapons, field artillery, tanks, machine-guns and ammunition, to which priority 1(b) was to be given.

duction Council put up a case to the War Cabinet for adopting the same solution as in the First World War. Instead of giving any department an overriding priority for certain products, each was to be allocated a guaranteed figure for materials (such as steel), for industrial capacity and for skilled manpower. The department had then to see that its production programme was carried out by the firms working for it within these allocations.

The change from one system to the other was not at all clear cut. The War Cabinet's ruling was anything but precise and as late as January 1941 Beaverbrook was insisting that he would cling to his labour priority, however regrettable the inconvenience to other departments. He was fighting a losing battle, however, and by the spring of 1941 (when the need for aircraft was no longer so desperate) it was settled in favour of allocation right across the board. Bevin had not waited so long. In November 1940 he instructed his local officers to treat priority lists as a general guide and use their common sense in filling labour vacancies. He did not in fact think much of allocation, regarding it as still too clumsy a method of dealing with manpower. His own preference was for identifying particular bottlenecks in production and then giving a labour preference to specific factories in order to remove them. However, the ending of priorities allowed his Ministry much greater freedom in controlling labour supply and, so far as materials and industrial capacity were concerned, Bevin needed no convincing that allocation was a better way of controlling scarce resources than priorities.

The Production Council's other important initiative followed a letter which Bevin wrote to its chairman in August 1940, pointing out that there was still no estimate of the labour which would be needed to carry out the programme of war production. The earlier figures submitted by the Wolfe Committee were out of date, but no one had yet sat down to add up the total demands on manpower likely to be made by the Services, the supply departments and civilian needs. Only when this had been done would it be possible to measure them against the total resources available and see how to make the two sets of figures match.

The Production Council adopted Bevin's proposal and handed over the job to a committee for which Bevin secured Sir William Beveridge as chairman. This was virtually its last decision of importance: between the middle of August and the end of the year the Council only met three times, and the last two occasions were wholly taken up with the report of Beveridge's committee. Criticism of the Council (with much of which Bevin privately agreed) was growing and it was clear that the Government would have to find some other means for carrying out the job which the Production Council had failed to do. In fact, however, before its disappearance it provided the Government, through Beveridge's report on manpower requirements, with what was to become the basis for all wartime planning, the first attempt at a manpower budget.

CHAPTER 2

Bevin's Labour Policy Takes Shape

I

To BE a minister, even the head of an important department, did not carry with it any right to a seat in the War Cabinet. This was a smaller body than the peacetime Cabinet and although it was reinforced, as need arose, by other ministers brought in for particular purposes, its six or eight permanent members constituted the inner circle of government. Much of its work was delegated to committees and the Prime Minister clearly exercised an authority of his own, of which Churchill made full use in his conduct of the war. None the less, the War Cabinet was the heart of the coalition and through its committees and secretariat the effective focus of power in wartime Britain.

In the autumn of 1940 Churchill felt his position strong enough to carry out a reconstruction of the War Cabinet, the way to which was opened by the resignation of Neville Chamberlain, who was mortally ill, both from the Government and from the leadership of the Conservative Party. The danger of a German invasion was at its height at the beginning of October and Churchill assembled in the War Cabinet the men on whom he most relied in what might well prove to be the climax of the war for Britain. The newcomers were Ernest Bevin, Sir John Anderson (Lord President of the Council in place of Chamberlain) and Sir Kingsley Wood (Chancellor of the Exchequer). Beaverbrook had already been brought in during August, and Eden replaced Halifax when the latter was sent as ambassador to Washington in December.[1]

Churchill's invitation surprised Bevin. "Of course, I'm very new at this game," he told Dalton, "and I didn't know what to say when

[1] The existing members who remained in the Cabinet were the Labour Party leaders, Attlee and Greenwood, and Churchill himself.

the P.M. asked me last night. But I thought it would help the prestige of the trade-union movement and the Ministry of Labour if I went in. No one has ever put the Ministry of Labour in the forefront like this before."[1]

For a man who had only been in politics five months it had been a rapid rise: once at the top, however, Bevin never stepped down again until his death more than ten years later. He enjoyed, to an unusual extent, the confidence of both the Prime Ministers under whom he served, and there were not many occasions when he really made up his mind to press hard for the adoption of a policy and failed to get it accepted. This was obvious enough in the Labour Cabinet of 1945–50: it was hardly less true of the wartime coalition. Bevin's influence, of course, depended on knowing when, and when not, to press: he did not make enough use of his position to satisfy the more militant members of the Labour Party. His interest and influence, however, were by no means restricted to matters which directly concerned his Ministry, but extended over the whole range of social and economic policy and, on occasion, into foreign policy as well. His membership of the War Cabinet gave him the chance to learn what was happening outside his own departmental field and the opportunity to raise anything that caught his attention in the Cabinet or one of its committees. He had direct access to the Prime Minister and to the Cabinet Secretariat, an access which was certainly not available to every departmental minister, and this fact, together with the standing which his membership of the War Cabinet gave him in Whitehall, also made it easier for him to carry out his policies as Minister of Labour.

Bevin soon found his feet as a Cabinet minister: the House of Commons was a different story. A seat had been found for him in the London constituency of Central Wandsworth. The sitting Labour member had just moved to the Upper House as Lord Nathan and the new Minister of Labour was returned unopposed. He took his seat at the end of June 1940 and made his maiden speech on 3 July, introducing the second reading of the Unemployment Insurance Bill in a sparsely filled House.

The first big debate in which he took part was in August when Greenwood made a statement on the Government's economic

Hugh Dalton: *The Fateful Years* (1957), p. 358.

policy.[1] Greenwood's speech irritated those critics who were already saying that the Government showed nothing like the grasp of the economic side of the war that it displayed in the conduct of military affairs. Shinwell opened the attack and other speakers followed in the same vein: there was no plan and no authority, too many committees and too little co-ordination.

At this point Bevin made an unexpected intervention. He had not meant to speak, but he was roused by the accusation that there was no planning of labour supply, and particularly by Shinwell's remarks about unemployment in the mining industry. When Shinwell tried to say that Bevin misunderstood him—he did not deny what he had achieved, but dealing with labour problems from hand to mouth was not a plan—Bevin turned on him:

"The suggestion was made that miners were being thrown out of work and nothing was being done...."

(Shinwell again tried to interrupt)

"The hon. gentleman may get annoyed with me. I am not used to this House but I am used to appreciating facts.... Let cynical bitterness and discontent get into the hearts of my army and we have lost the war.... I will not allow the miners of this country to feel for one moment that I am neglecting their welfare. In this war I have been as strong as most men on the platform and as bitter in debate as any man could be, but the time for that is past and I suggest that others might use caution in their language."[2]

Bevin's intervention had an effect. Clem Davies spoke of the "real pleasure to see him standing four square at that Box and dealing with the situation". But no one thought he had answered Shinwell's criticisms. Charles Brown, the member for the mining constituency of Mansfield, summed up the feeling in the House when he said:

"The Minister of Labour is a very skilful debater and yet all he did was to beg the question.... The Minister was very much concerned to show that he had what he called a 'plan' for labour. But it was not a plan at all. It was only a method of moving labour about when difficulties arise in particular places. It had nothing to do with the economic planning of labour resources.... Of course, when the Minister of Labour speaks, one feels that things cannot go wrong. He is very confident and he has every right to be confident. Few people know labour conditions in this country better than he does. When he

[1] House of Commons Debate on the Consolidated Fund (Appropriation) Bill 7 August 1940. Hansard (5th series), Vol. 364.
[2] Ibid., cols. 304–5.

sits down one feels that the last word has been said on the problem and that things cannot go wrong while they are in his hands. But, unfortunately, they do."[1]

This August debate was the first open expression of dissatisfaction with the Government's handling of economic questions and of Bevin's refusal to use his powers of compulsion. And Bevin's defence of his policy in the House, then as later, failed to convince his critics that they were wrong and he was right.

Many members, without distinction of party, admired his strength of personality and were impressed by the confidence which he conveyed; most of them were ready enough to be convinced, but they were disappointed by his parliamentary performance. The natural ease with which he spoke to audiences outside the House deserted him in the Chamber. There he was awkward, either reading out an official brief in pedestrian fashion, or when he spoke impromptu, often losing the thread of his argument and failing to catch the mood of the House. The formalities and procedure of parliamentary debate irked him. Criticism which a practised parliamentarian accepted as the common form of debate, without allowing it to ruffle his temper, goaded Bevin to anger. He took it too personally and was too fierce in rebutting it. As Dick Stokes remarked on this occasion: "The Minister was a little hard on the member for Seahaven (Shinwell) in flaying him alive for saying there was no plan." These were faults which Bevin was slow to overcome: it was a long time before he felt at home in the House or spoke with the authority he soon acquired outside it.

2

While the Prime Minister's and Foreign Secretary's wartime journeys took them to the United States, Russia and the Middle East, Bevin's took him to industrial Britain, to the Midlands, Lancashire, the North-East, South Wales and Clydeside. He visited all these in 1940, some of them more than once. They were grim, ugly towns, most of them, the source of Britain's industrial strength, where the war could be as surely lost as on the battlefields if production fell, if relations between managements and men broke down or the Govern-

[1] Ibid., col. 338.

ment failed to keep both sides' confidence. This was Bevin's responsibility, and he had no illusions about the troubles it would involve him in before the war was won.

Wherever he went he met groups of industrialists and talked to as many trade unionists as could be packed into the largest local hall. Both were critical audiences with plenty of awkward questions to ask and no patience with smooth answers. They got none from Bevin. He talked to them as bluntly as they did to him, impressing on them that if things were wrong it was no good expecting the Government to put them right: they had to think and act for themselves. At a lunch in Cardiff in November 1940 he told his audience of business men that the country could not wait four years, as it had in the earlier war, to reach its maximum production.

"Do not worry about what it costs. . . . You can easily rebuild wealth, but you cannot create liberty when it has gone. Once a nation is put under another, it takes years and generations of struggle to get liberty back. . . .
"Let me mention your ship-repairing, and I thank you for the improvement that has been shown. You people are losing too many man-hours—not because of the men. I hear firms saying that they will change men only within the firm. You are not firms, you are merely a branch of the national effort. You have your joint managements. Pool your brains and capacity, and utilise these men. . . .
"You cannot afford to lose a riveter or a boilermaker for a day when the Navy is struggling to bring food in. You must co-operate and you must find a way out. The same with the docks. That is the problem and you cannot solve it by regulations drafted in London."[1]

Bevin was well aware that he had suspicions and prejudice to overcome in his new role of Minister: many industrialists did not take easily to the idea of a trade-union leader telling them, in the name of the Government, what they ought to do. There were plenty of critics ready to say that he was favouring the workers at the expense of the employers; this was one reason why he attached so much importance to joint consultation with the representatives of both interests before issuing any order. On the other hand he was not prepared to repudiate either his origins or his beliefs in order to appease opposition: no one would have believed him if he had.

In fact, he saw no contradiction between the views he had expressed before 1940 and the part he had now to play as a minister. It was the

[1] Speech at the Park Lane Hotel, Cardiff, 23 November 1940, reprinted as a pamphlet.

Government's view of national interests which had changed, he argued, not his. War forced the State to remedy the waste of national resources, above all of human resources, which had roused him to protest in the 1930s. As he remarked to the Society of Labour Candidates:

"Immediately a nation is involved in a great crisis of this character it has to act collectively, and that brings into play great social forces. Individualism is bound to give place to social action, competition and scramble to order. . . . There is no other way by which a nation can save itself. . . . Thus will we defeat Hitler. We can at the same time create conditions upon which a new advance is possible."[1]

Bevin accepted as the basis of the coalition that major measures of a controversial character like nationalisation must be excluded during the war. But in joining the coalition he had not surrendered his conviction that there could be no more question of returning to the social and economic views of the 1930s after the war than to the policy of appeasement; both were equally discredited. The first task in a future peace settlement must be to lay firm political foundations for democracy. But the political foundations would never be secure, if there was a failure to recognise the social implications of democracy. The expansion of freedom could not end with political liberty: unless an attack was made on unemployment and poverty, liberty would be a precarious attainment. "Remember, unemployment has been the devil that has driven the masses in large areas of the world to turn to dictators. . . . What were they striving for but to find in the world, somehow, social security?"

How was this hunger for security to be satisfied? Bevin's answer, put to a Rotary Club audience in Bristol in exactly the same terms he used in addressing the T.U.C., was to make service to the community and social security, in place of individual profit, "the main motive of all our national life".

"I am afraid at the end of this war," he added, "unless the community is seized with the importance of this, you may well slip into revolutionary action—I do not mind a revolution if it is well directed and to achieve a great purpose, but what I am terrified of is a blind revolution of starving men, undirected, with no objective.

[1] Speech at the Annual Conference of Labour Candidates, Euston Hotel, London, 20 October 1940, reprinted as a pamphlet.

"You will have at the end of this struggle men demobilised from the forces, men taken from industry; tremendous readjustments will have to be made, and you will not solve it this time by putting a few shillings on the dole for a short period and tiding over by inflating your money for a few years and then crashing down as you did after the last war. The very training of your airmen today and those in the other Forces—remember it is a mechanised Army—is producing an outlook and an intensification of mentality that will never submit to the neglect that the untutored masses of the past had to under-go. . . ."[1]

Hitler had solved the problem of unemployment by rearmament and war. The only effective answer to Hitler was to show that a new economic order could be built up by peaceful construction. "You can spend millions of money now on the weapons of destruction; then may I not ask that the creative brains of the scientists, the works managers, the ability of the operatives be directed to this great objective?"

3

It was a mark of the changes which were already taking place in Britain that a member of the War Cabinet could talk like this and not bring down a storm of criticism round his head. The Cabinet, and Bevin in particular, were more generally criticised at the end of 1940 for not being radical enough.

"Mr. Bevin," *The Economist* wrote on 23 November 1940, "does not wish to 'coerce' employers; he certainly does not wish to 'coerce' workers; so he leaves to employers and workers the impossible task of marshalling and deploying themselves for victory. While this is so, the Minister has no plan."

In the House of Commons, one speaker after another got up to give instances of failure to mobilise manpower and industrial capacity effectively.

"When are these compulsory powers going to be used?" Clem Davies asked. "The complaint we have to make is that there are too many appeals instead of saying 'We have taken compulsory power and we are going to use it.' "[2]

Bevin's policy rested upon an accumulation of experience which he found it hard to put into words.

[1] Speech at the Bristol Rotary Club, 20 November 1940.
[2] House of Commons 27 November 1940, Hansard, Vol. 307, col. 253.

"Whatever may be my other weaknesses," he declared, "I think I can claim that I understand the working classes of this country. I had to determine whether I would be a leader or a dictator. I preferred and still prefer to be a leader."[1]

Quoting from an official German circular which showed that the enemy was having the same troubles and doing no better in overcoming them, he concluded:

"That leads me to the belief that working men are very much the same all over the world and that if you apply compulsion, you do not get good results. . . . One of the big contributions to the French disaster were the industrial decrees which failed to get enthusiasm in the workshops behind the production effort."[2]

The Economist dismissed the Minister's preference for leadership rather than dictation as "an almost perfect example of the politician's technique of diverting discussion from real issues to emotional prejudices".

"Let us grant Mr. Bevin all his grand generalities; let us give him his reading of British psychology; let us admit he understands the working man. And when we have done all that, let us ask him if he would kindly address his mind to the problem of manpower."[3]

When the debate was resumed early in December, Sir John Wardlaw-Milne demanded:

"Do the Government really believe that this tremendous change-over in the lives and occupations of the people of this country can be carried through on a voluntary basis? . . . I believe that the Minister of Labour is probably better fitted than any other man in the House or out of it to secure what we want by a voluntary system, but the test for him, I fear . . . is whether he is successful. . . . If we can get results on that basis, well and good, but we are not getting them now. We are not fully using our industrial strength and we must use it if we are to win the war."[4]

After the November debate, an informal alliance was formed between Emanuel Shinwell on the Labour back benches and Earl

---

[1] Ibid., col. 284.
[2] Ibid., cols. 287-8.
[3] *The Economist,* 30 November 1940.
[4] House of Commons, 4 December 1940, Hansard, Vol. 367, col. 610.

Winterton on the Tory,[1] in order to press the Government, and the Minister of Labour in particular, on the points which had been raised but not answered. Wardlaw-Milne, Hore-Belisha and Clement Davies, all experienced House of Commons men, gave them support and between them they subjected Bevin to sharp criticism in the next twelve months.

Bevin's handicap was not only his unfamiliarity with the procedure of the House and his awkwardness in striking the right note, but his difficulty in putting into words the philosophy which lay behind his actions at the Ministry of Labour.

When Bevin talked about "voluntaryism" (his own word for it) he meant something more than the traditional trade-union opposition to industrial conscription. He started with the question: how could a country with the democratic institutions of Britain hope to match the degree of organisation already achieved in Germany? Not, Bevin anwered himself, by discarding its own traditions and trying to copy the totalitarian methods it was fighting against: this was the mistake of those who wanted to treat the whole nation in wartime as if it were an army and organise it on military lines. The right way was to stick to the basic principle of democracy, government by consent, and rely on the willingness of people in an emergency to make greater sacrifices willingly than they could be dragooned into making by compulsion. This, he believed, could be more effective than dictatorship, provided that besides appealing to people, you took practical steps to remove the obstacles which inhibited or impaired consent.

What were these obstacles? So far as industrial relations were concerned, Bevin saw three. One was inadequate communication, as a result of which industrialists and workpeople alike complained that they did not know what the Government's plans were or what was expected of them. The answer to this, of which Bevin was to give plenty of practical examples, was for ministers not to sit in Whitehall but to get out and tell people what the Government was about and what it wanted them to do.

A second obstacle was inadequate participation, the failure to give people a chance to contribute to and feel involved in decisions. The answer to this, he believed, was to be found in extending, at every level from central government to shop floor, the practice of con-

[1] This incongruous but effective alliance was neatly christened "Arsenic and Old Lace", after the play of that title which was running in London at the time.

sultation, consultation on equal terms, with all the interested parties represented; consultation about the questions that really mattered, not trivialities; and consultation before the decisions were made, not afterwards.

The third obstacle was the most difficult of all to overcome, the legacy of an industrial outlook which treated workpeople as so many hands, indifferent to their well-being or to any obligation beyond paying them the lowest wages possible, an outlook which had produced a split society and an alienated labouring class. To hope that this could be overcome quickly, if at all, would indeed have been Utopian, but Bevin was determined to make a start by insisting to government and employers alike that, far from war being a good reason for bothering less about conditions under which people worked and lived, if they wanted to get the most out of them, they must respect their right, whatever was demanded of them, to be treated as human beings.

This was the background to Bevin's policy of "voluntaryism". It was a policy which he certainly did not think of applying to one side of industry only. He attached as much importance to securing the voluntary co-operation of the employers as of the unions, and perhaps most of all, to continuing and extending the voluntary co-operation of the two together in settling matters of dispute between them by collective bargaining. Bevin applied to the practice of voluntary negotiation in industry the same argument as he did to government by consent in society as a whole: the emergency of war required not its curtailment but its development as a source of strength, not (as the admirers of totalitarianism believed) a weakness.

Bevin had too much experience of organising men to suppose that the country would get through the war without having to apply some degree of compulsion to labour. According to Sir Godfrey Ince[1], he believed from an early date that this would have to be applied to women as well as men. "Voluntaryism" did not rule this out, did not mean that the Government had to rely solely on appeals for volunteers: what it did mean was that, instead of starting with a full-blown system of industrial conscription, you began from the opposite end, demonstrating to people that it was not only necessary but fairer to employ compulsory powers and keeping their use to a minimum. When the time came, Bevin proved that he was quite prepared to

[1] See below, p. 126.

issue orders if he thought this necessary; but when he did, it is not playing with words to say that it was upon a basis of consent, and that consent was more willingly given because he had plainly exhausted the possibilities of purely voluntary methods first.

4

If Bevin had succeeded in explaining what he meant by "voluntaryism", he would not have silenced all his critics in the House—it would only have provoked some all the more—but he would have reassured those members who felt that he was doing no more than deal with problems hand-to-mouth without a clear idea of what he needed to do or how to do it.

The crux of the matter was timing: how soon should he begin to tighten up the controls on labour supply, how long could the Government afford to wait? Whether the criticisms in Parliament had any effect or not, during the Christmas (1940) recess Bevin did some hard thinking about his problems. The appeal which the Government had made for an extra effort from all workers during the summer had produced remarkable results. To take the single most important figure, the output of fighter aircraft: in March 1940, the number delivered to the R.A.F had been 177, in April 256; in June, it shot up to 446, in July touched a peak of 496, and for the next four months ran at 476, 467, 469, 456. R.A.F. Fighter Command actually had more planes in the autumn, after its heavy losses of the summer, than it had at the beginning of the Battle of Britain. If much of the credit for this must go to Beaverbrook, it showed that Bevin's faith in the willingness of the industrial worker to rise to the occasion had not been misplaced: it was not compulsion which got men to work the hours they did in the summer of 1940.

But there were other sides to the story, in particular the high rate of turnover, the number of men leaving their jobs to look for better pay or to get away from the bombing. This could not be allowed to continue. What finally convinced Bevin that he would have to make more use of his powers as a minister was the report of the Manpower Requirements Committee which he had persuaded the Production Council to appoint. Its chairman, Sir William Beveridge, assisted by a young Oxford economist, Harold Wilson, who acted as secretary of the committee, cross-examined each government department in turn on its programme, converted the answers into terms of man-power and added up the total to present, for the first time, an estimate of what was going to be needed. The conclusions were startling.

If the three Services (excluding Civil Defence) were to reach their targets, they would have to take in an additional 1,700,000 men and 84,000 women in the sixteen months between 1 September 1940 and the end of 1941. This would bring them to a total strength (allowing for normal wastage but not for heavy casualties) of just under 4,400,000 men and 140,000 women. Such a target could only be reached by withdrawing close on half a million men from reserved occupations in the munitions industries.

But the men and women to be drafted into the Forces by the end of December 1941 could only be equipped if, before that, by the end of August 1941, the munitions industries themselves had expanded their labour force by an additional million and a half above the three and a half million already employed.

How could this paradox be resolved?

The only possible answer, Beveridge argued, was by bringing women into industrial employment on a scale which no one had so far contemplated. In munitions alone they would have to provide more than half the numbers (to be precise, 859,000) in the total increase which was called for. Another three-quarters of a million would have to take the place in other industries and services of the men who were either called up for military service or transferred to munitions. All in all, well over one and a half million women would have to be found for the Auxiliary Services and industry from among those who were either occupied in their own houses or in such employment as catering and domestic service. This was a formidable proposition in itself—quite apart from the obstacles to be overcome in persuading employers and unions to agree to the employment of women on such a scale.

Moreover, as the Committee pointed out, the totals of men and women necessary, startling though they were, concealed many of the real difficulties that would have to be overcome. What had to be done was to fit individual men and women in their millions, with varying skills and living in different places, into particular jobs in particular factories scattered up and down the country, often far away from their homes.

To complicate the problem still further, the total numbers needed could only be absorbed if the Ministry of Labour was able to maintain the right proportion of skilled to unskilled labour. This was true of the Services as well as industry: without a nucleus of skilled men among their recruits they could not hope to achieve the planned expansion of highly mechanised forces. Training would increase the supply, but it would also take time. The immediate shortage could only be met, the Committee reported, if dilution in industry was carried to the point where three skilled men were able to do the work of four.

For the first time, Bevin and his advisers were confronted with the magnitude of the task before them: if there was no general shortage of manpower yet (as distinct from a lack of skilled workers) there could be no doubt, Beveridge concluded, that it would appear in the course of 1941, or that it could only be met by a greater use of the Minister's powers of direction.

Bevin was not prepared to accept all the suggestions Beveridge made for dealing with the problem, but he recognised the force of the evidence he had assembled. By the middle of January (1941) he had made up his mind: under the title Heads of the Labour Policy he presented the Cabinet with a paper which showed the impression Beveridge's report had made on him.

Reduced to its essentials, Bevin's paper contained three proposals.

The first was to provide the Minister of Labour with the power to schedule any factory or undertaking as engaged on "national work". Once this was done, no one employed on such work could either leave or be dismissed without the Minister's consent. In this way Bevin hoped to provide much-needed protection for the labour force of industries essential to the war effort and to put a stop to the turnover which had been disturbing it. His previous objections to re-introducing the Leaving Certificate of the First World War were met by making the Minister's consent to registering an undertaking dependent on three conditions: he had to be satisfied that the terms and conditions of employment were not below recognised standards, that there were suitable arrangements for welfare and that proper training facilities were provided.

The second proposal was to start on the registration of the population by age groups. All men over the age of military service (forty-one) and women over the age of twenty were to be required to register as a first step towards directing them into war work of one kind or another.

Finally, Bevin proposed to make more selective the system of reserved occupations in which men were held back from military service. Henceforward it would not be enough to be employed in a reserved occupation: the question now to be asked was *where* a man was pursuing his occupation. A Register of Protected Establishments was to be drawn up and skilled men employed in these reserved at a much lower age than those in unprotected firms. This was an ingenious device which promised at one and the same time to release more men for the Forces, provide additional protection for the labour force of firms engaged on essential work and put a premium on transferring to such work.

The proposals contained in the Heads of Labour Policy paper were approved without modification by the War Cabinet on 20 January (1941) and presented to Parliament the following day. They still fell short of what Beveridge and others thought necessary. None the less, with the first, translated into the Essential Work Order, and the second, translated into the Registration for Employment Order, the Minister of Labour had armed himself with the two essential controls on which his Ministry was to rely in carrying mobilisation further than anyone yet believed possible.

5

At the same time that Bevin was turning over in his mind the new steps to be taken in labour policy, he was involved in discussions with Churchill and the other members of the War Cabinet about possible changes in the machinery of government on the economic side. The Production Council had plainly failed. Responsibility was too divided to be effective. What was needed was the concentration of the power to take decisions—a view as strongly held by Bevin and other ministers as by the critics.

The Prime Minister, however, was not prepared to accept either of the two solutions widely canvassed at the time: the appointment of a minister of war production with the same power to deal with economic questions that the Minister of Defence had in the conduct of military operations, or the creation of a small War Cabinet relieved of departmental responsibilities. A minister of production, he argued, could no more exercise the authority required than could

49

the minister of defence unless he was at the same time Prime Minister. As for a War Cabinet whose members were relieved of departmental responsibilities, this, he maintained, would lose the power of effective action.

Churchill proposed instead to create two new Cabinet committees—a Production Executive and an Import Executive—with membership restricted to the ministers who were most immediately concerned and who had the power to act. A third committee, under the Lord President of the Council (Anderson), was to keep watch on the large questions of economic policy which, as the Prime Minister recognised, "raise the most difficult and dangerous political issues. These issues were not solved in the last war and I cannot pretend they have been solved in this."[1]

Bevin was made a member of the Lord President's Committee, and chairman, in place of Greenwood, of the Production Executive. This took over the responsibility for war production which the Production Council had failed to exercise, and with it responsibility for the allocation of labour and materials.

These changes were announced on 7 January 1941. No mention, however, was made of Bevin's proposals in his Heads of Labour Policy paper which had not yet been considered by the Cabinet. The result was to exasperate rather than mollify the critics and a demand was at once made for a debate in the House on the Government's handling of the economic side of the war.

By the time this took place (21–22 January) the Cabinet had approved Bevin's proposals on labour policy, but their case was again badly presented. Bevin led off the debate with a lengthy account of the Government's difficulties and achievements in raising production. His speech, read from an official brief and full of vague assurances without supporting figures, made the House restive. A member interrupted to ask if it would not save the Minister trouble to let the Clerk of the House read his speech for him. Bevin gave an angry growl in reply but stuck to his brief. It was only after he had been speaking for three-quarters of an hour that he referred to the new measures for controlling labour and then introduced them almost casually without making clear that they represented an important change in government policy.

The result was a confused debate in which the chief opposition

[1] Quoted in W. K. Hancock and M. M. Gowing: *British War Economy* (1949) p. 219.

speaker, Earl Winterton, had clearly had no time to take account of Bevin's announcement before launching into the set-piece attack which he had prepared. The question to which Shinwell and Clement Davies wanted an answer was, why had Bevin and the Government taken so long to accept the measures they had been asking for all along, and why did they still refuse to admit how much further they would have to go?

The critics were eloquent, far more eloquent than Bevin had been in his opening speech, but the changes he announced weakened their case and the House was less sympathetic to the opposition than it had been on earlier occasions. Flight-Lieutenant McCorquodale told Earl Winterton that people were getting tired of his "self-sought position of Jeremiah to this country". "I would sooner trust the Rt. Hon. Gentleman's judgment of the British working class," he added, "than that of the Noble Lord."[1]

To Bevin's satisfaction he found support for his manpower policy from Labour rank-and-file members. Jim Griffiths, a miners' representative from South Wales, attacked the suggestion that the conscription of labour was a magic wand with which to solve the problems of production.

"We are paying the price for the last twenty years in allowing our industrial output to rust and rot. For twenty years we lived in a period when coalmines, workshops and ship-building yards were being closed down. For twenty years we have pursued a policy of restricting and cutting down production. . . . The Minister of Labour is expected in eight months to make up for losses in industrial capacity and labour that this House permitted to go on for twenty years."[2]

Woodburn, another miners' M.P., this time from Scotland, put his finger on the real question. There was only one good reason for compulsion he told the House: not wrong-headed comparisons with the Army or the cry for equal sacrifice, but proof that it would lead to greater production. After sitting on the Committee on National Expenditure—whose latest report had been highly critical of the Government—he had reached the conclusion that compulsion, like working over-long hours, would reduce production and that Bevin had been right to use his powers sparingly.

[1] House of Commons, 22 January 1941, Hansard, Vol. 368, cols. 235-6.
[2] Ibid., 21 January 1941, Vol. 368, col. 144.

C

Churchill wound up the debate and used all his powers of advocacy to persuade the House that the Production Executive was a better way of dealing with the problems of war production than any other proposed. There was all the difference in the world, he argued, between inter-departmental committees designed for consultation or advice and a Cabinet Committee like the Production Executive of which the members were ministers with the power to act. "They have the strongest incentive to agree . . . and if they do agree, they can make their departments carry out their decisions. . . . The way to help busy men is to help them come to a decision."[1]

Churchill reminded the House that he had many times warned them that the build-up of munitions production would be slow. Only now were the new factories coming into operation, and the process could not be rushed, however impatient they might be for results.

"In the next six months we shall have for the first time an intense demand upon our manpower and womanpower . . . because for the first time we are going to have the apparatus and the lay-out which this manpower and womanpower will be required to handle. That is the reason for the very far-reaching declaration of which the Minister of Labour thought it necessary to apprise the House and the country."[2]

Once the impression left by Churchill's speech had evaporated, it was clear that the House remained sceptical about the reformed structure of committees but hoped that the new measures for controlling labour might prove effective. *The Times* even found something kind to say about a speech which had been one of Bevin's least inspired performances in the House. "It was the note of firmness and resolution which gave character to his pronouncements. The country," it added, however, "will expect a train of energetic action to follow."[3]

6

Churchill later wrote in his memoirs:

"Looking back upon the unceasing tumult of the war, I cannot recall any period when its stresses and the onset of so many problems all at once or in

---

[1] Ibid., 22 January 1941, Vol. 368, cols. 261–3.
[2] Ibid., col. 267.
[3] *The Times*, 22 January 1941.

rapid succession bore more directly on me and my colleagues than the first half of 1941."[1]

Having taken the difficult decision to come to the aid of the Greeks, the War Cabinet saw the forces they were able to send overwhelmed, Crete captured, all the gains of Wavell's campaign in North Africa lost, and Rommel's advance halted only on the Egyptian frontier. Yugoslav as well as Greek resistance was beaten to the ground and the whole of the Balkans occupied. The German Air Force kept up its attacks on British cities and ports throughout the winter, and shipping losses in the first half of 1941 were the worst of the war, 2,200,000 tons of British shipping alone being sunk.

It is all too easy now, in writing about the problems of labour supply and economic organisation, to forget what was so obvious at the time, the situation in which these discussions took place, with bombs falling and fires burning in the streets of London, and Dover under shell-fire from German batteries on the French coast. No one could foresee in the winter of 1940–41 that before the summer was out, Hitler would swing his armies east and commit everything he had gained to the gamble of an attack on Russia. What weighed upon the members of the War Cabinet in the early months of 1941 was that, when they had mastered the emergencies and made good the losses, the most they were doing was to keep Britain in the war. It was hardly more than that: in June 1941, the British were no nearer to *defeating* Hitler than they had been in June 1940.

Bevin's concern was with the world of Whitehall, statistics and manpower, but he never lost sight of why he was in the Government at all—to fight and defeat Hitler. He was Minister of National Service as well as of Labour, responsible not only for holding back the men needed to make the weapons but for calling up those who were needed to use them against the enemy.

This is a point that needs stressing. Because the machinery for call-up worked smoothly and threw up few of the problems of labour supply to the factories, it would be simple to assume that Bevin attached far more importance to his responsibilities as Minister of Labour than to those he exercised as Minister of National Service. The reverse is the truth. In May 1945, he told the House of Commons:

"During the war, I had to a very large extent to fight for the Services. I think

[1] Churchill, Vol. III: *The Grand Alliance* (1950), p. 3.

that I can say without boasting that, in the five years during which I have held office, at no point after we got going, were the Forces short of the numbers allocated by the Cabinet . . .

"Subject to keeping production going, I have been obsessed during the whole of the war with the idea, in order to keep casualties down, of not leaving the Services short of the necessary manpower."[1]

This did not mean that Bevin was in favour of giving the Army all the men it asked for. But once the allocations were settled, he never failed to carry out what he had undertaken and to see that, whoever else had to go short, it was not the fighting services, a fact which accounts for the high regard in which he was held by the Chief of the Imperial General Staff, Sir Alan Brooke, and the other Service chiefs.

None the less if British fighting men were ever to meet the enemy on equal terms in equipment, what mattered more than anything else in 1941 was to draw out of the nation an *industrial* effort sufficient not only to meet the needs of the moment but to build up sufficient stocks of weapons and munitions to mount an offensive in the future. This was the vital "front" in 1941 and Bevin was deeply involved at every point of it.

He had, in fact, to provide for three major needs in the first six months of that year. He had to find more men for the Armed Forces; he had to prepare for the general shortage of labour, which the Committee on Manpower Requirements had forecast; and he had to continue his efforts to meet the shortage of skilled labour which was still acute.

First, the Armed Forces. Beveridge's Committee had accepted the current estimates of Service needs as they stood, having no authority to revise them. At a meeting of the War Cabinet, however, as early as September 1940, Bevin had expressed doubts whether the country could afford to meet the demand of the Army for a million new recruits in the next twelve months and another million and a half in the following year. The number of divisions was not in question: it was in fact raised from 55 to 57, thirty-six of them recruited from the United Kingdom. But the field army accounted for less than half the number (1,800,000) already on the ration-strength of the Army. Over a million more were employed at the various headquarters, in air defence, training depots and rear services. If industry had to

[1] House of Commons, 16 May 1945, Hansard, Vol. 410, cols. 2538-9.

manage with fewer men than it wanted, Bevin argued, the Army must do the same and make more economical use of the men which it already had.

On 6 March 1941, the Prime Minister issued a Directive on Army Scales which fixed a ceiling for the largest of all claimants on manpower. Churchill ruled that the Army could rely on being kept up to a strength of "about two million British"—later fixed after much argument at 2,374,800—"and they will be judged by the effective fighting use they make of it."[1] This decision was important not only because of the limit it placed on the Army's demands (which far exceeded those of the other two services put together), but because of the principle which it established that everyone, including the Armed Forces, would have to accept a limited allocation of manpower and exercise economy in their employment of it.

Having secured a limit, Bevin did everything he could to see that the Army got the men allotted to it. So far as possible he regulated the rate of call-up to keep pace with the rate at which equipment could be delivered, but the Ministry was so well ahead with recruiting that Churchill's original figure for the Army's ceiling (2,195,000) was found to be already exceeded by 10,000. The new recruiting programme could thus start without any arrears. By the middle of 1941 all men up to the age of 41 had been registered for military service and the Ministry began to call up the nineteen-year-olds in July.

The second of Bevin's problems, the *general* shortage of manpower, did not develop as rapidly as Beveridge's committee had suggested. One reason for this, as Bevin frequently pointed out to the supply ministries, was the fact that industry could only absorb labour which it had the supplies and equipment to employ. These other shortages—of machine tools, steel, fabricated alloys, drop forgings, and other key items—continued to act as a brake on the full expansion of production far into 1941. The worst of the "bottlenecks", however, was still the shortage of skilled manpower, a nucleus of which was needed to start up the new factories now being completed. Ironically, it was not until Bevin made more progress with this third problem that he needed to expect the second, the general shortage of manpower, to become pressing.

In the meanwhile, a number of preparatory measures could be

[1] Churchill's directive is printed in Vol. III of *The Second World War, The Grand Alliance*, pp. 705–7.

taken. The most important was to start on the registration of men over military age and of women. Until the Ministry knew what each man over forty was doing and what his qualifications were, what were the domestic circumstances of the women and their freedom to transfer, it was impossible to direct them into war work. The Registration for Employment Order was accordingly published in March 1941 and a start made with interviewing men aged 41–43 and women of twenty and twenty-one.

Bevin was very much alive to the social problems which the conscription of women would involve and the political repercussions which it could create if clumsily handled. Women officers were appointed to conduct the interviews and to sit on the boards to which appeal could be made; a special section was created in the Ministry and a Women's Consultative Committee set up with which he discussed all major questions of policy affecting women.

A second measure was to push ahead with the schemes for contracting the less essential industries begun by the Limitation of Supplies Order. In January 1941, Bevin pressed the Board of Trade to reconsider its policy in order to release the maximum amount of labour. The Board's response was the White Paper on the Concentration of Industry (March) which proposed to meet Bevin's needs by concentrating the reduced volume of civilian production in a few "nucleus" factories instead of spreading it thinly over as many as possible. The scheme encountered many difficulties when put into practice, but it was one more means of putting pressure on workers to move into the essential industries producing for the war. The same result was often achieved in another way when firms, rather than lose staff or risk their factories being closed down, turned over to producing war materials themselves.

In the early months of 1941, semi-skilled and unskilled labour was already short in one or two areas (Birmingham and Sheffield, for instance) and in a limited number of industries in which the work was heavy or dangerous, conditions poor or wages low—iron-ore mining, drop forging, building and the shell-filling factories, the last of which needed large numbers of women workers. But the general shortage of all kinds of labour did not become serious until the summer. Bevin's major problem in the first half of 1941 remained the same as it had been since he took office: the supply of skilled labour in the engineering industries. Until this had been overcome, the demand of many factories for semi-skilled or unskilled workers was held in check.

By March 1941 Bevin had ready the Essential Work Order[1] which gave him the power, proposed in his Heads of Labour Policy paper, to schedule a factory or undertaking as engaged on essential national work. Once this was done, under the new Order, no man employed there could either leave or be dismissed except with the consent of the Ministry's local National Service Officer. By this means he hoped to stabilise the labour force of firms engaged on war work. The first to benefit from the Order were the engineering and aircraft industries, and separate Orders rapidly extended the system to other industries, coal-mining, shipbuilding, the railways and building. Taken together these Orders represented a code of industrial employment matching in importance the Order (No. 1305) which governed collective bargaining and arbitration. By the end of 1941 30,000 undertakings had been scheduled and close on six million workers brought under their provisions.

The working of the Essential Work Orders came in for criticism from both sides of industry, and the A.E.U. denounced Bevin for abandoning the principle of voluntaryism. But the need for such a scheme could hardly be denied. What pleased Bevin was to have found an answer to the problem of labour turnover which provided him with a powerful lever for improving working conditions. For no undertaking was to be scheduled until the Minister was satisfied with the terms and conditions of employment and with the arrangements for welfare and training. With Bevin as Minister these clauses in the Order were anything but formalities. And if workers under the Orders could now be disciplined for absenteeism or persistent lateness, the employers on their side had to pay a guaranteed weekly wage, a pattern of mutual obligation which Bevin regarded as the only satisfactory basis for industrial relations. The introduction of a guaranteed wage was a major reform in occupations like building, the merchant marine and the docks where labour had hitherto been taken on casually. "If I have been justified by nothing else in coming into the Government," Bevin told the House of Commons, "I feel rather pleased about that."[2]

---

[1] S.R. & O. 1941 No. 302, 5 March 1941. For a list of the subsequent orders amending and supplementing the Principal Order, see the Ministry of Labour Report for the years 1939–46 (Cmd. 7225, 1947), pp. xix–xxx.

[2] House of Commons, 4 December 1941. Hansard, Vol. 376, col. 1350.

Bevin believed that it was a reform which would improve efficiency as much as conditions of work. He had long been convinced that those industries which were able to get away with low wages and poor conditions through their ability to draw on a large pool of un-employed casual labour were also, as a result, amongst the worst organised and least efficient. Full employment in wartime effectively ended casual labour, and the industries of which it had been char-acteristic had to accept other changes besides the guaranteed wage and to take steps to organise themselves more efficiently in return for the protection of their labour force provided by the Essential Work Orders.

Every industry represented a special case and required separate treatment. In the case of the building industry, for instance, the Order was used to strengthen the system of government control over building and to introduce payment by results to which the building unions had been traditionally opposed. Two other cases are of special interest in the light of Bevin's earlier career.

Under the Merchant Navy Order, officers and men of the mer-cantile marine were prevented from taking other employment ashore, but in return were given adequate leave and a guaranteed wage between voyages. The approval of a man's union had to be given before he could be ordered to take a job or be dismissed, and the right of the employers and unions to negotiate wages and conditions of employment through the machinery of the National Maritime Board was left unimpaired.

For years Bevin had led a dogged campaign on behalf of the seamen, both in Britain, where he had been a great friend of the National Union of Seamen, and in the I.L.O. Nothing gave him more satisfaction than to use his powers as Minister of Labour to improve the conditions of their calling.[1] The Dock Labour Order, which followed in the autumn, gave him the opportunity to do the same for the cause which had first won him a reputation as a trade-union leader.

The new Order did not follow the same pattern as Bevin's earlier

[1] An able-bodied seaman at the beginning of the war got £9–12–6 a month (including danger money). By February 1943, this had been raised to £24 a month. According to the official war history, it is likely that a quarter of the men in the Merchant Marine when the war began, were killed or permanently disabled before it ended. (C. B. A. Behrens: *Merchant Shipping and the Demands of War* (1955), c. 7.)

scheme for the Clyde and Merseyside. The responsibility for dock labour was placed, not on the Regional Port Controllers, but on an independent body created for the purpose, the National Dock Labour Corporation. The employers were left to engage their own men twice a day, as they had always done, but a reserve pool of dock labour was created. Each docker had to attend eleven calls a week. If he was not taken on by an employer, the pool guaranteed him a payment for attendance: on the other hand, he had to be prepared to accept any work of which he was capable and, if required, to transfer to another port.

In the engineering and other war industries, the Essential Work Orders prevented the loss of the skilled manpower on which they depended. Two other steps promoted its better distribution. In March 1941 a Labour Co-ordinating Committee was set up at which representatives of the supply ministries and the Board of Trade met under the chairmanship of the Permanent Secretary of the Ministry of Labour to watch over the working of the Essential Work Orders and establish a clearing house for labour problems. The second step followed the gradual replacement of priorities for particular industries by a system of labour preferences granted to individual factories. The job of sifting applications and granting these preferences was handed over to another inter-departmental body, the Headquarters Preference Committee. By the later months of 1941 this new procedure was working well and Bevin at last found in this idea of "preferences" a way of dealing with delays in production due to difficulties in labour supply, which was more flexible than priorities or allocations and had the additional advantage of not letting control of manpower out of the hands of his Ministry.

7

Bevin's main hope of meeting the shortage of skilled labour, however, was still to push on with the schemes for dilution, training and the wider employment of women which have already been described. There is no better way of illustrating the difficulties with which Bevin and the Ministry had to contend than by examining the case of one of the most important war industries, shipbuilding and ship-repairing.

The urgency of raising output was obvious in view of the heavy rate of sinkings and damage inflicted by the U-boats and air attack. But the industry, employers as well as unions, was individualistic and highly conservative. To take one example: plans to establish a training scheme for apprentice riveters on the Clyde were held up for months largely because of the Boilermakers' Society who demanded a promise, which the employers refused to give, that the ratio of apprentices to journeymen would not exceed 1:5. This ratio had been in dispute since the nineteenth century and in the end the Minister of Labour went ahead with the scheme without any agreement.

These attitudes had been hardened by the experience of the years between the wars. No other industry, not even coal, had been as hard hit by continued unemployment. In 1918 the annual launching capacity of British shipyards was some three million tons and this rose by simple technological progress to three and a half million in the 1930s. In the best year, 1929, only half of this was used and a total of a million and a half tons launched; in the worst years, 1932 and 1933, not more than 188,000 and 133,000 tons were launched. A hundred thousand of the casual workers drifted away from the yards between the wars, yet of the 176,000 men left, 20 per cent were still unemployed as late as July 1939. Quite apart from the material legacy of the Depression in old plant, poor physical conditions of work and housing, this experience left both management and men with a fear of recurrent unemployment after the war which no promises or exhortations could remove.

The labour force was divided in roughly the proportion of 1:4 between the five royal dockyards and the fifty-two principal private shipbuilding firms, with the largest concentrations on Tyneside, on Clydeside, at Barrow, Birkenhead, Belfast, London and Southampton. Bevin, having failed to secure the state control of the munitions industries which he had proposed in 1940, made two further attempts, in 1941 and 1942, to transfer control of the shipbuilding industry to the Admiralty, so that the work could be "carried on as a great public service and not limited by the pre-war conceptions of private interest and limited individualism."[1] Both attempts failed. In effect, although the responsibility for merchant as well as naval shipbuilding and repair was placed in the hands of the Admiralty, the greater part of the work continued to be done by private firms and the Admiralty's

[1] Quoted in Inman, p. 103.

Department of Merchant Shipping and Repairs was staffed by men drawn from the shipbuilding industry, chief among them the Controller, Sir James Lithgow, a former president of the Shipbuilding Employers' Federation and an old opponent of Bevin's.

When the Essential Work Order for the shipbuilding industry was made in March 1941, responsibility for labour supply was shared between the Ministry of Labour and the Admiralty. The Ministry's efforts to increase the total labour force in the shipyards were not unsuccessful: by June 1941 it had been pushed up from 144,700 men two years before, to 232,400, roughly half of whom were skilled workers. The difficulties arose from the resistance of men and management alike to any proposal for making a more economical use of the skilled labour available.

Shipbuilding was a complicated business in which seventeen major trades organised by seventeen separate unions were involved. The demand for the different trades in different yards fluctuated widely from week to week and there was an obvious case for making labour as mobile as possible between different yards and interchangeable between different trades. Such schemes, especially interchangeability, encountered strong opposition from unions such as the Boilermakers', each of which vigorously defended the interests of its members and insisted on traditional practices and lines of demarcation. Dilution, whether by upgrading semi-skilled labour, breaking down jobs or introducing women, was regarded with the same suspicion. Nothing could shake the men's conviction that dilution would depress wages and create unemployment again after the war.

After long negotiations, Bevin succeeded in getting agreements made at national level with the unions concerned, but their implementation depended upon the willingness of local union officials and shop stewards to put them into effect. This was grudgingly extracted only after every possible alternative had been examined which could provide better pay or conditions for members of the union already at work in the yards, or once employed there. "Whenever dilution is raised," one Ministry of Labour official complained in exasperation, "we seem to be brought up short against a ghostly army of unemployed boilermakers." Nor was there much support from the shipbuilding firms, each of which was more interested in preserving its own skilled labour force at maximum strength than in providing surplus labour to be transferred to its rivals, and all of which (like the Admiralty itself) preferred to put up with existing practices rather than risk trouble with the unions.

One example must do duty for a hundred which might be quoted. At Newport, fitters worked in pairs instead of one skilled man working with a mate which was the usual practice elsewhere. This arrangement was first questioned in October 1940 and eventually an independent inquiry reported that there was no justification for the Newport custom, which was wasteful of skilled manpower. The national executive of the A.E.U. however failed to persuade its members in South Wales to alter their practice. In June 1942 the Ministry issued directions, but the men refused to obey them and threatened to strike. The A.E.U. objected that it had not been given sufficient opportunity to discipline its members and it was decided not to prosecute. Negotiations dragged on, without effect, and at the end of the war the Newport fitters were still working in pairs.

By December 1943, the number of women employed in shipbuilding had risen to 13,000, less than 6 per cent of the labour force. At the same date, 15 per cent of the electricians, 10 per cent of the fitters and turners were "dilutees"—an average of 7 per cent of the skilled labour force over all trades in the 52 principal firms. In ship-repairing the figures were lower: not until 1943 was the A.E.U. willing to agree to the employment of women on repair work and the percentage of dilutees was only 4 instead of 7 per cent. This was the meagre result of three years' uphill work. In 1940 50 per cent of the total labour force in the industry had been rated as skilled: by 1942–3 this extravagantly high figure had been marginally reduced to 47 per cent, only to rise again to 48 per cent by the end of the war.

With the exception of coal-mining, shipbuilding remained the industry in which the Government's labour policy achieved least success. Nothing Bevin or anyone else could do succeeded in breaking down the mixture of suspicion, prejudice and conservatism which was the bitter fruit of the industry's history. Shipbuilding like coal-mining was haunted by the past and neither promises nor threats could exorcise its influence on men's behaviour.

No one has succeeded in producing a satisfactory way of measuring the progress of dilution in the engineering industries, but one figure may be cited for comparison with the shipbuilding percentages. Between June 1939 and December 1943 the number of women employed in engineering and allied industries had risen from 411,000 to over a million and a half, from 18 to 30 per cent of the total labour force, and four out of five of these were on semi-skilled or skilled work. This is enough to show that shipbuilding was fortunately not typical of the rest of British industry: if it had been, war production could never have reached the levels it did.

To keep the picture in balance, it is worth referring to a labour problem of a different sort with which Bevin was much occupied in the early months of 1941. At the beginning of the war the Ministry of Supply planned a big expansion of the Royal Ordnance filling factories to meet the demand for ammunition. In September 1939 there were only 8,000 workers employed on shell filling and although the number had been raised to 49,000 by the end of 1940 this was still well below what was needed. The R.O.F. at Chorley alone was planned to employ 28,000 and in the winter of 1940–41 was actually losing workpeople. At the beginning of 1941 the Prime Minister made sharp inquiries about what was being done to remedy the position.

Bevin was convinced that the key lay in the improvement of working conditions. Hours in the filling factories were long, transport bad, amenities poor. Although many women were employed, there were no day nurseries, inadequate canteens and no women welfare officers. Workers often spent up to two or three hours a day travelling to get to work and then had to walk long distances over a construction site to reach their shop.

While, therefore, he instructed the employment exchanges to give a high priority to R.O.F. vacancies, at the same time Bevin took energetic steps to make the conditions better. Hours of work were cut, the canteens reorganised, transport improved, medical and welfare services provided, hostels built. As a result, by the middle of May, the Ministry of Supply reported that six of the seven filling factories had all the labour they could absorb and by June the numbers had risen to 103,000, more than double the figure in December 1940. By the end of 1941 the labour force stood at 144,000. No less important, improved working conditions led to a rapid rise in productivity. Here was a case where action by Bevin's ministry transformed the labour position in a few months.

CHAPTER 3

Critics, Welfare and Wages

I

IN THE summer of 1941 Bevin completed his first year in office. The twelve months from June 1940 to June 1941 had seen big changes in the distribution of manpower, the most important of which can be summarised as a rise of a million men and fifty thousand women in the strength of the Armed Forces and a rise of a quarter of a million men and 426,000 women in the munitions industries.[1] The Forces and munitions industries together now accounted for eight million men and women, 37 per cent of the total working population against just under 30 per cent a year before.[2] There were still troublesome shortages of skilled workers but the number of unfilled vacancies had been brought down to 10,000, a much less serious shortage than six months before.[3]

Altogether, Bevin thought it a satisfactory balance sheet. The changes it represented had been accomplished by consent; over industry as a whole labour supply had roughly kept pace with industrial expansion, and a start had been made with the more effective use of the manpower available. His critics, however, regarded such claims as complacent and renewed their demand for more drastic measures.

After the January debate these criticisms had died away for a time. Even *The Economist* was prepared to say in March:

"Slowly and painfully but with sureness, Mr. Bevin has produced the outlines of a flexible plan for the effective use of labour as requirements grow."[4]

[1] The detailed figures are given in Parker, *Manpower*, Statistical Appendix.
[2] This figure includes Civil Defence but not mining, transport, shipping or the other important Group II industries. The comparable figure for June 1943, the peak year of mobilisation, was 10.3 million, 46 per cent.
[3] W. K. Hancock and M. M. Gowing, *British War Economy* (1949), p. 292.
[4] *The Economist*, 22 March 1941.

At the end of April, Bevin was able to present to the House of Commons regulations abolishing a practice which for years had angered the Labour Movement. Whenever anyone had applied for unemployment or some other form of social assistance, the resources of the whole household had been brought into account in determining the applicant's means. The new regulations which Bevin introduced to the House on 29 April abolished this hated household means test in favour of a test of personal need. "With this," Bevin told the House, "the Poor Law is now buried."

The lull in criticism of the Government, however, did not last more than a few weeks. The setbacks in Greece, North Africa and Crete roused a storm of anger in Parliament and the press and many of those who spoke or wrote about the war gave vent to their anxiety by finding fault with Churchill, Bevin and any other minister in sight for their slowness in mobilising the country's resources.

On 6 May, the Government asked the House of Commons for a vote of confidence. Most of the debate was devoted to the defeats in Greece and North Africa, but Lees-Smith, who spoke after Eden, remarked that there was more criticism in the country of production than of strategy. Other speakers said the same, and Hore-Belisha, a former War Minister, after a stinging attack on British strategy in Greece, spoke with scorn of the Government's lack of preparation:

"The Germans used the winter thoroughly. We spent the winter arguing—at least the Rt. Hon. gentleman the Minister of Labour was arguing with himself—whether the compulsory or the voluntary principle was better in industry. It was an irrelevant discussion."[1]

The leader writers echoed the opposition in the House.

"The most flagrant example, of course," *The Economist* declared, "is Mr. Bevin whose course of action over the conscription of labour . . . is not very far removed from deception of the people. It is the spectacle of Mr. Bevin talking hot and acting cold which, perhaps more than any other single factor, accounts for the undercurrent of uneasiness in the country."[2]

These feelings, further sharpened by the loss of Crete, found acrimonious expression in the Commons debate on production on 9–10 July. Members were irritated rather than appeased by the

[1] House of Commons, 6 May 1941, Hansard, Vol. 371, cols. 779–80.
[2] *The Economist*, 10 May 1941.

announcement shortly before that Beaverbrook was to go to the Ministry of Supply (the fourth Minister of Supply since the beginning of the war) and that changes were to be made in the procedure of the Production Executive. It was with the assurance of support from all quarters of the House that Horabin summed up the first day's debate with the question:

"Is the Prime Minister at long last going to face what almost every hon. member believes to be necessary at the present time? Is he going to appoint a Minister of Munitions with full power over the whole field of war production?"[1]

Members vied with each other in citing examples of inefficiency and breakdowns in production. A maiden speech by a serving officer just back from Egypt attracted particular attention: 70 to 80 per cent of the tanks supplied to the Army in North Africa, he claimed, were of such poor quality that they broke down before engaging the enemy.

These charges were supported by the Conservative chairman of the Select Committee on National Expenditure, Sir John Wardlaw-Milne:

"I said in the House not long ago that I did not believe that our people were working up to more than 75 per cent of our possible efficiency and I cannot alter that opinion yet."[2]

No member of the Government came more frequently under attack than Bevin, both as chairman of the Production Executive and as Minister of Labour. Each time he attended he was provoked into intervening, particularly by Wardlaw-Milne's censure of the lack of a wages policy and the working of the Essential Work Orders. More than one speech came close to a personal attack on him and one member referred to him contemptuously as an "unskilled labourer", a remark which Bevin turned to his advantage by welcoming the epithet as a compliment. The Government, however, made the serious mistake of not putting up a senior minister to answer the debate. "In the Government's view," *The Times* commented the following morning, "it was a second-class occasion. . . . But the Government surely cannot ignore the body and the weight of

[1] House of Commons, 9 July 1941, Hansard, Vol. 373, col. 253.
[2] Ibid., 10 July 1941, col. 336.

complaints which come from employers and labour representatives alike."[1]

Recognising the mistake and the damage that had been done by failing to answer criticisms widely reported abroad, Churchill proposed a third day's debate at the end of the month and opened it himself. This time the Prime Minister did not wait for anyone else to start the attack, turning his scorn on those members "who feel, no doubt quite sincerely, that their war work should be to belabour the Government and portray everything at its worst, in order to produce a higher efficiency," and their allies in the press who "cry a dismal, cacophonous chorus of stinking fish all round the world".

"If we are depicted by our friends and countrymen as slack, rotten and incompetent, we are entitled, nay it becomes a pressing duty to restore the balance by presenting the truth."
"We are not a totalitarian state," he reminded the House, "but we are steadily and, I believe, as fast as possible working ourselves into total war organisation. When we are given vivid instances of lack of organisation or of inter-departmental rivalry, and when these are all bunched together to make an ill-smelling posy, it is just as well to remember that the area of disputation is limited, circumscribed and constantly narrowing."[2]

The Prime Minister did not leave Bevin to defend himself against the opposition:

"It is the fashion nowadays," he said, "to abuse the Minister of Labour. He is a working man, a trade-union leader. He is taunted with being an unskilled labourer representing an unskilled union. I daresay he gives offence in some quarters; he has his own methods of speech and action. He has a frightful load to carry; he has a job to do which none would envy. He makes mistakes, like I do, though not so many or so serious—he has not got the same opportunities. At any rate he is producing, at this moment, though perhaps on rather expensive terms, a vast and steady volume of faithful effort, the like of which has not been seen before. And if you tell me that the results he produces do not compare with those of totalitarian systems of government and society, I reply by saying 'We shall know more about that when we get to the end of the story'."[3]

Bevin wound up for the Government and despite the harsh things that had been said about him earlier in the debate made a better speech than any the House had yet heard from him. Instead of getting

[1] *The Times*, 11 July 1941.
[2] House of Commons, 29 July 1941, Hansard, Vol. 373, cols. 1300–1, 1278–9.
[3] Ibid., col. 1300.

angry with the opposition, he showed that he was as much alive to the shortcomings and as concerned to find solutions as they were. Defending the decision not to create a super Ministry of Munitions he made the obvious point that the practical effect of such a decision at this stage of the war might well be to throw plans into confusion and hold back production. He also made the less obvious but shrewd point that the real need was to place the responsibility for increasing production lower down, not to pass it up and so create more of a bottleneck than ever. To Wardlaw-Milne's criticism that production was only running at 75 per cent he gave the answer:

"I want to have more than a 25 per cent capacity in the kitty all the time, so that when the last emergencies have to be met, there is a last reserve of production to carry them through, and therefore, I want to keep that in reserve, in advance of the defence and production plans, so as to avoid waiting for supplies."[1]

The opposition was not convinced, either by Churchill or Bevin. Their case, they complained, had not been answered. "Facts are facts," Shinwell retorted, "whatever the Government may say." But by standing firm and refusing to make any concessions, Churchill forced the critics out into the open. They had either to challenge the Government in the lobbies—and with no alternative government in sight, this meant certain defeat and much censure for harming national unity—or to watch their protests fizzle out without a sequel. This was exactly what happened. The press continued to grumble for a few days more, then as the sting of defeat (the real source of the trouble) was overlaid by the events of the Russian campaign, better news of the U-boat war and a local success in Iran, the mood of indignation changed to one of confidence. When the next debate on manpower took place, in October, it was in a different atmosphere and for almost the first time Bevin was able to discuss his difficulties with a friendly House.

Looked at from this distance of time, this second half of 1941 stands out as the period in which the whole prospect for the British was changed by the entry into the war, first, of the Soviet Union, then of the United States. It was hard to grasp this at the time, however. Hitler's invasion of Russia (22 June 1941) was followed by his most spectacular successes: not until December, when the Germans were

[1] Ibid., col. 1371.

already at the approaches to Moscow, was there any indication—in the severity of the Russian winter and the violence of the Russian counter-attacks—of the price which they might have to pay for Hitler's gamble. In the autumn of 1941 it still appeared possible that the gamble might come off and the Germans overrun Russia as they had every other country they invaded. As for America, although Roosevelt went to the limit of neutrality in helping Britain, she remained outside the war until the Japanese attack on Pearl Harbor and the German declaration of war in December. When Churchill crossed the Atlantic for his first meeting with the President, off the coast of Newfoundland, in August 1941, the United States was still officially neutral.

It was at this meeting that the President urged on the Prime Minister the publication of a joint declaration of principles. The draft of the Atlantic Charter was drawn up by Churchill, revised in discussions with Roosevelt and Sumner Welles and cabled to London for approval on 11 August. The members of the War Cabinet were summoned to a special meeting at quarter to two in the morning in order to consider the text. They suggested amendments to the controversial clause 4 (non-discrimination in world trade), which were subsequently abandoned in deference to the President's wishes, and then added an entirely new clause proposed and drafted by Bevin. This read:

"They support the fullest collaboration in the economic field with the object of improving labour standards, abolishing unemployment and want, securing economic advancement and social security for all people."

The text was altered in the final version,[1] and the reference to the abolition of unemployment and want omitted, but with this change the President and the Prime Minister agreed to the new clause which appeared as the fifth point of the declaration immediately after "access on equal terms to the trade and raw materials of the world".

It is impossible to read the Atlantic Charter now without a sense of disillusion, but at the time it raised great hopes. To Bevin it was a cause of satisfaction that he had succeeded in getting both the British and American Governments to accept in principle the concern with

[1] The final version ran: "Fifth, they desire to bring about the fullest collaboration between all nations in the economic field, with the object of securing for all improved labour standards, economic advancement and social security."

economic and social problems which he believed ought to figure as prominently as political factors in the post-war settlement.

2

Churchill's answer to the criticisms of the production programme had been the Production Executive,[1] and Bevin put all his energy into the task of making it more effective than the Production Council it replaced. The permanent members were still the same, the three supply ministers, the President of the Board of Trade and the Minister of Labour; the biggest difference was Bevin instead of Greenwood as chairman. The Executive not only met more regularly (thirty meetings in the course of 1941) but it was the ministers themselves and not their deputies who took part in the proceedings. Out of a total of 261 attendances spread over the year, the five senior ministers accounted for 205, other ministers for 48 and officials for no more than eight. To make sure that Beaverbrook did not treat the new Executive with the same contempt as he had the earlier Council, Bevin held all the meetings in Beaverbrook's room as long as he was Minister of Aircraft Production.[2]

Between them the five members of the Executive were responsible for the whole production programme and, if they were agreed, could put any decision into effect at once. Why then did it incur so much criticism, much of which inevitably fell on Bevin as its chairman? To answer this, it is necessary first to ask what was the job it did.

The Executive's agenda refer to a great many miscellaneous items —more equipment for the electricity and gas industries, the quicker turn-round of railway wagons, the dispersal of timber from the docks, improving rail links with South Wales. These appeared as matters of great urgency for two or three meetings and then once they had been dealt with, disappeared altogether.

This was the emergency side of the Executive's activities, on which it could often act with great effect. But much more important was its regular business.

Bevin's position as chairman was reflected in the extent to which labour-supply questions figured on its agenda. Some of these dealt

[1] See Chapter II, 5 above.

[2] Beaverbrook continued to work at his desk while the committee met, but he could not avoid getting drawn into the discussion.

with specific shortages: scientific-instrument makers, electricians for Admiralty work or the shortage of labour for the Royal Ordnance filling factories which was made the subject of fortnightly reports. Other items were more general, less a matter for action by the Executive itself than for discussion: the provisions of the Essential Work Orders, training schemes, the schedule of reserved occupations, and a host of related questions, from hours of work and holidays to joint consultation in industry.

Another range of questions in which Bevin took great interest was connected with building. The construction industry, with a total capacity of £350 million p.a., was still trying to carry out a Government building programme of twice that figure. In an effort to close the gap, the new Ministry of Buildings and Works had devised a system of allocations between departments, at first based on money, later on labour. But the chances of the newest ministry of all enforcing such a system when it might mean a reduction in the Admiralty's or the War Office's building plans were not great. Every decision taken in the Ministry of Works, therefore, went as a recommendation to the Production Executive where Bevin used his position as chairman to bolster its powers. Under the authority of the Executive, Bevin and Reith, the first Minister of Buildings and Works, carried through a reorganisation of the building industry, and in July Bevin succeeded in getting Reith added as a permanent member of the Executive.

This side of the Executive's interests grew out of the Works and Building Committee which it had taken over from the Production Council. The Materials Committee under Colonel Llewellin dealt with the allocation of scarce materials: steel, drop forgings, tinplate, timber and cotton. The Industrial Capacity Committee, under Harold Macmillan, hunted out additional capacity which might be employed on arms production (such as the railway shops, and firms in Northern Ireland) and sought to relieve the shortage of machine tools. A Manpower Committee supervised the revision of the schedule of reserved occupations and a Central Priority Committee dealt with such priority questions as remained now that the system of allocations was becoming established. Much of the Executive's time at its fortnightly meetings was spent on the reports and recommendations of these committees which, if accepted, were put into effect under the Executive's authority.

All this represented useful and necessary work without which there would have been many delays in production. By bringing together the ministers most concerned with industry the Production Executive gave them the opportunity and the incentive to settle the issues that were bound to arise between them. *The Economist's* original comment: "New Executive may be Old Council writ rather differently, but Mr. Bevin is likely to prove an abler and more decisive chairman than his translated party colleague",[1] proved to be an accurate forecast—in both its parts.

But its role remained the same: it was a co-ordinating committee, not a directing authority, "Old Council writ rather differently". And it was from this fact, dissatisfaction with the design rather than the execution of its functions, that the criticisms sprang. Whether the critics were right and the appointment of a Minister of Munitions or a Minister of Economics would have speeded up the expansion of production in 1941 is a hard question to answer, as anyone will find who reads the discussion of the evidence in Professor Postan's *British War Production*.[2] But it is not a question that needs to be answered here, for the decision was Churchill's not Bevin's. Bevin was inclined to favour a single minister and his name was mentioned from time to time as a strong candidate for the post.[3] But Churchill had made it clear that he was opposed to the idea, that he meant to keep the crucial decisions on production in his own hands, and Bevin accepted his view. Whatever his doubts may have been, he kept them to himself and did his best to make a success of the Executive within the limited sphere and with the limited powers allotted to it.

The Production Executive, in fact, never took charge of the main production plans, which continued to be determined by the Defence Committee of the War Cabinet in its supply meetings. This was a body of no fixed constitution and a fluctuating membership. It never, however, included the chairman of the Production Executive and was presided over, or, more truly, dominated by the Prime Minister in his capacity as Minister of Defence, with the expert advice of Lord Cherwell's Statistical Section. Few of those who criticised the administration of war production at the time seem to have grasped the importance of the Defence Committee (Supply), if it can really

[1] *The Economist*, 11 January 1941.
[2] Postan, pp. 269–74. See also, in the same series, J. D. Scott and Richard Hughes: *The Administration of War Production* (1955), Part V.
[3] E.g. by *The Economist*.

be called a committee, or of the Statistical Section. No doubt the Prime Minister did not mean them to. Here was one of the keys to what Professor Postan calls "the miracle of Britain's Government in the war . . . a Government which was largely personal and yet free from the intellectual limitations of an autocracy".[1]

If the Production Executive did not take the decisions, no more was it (despite its name) the body to which they were handed for execution. The links between it and the Defence Committee (Supply) are described by the official war history as "tenuous and occasional".[2] The Prime Minister's directives went straight to the supply ministries, the heads of which retained their independent powers and constitutional responsibility for carrying out their own programmes. Finally, where production matters touched upon questions of general economic policy—for example, wages, manpower budgeting, and even the heavy bomber programme—these were sent to the Lord President's Committee, which took over, with great success, an increasing part of the Cabinet's responsibilities for home affairs. Bevin, it is true, was a member of the Lord President's Committee and played a big part in its discussions, but he would have done so in any case, as a member of the War Cabinet and Minister of Labour, not because he was chairman of the Production Executive.

The Production Executive was left, therefore, to deal with the peripheral rather than the central issues of war production. Its powers reflected this view of its functions: it had no authority apart from the collective authority of its members—which meant no power to act in the absence of agreement among the the ministers who composed it— and it could take no initiative except on matters brought before it by one of its members.

This initiative was exercised almost entirely by its chairman: it was thanks to Bevin for instance that matters of real importance in the field of labour supply were discussed by the Production Executive and it was again largely thanks to his personal authority that its decisions were acted on and not ignored. Granted its limited role and powers, Bevin made as much of the Production Executive as any chairman could have done. But the chairman of the Production Executive was never, and was never intended to be, a Minister of Production. Even to Bevin, who made more use of it than any other

[1] Postan, p. 144.
[2] Scott and Hughes, p. 429.

of its members, its value lay in enabling him to deal more successfully with problems with which he would in any case have been concerned as Minister of Labour.

3

Whenever he could, Bevin still tried to get away from London to meet and talk about the war with people outside the closed circle of ministers and officials. The first of his 1941 tours took him to the North-East in March and was characteristic of many others. In the space of two days, he visited a number of works, met and talked to the local staff of the Ministry, picked up old trade-union contacts, spoke to a Durham miners' conference, was guest at a civic dinner in Newcastle and addressed crowded public meetings in Newcastle City Hall and in a cinema at Ashington.

In May 1941 he visited Yorkshire, with big meetings at Hull, Sheffield and Leeds; in June, the Midlands, speaking on successive days at Leicester, Nottingham and Derby. In July he went to Lancashire, in August to Llandudno for the Biennial Conference of his old union, then travelled across to East Anglia, to speak at the Theatre Royal in Norwich. In September he spent a weekend at Southampton and Reading, and later in the autumn fitted in a two-day tour of South Wales between a visit to the Potteries in October and another to Middlesbrough in November.

This was a heavy programme for a man of sixty on top of his work as a minister. But Bevin was restless if he was shut up too long in the atmosphere of Whitehall and Westminster. It was on such visits that he formed the impressions of public opinion on which he relied heavily when it came to issuing orders and regulations. And it was still the case, as it had been when he was a trade-union leader, that to go out and meet the people he was trying to lead restored his confidence in them and in himself.

Churchill was supreme in the House of Commons or as a broadcaster: Bevin had none of his magnificent command of English. But, put him in front of an audience of a thousand or two thousand in a works canteen or one of the big provincial halls, and then the directness, the force of conviction with which he spoke left no less an impression than Churchill. On occasion he talked with the authority of a

member of the War Cabinet; at other times he spoke as one of themselves, finding at once the arguments and homely metaphors that
would relate the issues and events with which the Cabinet had to deal
to the everyday world of his audience.

Almost invariably in his 1941 speeches Bevin went back to the
situation in which the Government had found itself in May and June
1940: the lack of preparations, the dramatic scene in the Prime
Minister's room when the Chiefs of Staff weighed the chances on the
eve of Dunkirk, the threat of invasion, the sudden realisation of how
close the country was to disaster. And now, twelve months later, to
the astonishment of their friends as well as their enemies, they were
still in the fight. Hitler had concentrated all his forces against them
and failed to break them. If ever a people had been put to the test and
proved, it was the British. The R.A.F. had not been swept out of the
skies; the people of London and the other bombed cities had not
cracked under the Blitz; the Navy and the Merchant Marine had not
allowed the U-boats and the bombers to drive them from the seas.

"You people in local government have a right to be proud, for most of these
young fellows came from your elementary, your secondary schools, sons of
our own people. Who then can say that Britain is decadent, that Britain is
down with such material as that, people who never expected fighting anybody, who never had war in their minds, who never looked upon it as a profession, sons from our own homes who went to battle. . . . It makes you feel that
notwithstanding our industrial depression, notwithstanding all we have gone
through, the spirit of the British people is marvellous and Britain is destined
to play a bigger part yet than ever she has done."[1]

From this Bevin went on to trace the progress of the war, to show
the relevance of the fighting in Africa (the East African campaign
and the restoration of Abyssinia had made a deep impression on him)
and the magnitude of the war at sea.

"There were a lot of things happening at that time that you could not understand. One day all your bananas went. . . . Then your meat went down to a
shilling a week and some of you good women said 'This is a bit of a puzzle'.
In the canteens it went down to a penny a day. You must have thought we
were mad. But there was a reason for it all. To save the Middle-East Army
we had to take away every fast ship we could get, and the main fast ships were
refrigerator ships. We had to put them on because they could dodge sub-
marines. The Government had to choose whether it would cut down food

[1] Speech at Ashington, Northumberland, 9 March 1941.

supplies or save the Middle-East Army. You will agree it made a wise decision. I mention this because when we have to do what may appear to you very odd things, there is a good reason. But you must trust us. We cannot always tell you until it is accomplished."[1]

To his audience at Newcastle in March he said:

"The next battle is to be the Battle of the Atlantic. . . . We know the preparations that have been made by the enemy to strangle us, and we know that the Navy has to meet the same or even greater intensity than existed at the end of 1917 or 1918. I put it to the employers and managements and to the men in these great shipyard districts to feel that they are a part of the Navy and that it is not right to expect that great force to carry all our burden. . . ."
"It is up to us to shorten the turn-round of the ships by co-operation between the dockers and the managements. . . . Ship management in ports—repairs, turning round and clearance—is not as good as it ought to be. In my opinion it is possible on every turn-round, by a combined effort of ship managements, repair yards, dockers and railway companies, to save eleven days on every trip. You can do it. . . ."[2]

In Durham he put the same argument to the miners, in Leicester and Derby to the building industry. The remaining months before the next winter were vital to the production programme. Whatever plans the Government made, they had to depend on the rank-and-file worker—the miner at the pit face, the building labourer, the women in the filling factories—making the extra effort to carry them out, and Bevin used every illustration he could think of to drive home the point that war was no longer a matter for cabinets and generals but affected directly the life of every man and woman listening to him. He made a special appeal to the women to join the auxiliary forces and to take on jobs in industry. "I saw one headline the other day," he remarked at Leicester, "which said: 'Bevin wants 100,000 women, the State to keep the children.' "[3] If they had husbands or sweethearts in the Forces, they could help to make the weapons to equip them: this was the quickest way to end the war and get their menfolk home again.

No one believed more strongly than Bevin that, even in the middle of the war, it was right to think about the world that was to come out of it and that men and women would more willingly meet the demand

[1] Speech at Middlesbrough, 16 November 1941.
[2] Speech at Newcastle on Tyne, 8 March 1941.
[3] 13 June 1941.

for sacrifice and effort if they felt that these would contribute to making a juster and more equal society. "Why," he asked his own Union delegates, "do you think the Russian peasant is fighting as he never fought in the First World War?"

"It's not the political system that has made him fight. He is not fighting for a landlord, it is his own, his roots are in it, he owns what he is fighting for. . . . It is not a particular theory of Marx and Engels, or something of that kind. It goes fundamentally deeper and, equally, civilisation cannot survive if it rests upon a propertyless proletariat. That is why I have urged that if our country is not big enough to solve our problem by means of the land, like the peasant countries can, you have got to find a substitute and the substitute is the vested interest of social security within your own state in which all shall participate."[1]

Nor should they be content to think only of their own country: there would be no security against future wars unless they carried out the promises of the Atlantic Charter, in particular those which dealt with freedom for the dependent peoples, access to raw materials, and the raising of living standards throughout the world.

"I deny Hitler's right," Bevin declared, "I deny any State's right, to put a limitation on the progress of the human mind. I believe that the opportunities that lie ahead of us are great."[2]

But all this, he reminded them, the chance to make a better world, depended on defeating Hitler, and that in turn on the effort they made now in the remaining months of 1941 to turn out the tanks, the aircraft and the guns. He set before them all the demands that had to be met from a working population of no more than 20 millions. The war could be won only if every man and woman did what was asked of them. This was no time to argue about wages or profits: unless they won the war and kept their freedom, there would be neither wages nor profits.

"Ah, then, my friends, give me six months resolute, urgent, persistent, consistent effort over the whole field of industry . . . and it may be it will not be months but years of human suffering we shall have saved."[3]

[1] Speech at the T.G.W.U. Biennial Delegates' Conference, Llandudno, 18 August 1941.
[2] Speech at Swansea, 1 November 1941.
[3] Speech at Llanelly, 2 November 1941.

4

Bevin saw nothing inconsistent in combining his appeal for an all-out effort in the workshops with the improvement of working conditions and welfare. On the contrary, it was the low value which the nation had hitherto attached to labour, he argued, which accounted for half the problems he was called upon to solve, from bad industrial relations to the shortage of skilled manpower. Instead of accepting the view that in wartime any conditions were good enough for people to work in, he was bent upon making up for past neglect and expanding the conception of industrial welfare. If anyone told him that this was no time to introduce far-reaching reforms, his answer was that this was the surest way in which to increase productivity and efficiency. To treat working people as human beings, as you would yourself expect to be treated, was not only a matter of principle: it was also, as he had been trying to persuade employers for twenty years, the commonsense way in which to get them to work willingly and well. By insisting that he would only schedule factories under the Essential Work Orders if he was satisfied that the conditions of work and arrangements for welfare as well as wages were satisfactory, Bevin provided himself with a powerful lever for getting his views listened to.

His first step had been to get the administration of the Factories Acts and the Factory Inspectorate removed from the Home Office where he felt that they were in danger of being identified with police, prisons and the control of vice, a totally wrong context for the developments he had in mind. He had always been interested in the safety side of the inspectors' work, and the rapid changes in industry during the war and the large numbers of inexperienced workers, especially women, brought into the factories made doubly important the prevention of accidents and the improvement of conditions, such as bad lighting and bad layout, which produced them. Bevin saw this, however, not as an extension of police functions into the factories but as the beginning of a comprehensive welfare service, set up a Factory and Welfare Division in the Ministry and strengthened his hand by the appointment of an Advisory Board with which he could discuss his ideas.

Bevin had long been concerned with health in relation to industry.[1]

[1] See Vol. I, pp. 602–3.

He wanted to enlarge the notion of "industrial health", to take account of the general health of those working in factories and mines, not just the diseases associated with industrial processes like anthrax or silicosis. He was certain that in a long war the health of the working population was going to be far more important than was yet realised; as early as July 1940 he secured an Order for the full or part-time appointment of doctors in large factories and persuaded the British Medical Association to set up a Committee on Industrial Health in Factories. There were so many other claims on doctors and nurses in wartime that Bevin had to fight a long battle with local medical committees to get a nucleus for what he planned to make a permanent service. By the end of 1944 the number of full-time appointments for doctors had been pushed up from 30 to 181, and of part-time from 50 to 890; the number of nurses had risen from 1,500 to 7,800.

Bevin had made up his mind that the administration of the Factories Acts should not return to the Home Office: he was equally determined to keep medical services in the factories out of the hands of the Ministry of Health and link it with the work of the medical branch of the factory inspectorate. He got this principle accepted in the 1942 White Paper on a National Health Service; in March 1943 set up an Industrial Health Advisory Committee, and in April of that year convened the first public conference to be held on industrial health problems.

The Order of July 1940 also gave the factory inspectors power to require the appointment of officers for the supervision of welfare where more than 250 people were employed in a factory. The Minister was ready to sponsor and subsidise training to provide such a service and under this official prompting the demand for welfare supervisors exceeded the supply, especially in factories where women were employed in growing numbers. Bevin's ambition was to see this development contribute to the establishment of a proper profession of personnel management, another object for which he continued to work with missionary zeal throughout his time at the Ministry.

Closely linked with both health and welfare were good feeding arrangements. Bevin's answer was to increase the number of works canteens. More than 200 had been set up before he became Minister, but he believed that firms—as in the case of medical officers—would act more rapidly on their own if they knew that the Ministry had powers to make them provide meals. The appropriate Order was issued in November 1940 and arrangements were made for the State to help with any capital expenditure involved. Food supplies were a matter for the Ministry of Food, but Bevin was dissatisfied with the complaints of poor meals and inefficient catering. His own solution was to set up a Catering Corporation. This did not find favour with the Minister of Food. So Bevin drew up a new Canteen Order[1] giving the Chief Inspector of Factories the power to serve notice on an employer whose canteens were being run unsatisfactorily and direct him to remedy the defects. This, and the appointment of catering advisers, helped to raise the standard of catering. By 1944 the number of canteens set up under the 1940 Order had passed the 5,000 mark and the effect of the Ministry's campaign had extended to many smaller firms which were not subject to compulsion: another 6,800 of them had established their own canteens by the end of the war.

Bevin was not content to press for communal feeding only in factories. At the beginning of 1941 he extended the provisions of the Order to building workers on construction sites, and yet another Order made in February 1941 placed an obligation on Port Authorities to provide hot meals for dockers. By the middle of 1943, 171 canteens had been set up in the docks, where the conditions of work in the open and in all weathers made them particularly necessary.

As early as the end of June 1940 Bevin became convinced that the long hours of work called for in the emergency of May would defeat the Government's own objective by undermining health and reducing rather than increasing output. At that time the war industries were commonly working twelve hours a day, often seven days a week. Bevin insisted on restoring the statutory limitation of the Factories Acts for the hours worked by women and young people—fixed at a 60-hour week of 6 days—and (although they were not covered by the Acts) strongly urged the same limits in the case of men, particularly a day off in every week. The supply departments, for understandable reasons, were reluctant to agree. As late as 1942–3, men were still working excessive hours in many factories and women in the aircraft industry were called on to stay at work for the maximum of 56–60 hours—in the face of conclusive evidence that such demands lowered efficiency and raised the rate of absenteeism. It was only in the last two years of the war that Bevin won his protracted battle with

[1] Factories (Canteen) Order, 1943: S.R. & O. 1943, No. 573 (7 April 1943).

managements and the supply departments to limit hours of work in the interests not only of the workers but of productivity.

The same arguments applied to holidays, but in this case met with less opposition. Paid holidays for factory workers was a reform which Bevin had played a great part in securing shortly before the war.[1] Although this went by the board in the critical summer of 1940, he succeeded in getting a week's paid holiday and a break at Whitsuntide restored in 1941 and for the remainder of the war.

5

It was not enough, in Bevin's view, to expand provision for welfare inside the factories; he wanted to extend Government initiative beyond the factory gates to living and travelling conditions and the recreation of the workers. This was a new departure and he showed himself aware of the obstacles it could encounter. As he told the House of Commons in 1942:

"I explained to employers and trade unions the need to remember that the worker inside the factory is a different person from the worker outside the factory: that we must not have a kind of industrial feudalism growing up in war, under which firms would take the responsibility of looking after their people even when they had left the factories. A person will accept discipline inside the factory, but immediately he is outside the door he becomes a free citizen. Therefore it is necessary to have a different organisation to deal with him after he leaves the factory gate."[2]

With this in mind, he went out of his way to enlist the support of the voluntary bodies already at work. He set up a Central Consultative Council of Voluntary Organisations, and the welfare officers appointed by the Ministry—a hundred and twenty by the end of the war—were instructed to work in the closest collaboration with the voluntary organisations and local authorities in their areas. This was a wise policy for it enabled him to draw on a fund of energy and local initiative which no government department could have supplied. The impetus, however, continued to come from the Ministry, and by the summer of 1941 a number of schemes had been started which

[1] See Vol. I, p. 601.
[2] House of Commons, 22 July 1942, Hansard, Vol. 382, cols. 52-3.

helped substantially to relieve the human problems of transferring men and women to unfamiliar work and unfamiliar surroundings in the straitened circumstances of wartime.

The first and most obvious of these was where to live, especially for those working in the shadow factories, on construction sites or in Royal Ordnance Factories situated in isolated areas. Billeting officers helped those transferred to find lodgings, but compulsory billeting was unpopular and difficult to enforce. In April 1941, Bevin secured agreement to the setting-up of a National Service Hostels Corporation under his Ministry's aegis and characteristically appointed as chairman of its board the managing director of the Dorchester Hotel, H. R. (later Sir Harry) Methven. At its peak, the Corporation was running 155 hostels and camps (most of which it had built), with a total capacity of 73,000 beds.

Travelling to a strange town and arriving there in the blackout, with no idea of where to go, was another problem which caught Bevin's attention. Many who agreed to transfer gave up and returned home when they could find nowhere to live and no one to help them overcome the initial difficulties. Through the Minister's efforts, reception offices were established at the railway stations in the main industrial towns and kept open, when necessary, round the clock. The local offices of the Ministry of Labour were used to provide a centre where the national registration, food and billeting officers were brought together with representatives of the employing firms, to meet new arrivals. By November 1941, 25 reception hostels had been opened and 27 more were planned in which newcomers could spend a night or two while lodgings were found for them. Escorts were provided for women on rail journeys and arrangements made for those who had to pass through London to be met, given a meal and put on their way. These arrangements had a big effect in overcoming working-class opposition to drafting young women away from home.

Even for those who continued to live at home, travelling to and from work every day became a major problem, especially in the winter months and the blackout. The time spent in waiting for over-crowded buses, and the lack of shelter from the weather, depressed morale and sent up the figures for absenteeism and illness. Bevin went to Bristol to see for himself and when he came back took the matter up urgently with the Minister of War Transport—if there was one subject he knew about, it was bus services. Using his own powers,

The War Cabinet in October 1941. *Seated*: Anderson, Churchill, Atlee, Eden. *Standing*: Greenwood, Bevin, Beaverbrook, Kingsley Wood.

The Minister of Labour in his element. *Above.* In a Midlands factory. *Below.* On a tour in Lancashire.

Bevin did everything he could to provide more bus crews, including the loan of drivers and even vehicles from the Services. He persuaded the Ministry to set up Regional Transport Consultative Committees to help with the regrouping and improvement of services; at the same time he pressed the bus companies through the Joint Industrial Council to improve wages and conditions. Bevin was strongly in favour of imposing on employers a general scheme for travelling allowances; he was persuaded to abandon this as an encroachment on the freedom of negotiation between workers and employers, but he succeeded in carrying out a modified proposal for such allowances to be paid in the engineering industries whenever conditions made longer journeys necessary.

Recreation and entertainment were just as important, he thought, as decent living conditions. He gave strong support—including financial assistance—to the National Association of Girls' Clubs, which by the end of the war had organised fifty-six clubs for women workers; he took a personal interest in such B.B.C. programmes as "Music While You Work" and "Workers' Playtime", and by the beginning of 1942 the welfare officers of his Ministry were arranging E.N.S.A. concert parties to visit over a thousand factories, most of them once a week. In April 1941 he made a grant to the first rest home for women workers at Colwyn Bay, the forerunner of another scheme in which he took great interest and which by 1944 was providing over 12,000 women a year with a break of one or two weeks.

Opening one of these rest homes, provided by American gifts for the working women of Bermondsey (at Matfield Court, near Tunbridge Wells), Bevin showed some of the feeling that lay behind his insistence on welfare:

"Bermondsey is specially near my heart because it is Dockland. It is there that the people have lived and reared their families, provided the labour that has helped to create the wealth of Britain and the world, and being part of Dockland it is one of the enemy's principal targets. The bombing has been terrific, homes have been destroyed and many have paid for our future freedom with their lives. Night after night the women's and girls' only rest has been in the shelters and yet with cheery hearts and magnificent courage they have returned to their factories to produce the necessities of life. . . .

"What people! They have never correctly estimated their own value and no one has put a correct value upon them, but this war has brought out a character and courage far beyond anything that was dreamt of before it began. . . . It is this that makes me feel that when victory is achieved and the grimness of war over, the resilience of our women folk and the indifference of our people to danger have been such that we are bound to have a renaissance. The creative genius of their brothers will be revealed and they will proceed to use all their resources and skill to wipe out the horrible mortgage left to us by the industrial revolution with all its narrow streets and insufficient space to live. We feel in our hearts that these conditions are not worthy of a great people and do not give the right opportunity for the expansion of the character and possibilities of the ordinary folk of these islands."[1]

There was another group of workers whose treatment Bevin had long felt was not worthy of a great nation—least of all of a great maritime nation—the seamen of the Merchant Marine. In addition to the Merchant Navy Order and the reforms he was able to carry out under this, he established a separate Seamen's Welfare Advisory Board (October 1940), encouraged the setting-up of Port Welfare Committees in all the main ports and appointed seamen's welfare officers to stimulate the work of voluntary bodies.

Thanks to this initiative, a number of seamen's hostels and clubs were opened, amongst them a merchant-navy hotel in London where seamen and their wives could stay. Nor were these benefits limited to seamen from the United Kingdom. Special hostels and clubs were provided for men from the colonies, India and China, and the Allied Governments followed the British example in making arrangements for their own merchant marines. No task he undertook gave Bevin greater satisfaction than improving the conditions of the men on whom Britain's seaborne supplies depended, and however busy he was he would always find time to go down to the ports and open a new seamen's hostel.

6

Better welfare arrangements could do much to improve human relations and increase efficiency in industry, but wages and hours remained the central issue of industrial relations. For the first time for many years there was full employment and competition for labour. This was bound to send wages up and in some cases it was not only fair but essential for them to rise. Although munitions work was often well paid, there were other industries equally important in wartime

---

[1] Speech at Matfield Court, 21 June 1941.

(coal and agriculture, for instance) where wages were low and had to be raised if the country was to get the coal and food it needed. Women's work was another case: it would have been impossible to get the large numbers of women required to take up work if something had not been done to improve women's pay.

At this point, however, industrial policy had got to be considered in relation to the Government's economic and financial policies. The great fear of the Treasury was a runaway wartime inflation which they rightly insisted would not only hamper the mobilisation of the nation's resources but produce social injustice and bitterness. The danger was a real one. The demands of the war meant, on the one hand, the economy running at full stretch, a boom in prosperity and more money being paid out in wages and profits; on the other hand, with the maximum possible concentration on war production, far fewer goods and services for people to buy.[1] More money in circulation with fewer goods and services was bound to produce higher prices; these in turn would lead to demands for higher wages, and so ad infinitum, unless something was done to check the process.

What could be done? Any increase in the supply of consumer goods was ruled out by the needs of the war. The problem therefore had to be tackled from the other end and the measures proposed fell into four main groups. The first was government control of wages and profits in order to check the growth in people's incomes. The second was increased taxation and a drive for increased savings (the National Savings Movement) in order to mop up at least part of the increased purchasing power arising from higher incomes. The third was to cut down people's expenditure by rationing and other restrictions on buying. The fourth was to hold down prices by government order and by subsidising the cost of food.

A start had been made with most of these measures in 1940, but it was not until 1941 that they were combined into a coherent policy which did not prevent inflation during the rest of the war but succeeded in keeping it within reasonable limits. This policy was thrashed out in a lengthy series of discussions in the Lord President's Committee, of which Bevin was a leading member, and put into effect in the April

[1] In the spring of 1941 it was calculated that the probable increase in incomes during the coming financial year would be £100 million and the probable reduction in consumer goods and services £400 million. (Alternative figures were £150 million and £350 million.) This faced the Government during 1941 with an inflationary gap of £500 million. (Hancock and Gowing, p. 325.)

1941 budget and a succession of government orders during the rest of the year. The standard rate of income tax was increased to 10/– in the £; tax allowances cut; purchase tax and Excess Profits Tax (at 100 per cent) continued; rationing extended by the introduction of the "points system" for food and clothing; prices controlled, food prices subsidised and a sytem of "utility" clothing at controlled prices introduced.

The obvious omission from this list was any provision for government control of wages. This had been a feature of the Treasury's original proposals in 1929; the case for it on purely economic grounds was hard to refute, and there were economists both inside and outside the Government service who argued that, without such control, it would prove impossible to prevent inflation. The trade unions, however, were strongly opposed to any such policy (as were many employers) and in the discussions in the Lord President's Committee Bevin was a powerful advocate of their views.

He was not against the State intervening in wage questions if necessary. The Essential Work Orders not only made a guaranteed weekly wage obligatory but gave the Minister of Labour the opportunity to insist on wage increases in badly paid industries before he would agree to schedule them. None the less he looked on these as exceptions, emergency action which did not invalidate the general principle that wages were best left to industry itself to settle by the normal process of collective bargaining, with the fall-back of compulsory arbitration if the two sides failed to agree. This was the policy he had adopted in June 1940 on the recommendation of both the employers' and the unions' representatives on his Joint Consultative Committee,[1] and he defended it with the same tenacity as the parallel policy of keeping the compulsory direction of labour to the minimum.

The Treasury, however, maintained that in wartime the State had a direct responsibility to see that wage increases did not lead to inflation and argued that Bevin's policy of leaving the employers and unions in each industry to act as "the trustees of the nation" provided no safeguard other than the negotiatiors' sense of responsibility.

Bevin believed it worth taking this risk, for several reasons. The

[1] See p. 22 above. On that occasion Bevin put up the alternative suggestion that wages should be stabilised at existing levels with four-monthly reviews by a national arbitration tribunal. This was rejected by both the T.U.C. and the Employers' Confederation in favour of retaining collective bargaining.

first was that any alternative to collective bargaining, such as the statutory regulation of wages, would undermine the network of joint negotiation on which depended something more important even than wages—the chance of preserving sufficient mutual understanding and goodwill in industry to maintain industrial peace. The second was that he had sufficient experience of collective bargaining to believe that the two sides' sense of responsibility was, in practice, a very good guarantee of restraint, particularly if it was voluntary, not imposed. In any case—and this was a reason that carried great weight with all trade-union leaders—he believed that the right way to prevent inflation was, not by holding down wages at the expense of one section of the nation, but by increasing taxation and by holding down prices with the aid of subsidies paid out of general taxation at the expense of the community as a whole.

If prices could be stabilised, Bevin argued, it would be possible to keep wage rates steady as well, since it was the rising cost of living which, far more than anything else, had been the chief stimulus to wage claims in the first eighteen months of the war.[1] In the spring of 1941, the Government accepted this view which was explicitly stated by the Chancellor in introducing his budget (7 April). Henceforward wages policy rested, as the official history puts it, on "a combination of faith and works—faith in the moderating influence of the trade unions and action to control the cost of living".[2]

7

Such a policy did not satisfy those, like *The Economist* or Sir William Beveridge, who were as critical of Bevin's "voluntary" policy on wages as they were on manpower and who continued to urge the Government to control wages as well as prices. This was not a view, as Bevin pointed out, which received much support from industry where the practical problems were better appreciated. But soon after the announcement of the stabilisation policy, the Lord President's

[1] In 1939 the wage rates of about one and a half million workers were directly tied to the cost of living index. This number was increased during the war to two and a half. Even in industries where wages were fixed by negotiation, applications for an increase had been almost wholly based on the rise in the cost of living. (Hancock and Gowing, pp. 337–8)

[2] Parker, *Manpower*, p. 428.

Committee again examined the matter and Bevin had to defend his opposition to wage control.

He did so on two grounds. The first was that for the Government to try and prevent wage claims being made and discussed would at once stir up widespread discontent throughout industry. In view of its overriding need for increased production the Government would not be able to enforce its policy; it would back down on its own pronouncements, as it had in the First World War, and would end by losing far more than it gained. His second reason was that to attempt to carry out such a policy would put an end to the policy of co-operation between trade unions and the Government on which much else besides wage restraint depended. Trade-union leaders and officers had shown themselves ready to go a long way in pressing their members to put the national interest before sectional advantages— for example in agreeing to set aside hard-won rules and regulations in many industries and in damping down wage claims—but they could only do this if they preserved their authority with the members who paid and employed them. This depended upon leaving them freedom to negotiate the day-to-day adjustment of wages and conditions of work. If the Government took away or impaired their freedom, instead of trusting to their sense of responsibility, this would either destroy their authority or drive the union leaders in self-defence to take up a much stiffer attitude. In either case, the Government would make it much more difficult for the unions to co-operate.

The other members of the Lord President's Committee accepted Bevin's argument. But, in view of criticism in Parliament and the press, they felt it necessary to justify their economic policy by a statement which would make some reference to the need for restraint in raising wages. Bevin warned them that if they did so they would stir up trade-union suspicion that the Government was giving a lead to employers and, more important, to arbitration tribunals to turn down wage claims. When the Committee persisted, however, he agreed to join the Chancellor (Kingsley Wood) in meeting the T.U.C. and putting the Government's anxieties to them.

The General Council reacted exactly as Bevin had foreseen. Although none of its members disagreed in practice with the Government's concern about wages and inflation, the fact that the Government should want to draw public attention to this at a time when there was a strong demand for a wage-stop in quarters not friendly to the trade-union movement met immediate opposition from the T.U.C. They had heard all this before, Citrine told the Ministers, when Simon was Chancellor in December 1939. "They expressed then the emphatic view that any attempt to control movements for increases of wages is impracticable and undesirable. . . . The Council just as emphatically affirms that conviction today."[1] The only result of the encounter was a revision of the text of the proposed White Paper[2] which neither removed the impression left on the T.U.C. nor satisfied those critics who thought the Government's policy weak.

Before the end of 1941 the question of wages again came up for discussion in the Lord President's Committee. One of Bevin's strongest arguments against a wage-stop had been that the Government itself would be the first to ignore it when it threatened to create either industrial or political difficulties. The truth of this was soon proved by the grant of a 12/- a week increase to land workers by the Agricultural Wages Board. This was twice as much as Bevin himself had recommended and it had the effect of throwing wages in other industries out of balance. Railwaymen, for instance, had always been paid a little more than agricultural labourers and soon put in a claim for an increase which the Lord President's Committee found it impossible to refuse.

The Treasury, however, still hankered after some stronger guarantee against inflation than leaving wages to be settled by the two sides of industry. On top of the rise in agricultural and railway wages, an increase of 5/- in the weekly rate for the comparatively well-paid workers in engineering and shipbuilding—turned down by the employers but granted by the Arbitration Tribunal in December—appeared to confirm fears of the situation getting out of hand and revived complaints that the Government had no wages policy. Bevin was unmoved either by the fears or the complaints. He took the lead in the renewed discussions in the Lord President's Committee (December 1941) and this time marshalled a wide range of arguments in support of his policy.

He began by explaining why he attached so much importance to the practice of settling wages by negotiation backed by arbitration. Collective bargaining had already been accepted over a wide field of

[1] T.U.C. Statement: *The Trade Unions and Wage Policy in War Time* (1941). See also Citrine's speech to the 1941 Trades Union Congress.

[2] Cmd. 6294 (July 1941): *Price Stabilisation and Industrial Policy.*

industry before the war, and if it could be extended and preserved would mean the minimum of industrial disturbance in settling wage claims. This was important enough in wartime, but the addition of compulsory arbitration—which, if it became established during the war, had a good chance of being continued afterwards—provided the best guarantee against a runaway inflation when the war was over. That, not wartime, Bevin argued, was the real danger period and he pointed to the inflationary experience of 1918–20 as support for his view. It was then, far more than during the war itself, that the damage had been done. Immediately the 1914–18 war was over, the unions had repudiated arbitration and struck for higher wages. At the same time, in the absence of an independent court to examine the facts objectively and guide opinion, the Treasury had resorted to a deflationary policy without proper understanding of its implications. This combination of the unions pressing for wage increases (which led to inflation) with a Treasury policy of deflation, Bevin maintained, had produced the bitter industrial strife of the early 1920s which ruined any prospect of the revival of British trade. It was to avoid a repetition of this experience and build up arbitration as an objective check on union demands and government policy, after as well as during the war, that Bevin argued for retaining and strengthening the procedure he established in June 1940.

What were the arguments against it? Had it in fact worked badly? He produced figures to show that the increase in wage rates between October 1938 and July 1941 did not amount to more than 18 per cent, well within the 28 per cent rise in the cost of living. Earnings, it was true, had risen much more steeply, by 43 per cent. But once they had cut out the anomaly of high Sunday pay (which Beaverbrook, he could not resist adding, had encouraged against his advice), higher earnings largely represented greater effort and increased production, exactly what they wanted. Provided it came as higher earnings and not on the wage rate, Bevin said flatly that he did not mind how much a man drew and had in fact encouraged some industries like building to go over to payment by results in order to get higher output. Far from breaking down, as the alarmists feared, the existing system was justified by its results.

"Now what," Bevin continued, "would be the alternative to the present machinery of adjustment? I am waiting for my colleagues to show me."[1]

If they meant to freeze wages, they would have to make adjustments to the lower rates. So the first result would be to raise wages. None the less that could be done and a national wage scale on the lines of the Civil Service worked out. But the old guide in collective bargaining—what an industry could afford to pay—would have to be abandoned and the payment of subsidies would have to be accepted not only for the war but for the post-war period. Quite apart from the loss of flexibility needed to recapture trade after the war, were his colleagues prepared to face the further consequences of such a policy? For the question which would immediately be raised was whether, if a system of fixed maximum wages was to be established, the private employer must not disappear and a fully socialised system of industry be introduced.

"I have no objection to facing this," Bevin added, "but I must resist to the uttermost and make it a real issue, that the State should say to one section of the community, you must work for another section at a maximum wage."

He was not prepared to accept the argument that the Excess Profits Tax represented a *quid pro quo* for the control of wages: the equivalent of the Excess Profits Tax was the Essential Work Order and the other Orders restricting the freedom of labour. If wage control was imposed by the State, Bevin made perfectly clear, then Labour would demand the socialisation of industry.

No doubt, in a Coalition Government, this was the decisive argument, but he added at least one other which carried weight. If the State took over responsibility for wage rates, not only would it be faced with an immediate demand to raise the lowest wages, but the claims of different industries would become a matter of parliamentary debate and political competition. Was it worth while to embark on these dangerous courses, stirring up vehement political and industrial controversy, when the existing system of adjustment was working satisfactorily, had preserved industrial peace, helped not hindered the mobilisation of the economy, and not produced more than a mild degree of inflation?

[1] The whole of this account is based on notes, dated December 1941, which Bevin dictated to guide him in the Cabinet discussions.

Wages policy continued to be a subject of controversy, particularly Service pay and allowances, until almost the end of the war.[1] But so far as the Cabinet was concerned, Bevin's argument settled the matter and the possibility of the direct control of wages was not raised again.

Bevin never pretended that the policy he advocated would prevent wage increases: he thought it impossible for the country to get through the war without substantial rises, certainly in the worst paid occupations. All he maintained was that, if wages were left as far as possible to be settled by negotiation, this was the best way of avoiding serious industrial trouble and, secondly, that if the Government succeeded in holding prices steady, the wage increases which had to be granted need not lead to a runaway inflation. The later history of the war confirmed his forecast: there were substantial wage increases and there would have been serious industrial trouble—in the mines, for instance—if these had been refused. But the policy of stabilisation worked at least well enough to hold inflation within bounds and the fears that wages would get out of hand proved exaggerated.[2]

[1] Reacting to renewed criticism in the press, Bevin dictated a note on 2 September 1942 in which he asked:
"What does *The Times* mean by absence of wage policy?
What is the policy they are advocating?
Are wages to be fixed for each Trade by legislation?
Is payment for overtime to be abolished?
Is payment by results to be wiped out?
If not, how is a level to be kept?
Is there to be grading?
At what level should State wages be fixed?
Would it not become as inelastic as the Public Service?
How would you deal with Management?
What is the position to be of the Trade Unions?
How would you deal with the Railway claim?
The State makes an offer which it thinks just; the men and their Union refuse. Should Parliament debate it. Or what?
If the men struck, how would you deal with it?
How would you readjust at the end of the war?"

[2] The evidence for this is set out in Parker, pp. 433–7. Briefly, there are three sets of figures which matter. First, in the group of industries that were most important in war production, until July 1943 the upward movement of *wage rates* was below the rise which had occurred in the cost of living even after the stabilisation of prices; at the end of the war the increase in wage rates above the level of October 1938 was still only 9.5 per cent more than the rise in the cost of living during the war period. Compared with October 1938, wage rates in these industries had risen by 43 per cent. Second, *weekly earnings* in the same group of industries: these went up by 60 per cent between October 1938 and July 1942, but this includes the

8

Bevin's belief in collective bargaining as the best means of settling wages echoed the conclusions of the Whitley Committee[1] which had been set up to advise on labour problems in the earlier war. They too had laid down as a general principle "the advisability of a continuance, as far as possible, of the present system whereby industries make their own agreements and settle their differences themselves", and had made their first recommendation the setting up of joint industrial councils, with equal representation of employers and workers, for all important industries.

There have been two productive periods in the formation of such councils, the years 1918–21 just after the Whitley Report came out and 1940–45 when Bevin was Minister of Labour. Of the 112 joint industrial councils[2] or similar bodies active in 1946, forty-six had been set up in the first of these two periods, another forty-six in the second. The latter group included the National Negotiating Committee for the mines, seven separate councils for different branches of retail distribution covering most of the shops in the country, and others for such varied occupations as rubber manufacture, plastics, road-mending, bacon-curing, inland waterways and the ophthalmic-optical industry. Taken together with the statutory Wages Boards, which Bevin extended in 1943 and 1945,[3] they were responsible at the end of 1946 for settling the wages and conditions of fifteen and a half out of a total of seventeen and a half million workers. And this time there was not the same falling away that there had been after the

---

period when very long hours were worked, after the fall of France. Between July 1942 and July 1945 the rate of increase was slowed down and added only a further 20 per cent. Finally *weekly wage rates* in industry as a whole. Surprisingly enough, the rise here was greater than that in the group of manufacturing industries, 50 to 51 per cent above September 1939 by the end of the war. The higher figure seems almost certainly to be due to the award of increased wages to workers in less well paid occupations, such as coal mining and agriculture, and can hardly have had serious inflationary consequences.

[1] Set up in October 1916 under the chairmanship of J. H. Whitley, then Speaker of the House of Commons, to inquire into the relations between employers and employees.

[2] Names and functions varied a good deal, but it is simpler to use the generic term.

[3] See below, pp. 351–4.

First World War. At the time Bevin died (1951) the percentage of workers covered in this way was still as high as 80 per cent and the number of joint industrial councils had been brought up to 130.

The Whitley Report had looked forward to the extension of collective bargaining into a form of industrial self-government based on the same joint procedure. In seeking for a permanent improvement in the relations between employers and employed the Committee reported, "What is wanted is that the workpeople should have a greater opportunity of participating in the discussions about and adjustment of those parts of industry by which they are most affected." Bevin fully shared this view. Between the wars he had shown what could be done by the part he played as chairman of the National Joint Industrial Council for the Flour-Milling Industry,[1] and as Minister of Labour he took every opportunity to urge the newly founded councils to extend their interests to other questions besides wages and hours, and to discuss such matters as the future prospects of their industries, training, redundancy, post-war reorganisation and anything else which might affect the lives of those who worked in them.

The most interesting development of the practice of joint consultation during the Second World War was at the factory level—the joint production committee. This, too, had been recommended by the Whitley Committee, but had failed to make much progress in face of the suspicion with which both management and the trade unions (for very different reasons) regarded Works Committees, as they had earlier been known. The war provided a much more favourable atmosphere for such experiments. There was a common interest in the expansion of production, to which even the most militant of shop stewards were converted after the attack on the Soviet Union. There were many matters arising from the change to new types of production, the application of dilution agreements and the operation of Essential Work Orders which could only be settled by workshop discussion, and the demand for this on equal terms between management and workpeople was strongly backed by the unions.

Woolwich Arsenal claims to have started the first joint production consultative committee, and the practice soon spread for obvious practical reasons. With Bevin at the Ministry of Labour there was no lack of official support. In February 1942 the unions concerned

[1] See Vol. I, pp. 380–1.

signed an agreement for the establishment of such committees in all undertakings operated by the Ministry of Supply and the next month the engineering unions reached agreement with the Engineering Employers' Federation to set up similar bodies in private under-takings "to consult and advise on matters relating to production and increased efficiency . . . in order that maximum output may be obtained from the factory". By July 1943, according to the Ministry of Production, there were 4,169 joint production committees or their equivalents operating in private engineering and allied firms with over two and a half million workers on their payrolls.[1] Under different names, the Yard Committees in shipbuilding, Pit Production Committees in virtually every coal mine in the country, Shop or Site Committees in the building industry were variants on the same principle. How far all this activity contributed to raise productivity it is impossible to calculate.[2] Experience varied widely: one survey concludes that "many committees were generally regarded as successful and perhaps almost as many as unsuccessful".

The difference between success and failure usually turned upon local factors, the quality of the representatives on both sides (one or two men with personality and enthusiasm could make the whole difference), the attitude of management and of the unions, the local tradition of labour relations, the nature of the processes involved in the work and so on. This seems to be equally true of experience after the war, when there was a sharp fall in the total of active committees, but a number continued with success as a permanent feature of in-dustrial relations in certain works. A generation later, opinion in industry on both sides still seems to be divided whether this is the sort of experiment which can only be made to work, and only has value, in the special circumstances of wartime, or whether, as the Whitley Committee and Bevin hoped, this offers one of the directions in which to look for an improvement in the relations between management and workers.

As Minister of Labour Bevin practised what he preached and made great use of joint consultative and advisory bodies in the belief that if

---

[1] According to K. G. J. C. Knowles, *Strikes* (1954), in 1944 this had risen to 4,500 and three and a half million workers.

[2] See Allan Flanders and H. A. Clegg (ed.), *The System of Industrial Relations in Great Britain* (1961), c. vi, and Inman, *op. cit.*, pp. 371–92, for a very interesting account of the development and problems of joint production committees.

you wanted people to co-operate it was common sense to discuss in advance what you proposed to do and get the views of those who would be most affected. His success was largely due to a clear conception of the purpose of consultation. He did not make the mistake some ministers have made, either of supposing that consultation could be a substitute for a policy or on the other hand of treating it as an exercise in public relations at the end of which the committee's job was to approve with the appropriate formalities what the minister proposed to do in any case. Particularly if Bevin himself was in the chair, there was no lack of ideas about policy, no abdication of the Minister's responsibility to give a lead, but he expected the members of a committee to express their own views and frequently modified proposals to take account of them. The Minister's Joint Consultative Committee with the Employers' Confederation and the T.U.C. was certainly no sham: all the major decisions on manpower and labour policy were discussed in advance with its members. The same can be said of two other bodies of which Bevin made the fullest possible use. One of these was the Factory and Welfare Advisory Board on which he tried out his ideas about the development of welfare services and which discussed the most varied range of subjects from industrial health and day nurseries to transport services and the rehabilitation of the disabled. In March 1941 he created a Women's Consultative Committee which met fortnightly under the chairmanship of one of his Parliamentary Secretaries and played a parallel role in framing the far-reaching proposals for the conscription and employment of women.

Similar machinery for consultation on a tripartite basis of ministry, employers and unions, was set up to deal with the problems of particular industries. Examples are the Engineering Industry Advisory Panel and the Building and Civil Engineering Labour Advisory Panel which produced the Government White Paper on Training for the Building Industry. Another is provided by the distributive trades: when it was decided to withdraw younger women from these for more essential work, Bevin gave the job of drafting a scheme to a Central Advisory Panel of employers and workers under a Ministry chairman, with the result that the operation, which gradually withdrew a majority of women up to the age of forty-one, was carried through without friction.

In three other major industries, shipbuilding, iron and steel, and chemicals, each of which was allowed to preserve its labour force intact within a "ring fence" but had to meet its labour requirements from its own resources, the transfer of workers on which the scheme depended was placed in the hands of local committees of employers and workers with a Ministry of Supply representative and a Ministry of Labour chairman.

What is interesting in this development is that Bevin as Minister of Labour not only established for the first time the regular practice of the Government calling in the trade unions for consultation,[1] but began to introduce a tripartite pattern of consultation and co-operation between Government, employers and unions as a way of dealing with industrial and economic problems. He thought it equally applicable in the international field and, whether or not he took over the idea from the I.L.O., one of the reasons for his un-wavering support of that body was the fact that this was its method of organising its activities.

In the later course of the war this practice brought him under sharp criticism (from such diverse quarters as *The Economist* and Aneurin Bevan) for undermining the parliamentary in favour of the corporate state.[2] He himself thought of it, not as an alternative to parliamentary democracy but as a practical way of dealing with problems for which traditional parliamentary institutions provided no answer. For the same reason both Labour and Conservative Governments have continued the practice since the war (e.g. the National Economic Development Council). Bevin, however, never supposed that consultation meant that Government had to play a neutral or passive role: on the contrary he thought that consultation would only succeed where the representatives of Government gave a positive lead, a condition which has too often been ignored on subsequent occasions.[3]

---

[1] See below, pp. 136–7.
[2] See below, pp. 191 & 275–6.
[3] See Andrew Shonfield: *Modern Capitalism* (1965), esp. c. 5, "Arm's Length Government".

CHAPTER 4

Character and Colleagues

I

ENOUGH HAS now been said to give an idea of the policy which Bevin pursued as Minister of Labour; it is time to say something about the man himself, about the effect on him of office and about his relations with his colleagues.

On the day Bevin's appointment as Minister of Labour was announced, the *Daily Express* described him as "a bad mixer, a good hater, respected by all". The *Express* had never been friendly to Bevin, but its snap judgment does not misrepresent the view which was held about him at this time by many people.

No trade-union leader can expect to be popular, and the press delighted to draw an unflattering picture of the trade-union boss dictating to the public as well as to the members of his own union. His rise in the trade-union movement to the point where his only rival was Citrine had left him with plenty of enemies who resented both his ability and the power he exercised as general secretary of the biggest union in the country. In politics, the episode most commonly recalled and always held against him was the attack he had made on Lansbury at the 1935 conference, an episode in which the standard adjective applied to Bevin's behaviour was "brutal". No one had expressed stronger criticism of politicians who believed in "tactics" instead of "playing straight", and his quarrel with the Left and "the intellectuals", a comprehensive term of abuse on Bevin's lips, had been revived by the attempt to create a Popular Front.

Bevin did nothing to placate his critics. Where he felt strongly about an issue, he expressed himself forcefully and refused to follow the parliamentary tradition of keeping what was said in a debate and the ordinary civilities of life separate. When he fought, he fought hard and to win. Conciliation was not in his nature, he was a "good hater" as the *Express* said, and he made no attempt to conceal his scorn for his opponents, particularly those in the Labour Movement. The fact that he was also "a bad mixer" put him at a further disadvantage. Even those who agreed with him or at least admired his independence and integrity often found him difficult to approach. He was reserved in private life, suspicious and slow to give his trust or admit anyone as a friend. His ability, his strength of character and determination were obvious and admitted, but he was respected or feared rather than loved and his position in the Labour Movement, although powerful, left him personally isolated.

In the circumstances of 1940, none of this appeared to matter—any more than it did in the case of Churchill—compared to the fact that Bevin had the qualities of toughness and courage needed to stand up to the crisis. This was what counted and almost overnight the picture of Bevin given by the press began to alter. But many who knew him wondered whether, once the crisis was over, he would ever make a success of politics, especially parliamentary and party politics for which he seemed temperamentally ill-suited.

These doubts were justified, at least in part. As this volume shows clearly enough, Bevin took a long time to adjust himself to the differences between the House of Commons and the T.U.C., even longer (if he ever did) to those between the trade-union movement and the Labour Party. But this is only part of the story. Ernie Bevin had not risen to the position he held as a trade-union leader, nor won the respect of many outside the Labour Movement without other qualities besides those of being tough, determined and a hard fighter. These were essential: without them Bevin would never have survived in a movement which was anything but a band of brothers and where power, to be effective, had to be exercised openly—whether it was the power of fists, voice or votes. What distinguished him from the run of trade-union leaders, however, was the other qualities he possessed besides these.

If it took time for these to be recognised in the House of Commons and, last of all, in the Labour Party, they were very soon recognised by those who worked most closely with him in the Ministry of Labour, in Whitehall, in the Cabinet and its committees. Reversing the normal order of events, Bevin made his mark as a minister before he made it as a politician. Recognition, however, when it did come, was complete and he ended the war with a greater reputation than any other member of the coalition except the Prime Minister and with his standing as a national leader accepted by all.

What Bevin lacked before he entered the Government was a framework big enough to develop his gifts in. His own union, which he had created nearly twenty years before, no longer presented the stimulus of a challenge and, although he found greater scope in the T.U.C., it was still too far from the centre of power in the 1930s to bring out all that was in him. Much of the reputation Bevin had before the war for being (in a phrase used of Sir Robert Peel) "a bad horse to go up to in the stable"[1] sprang from frustration.

Office changed this. In the first two years (1940–42) he had a rough time in the House of Commons and came in for more criticism than any other member of the War Cabinet. But he was no longer frustrated. At last he had a job to do big enough to tax all his abilities to the full, a job which he believed he was the one man in the country capable of carrying out successfully. From the day he walked into the Ministry of Labour he acquired a new access of energy and self-confidence.

Like Churchill, Bevin enjoyed being a minister and exercising power with that uninhibited enjoyment of his own performance which shocked the conventional but was one of the hallmarks of both men's strength. He only needed to walk into a room or start speaking for his confidence in himself to make an immediate impact. Where he got this confidence from is a mystery. His early life might have been expected to produce exactly the opposite. But from the time anyone began to take notice of him, in Bristol as a young man, it was this characteristic which attracted attention.

Bevin might be a newcomer to parliamentary politics, but when it came to the contest for power between ministers and departments, he had little to learn from anyone. He had been playing this sort of game since he first became a trade-union official; he took it for granted as part of any power structure and assumed that this was a large part of what politics was about.

Attlee wrote of him:

"Because of his own genius for organisation and his confidence in his own strength, he did not fear—he embraced—power. Lord Acton's famous

[1] This phrase was used by the *Manchester Guardian* in its obituary notice of Bevin, 16 April 1951.

dictum on power probably never occurred to him. If he agreed that power corrupts, he would have said that it corrupted only the men not big enough to use it. And power was given to Ernest. Men recognised in him a national leader, someone to lean on. He attracted power. At a time when the Labour Movement had all the hopes, aspirations, ideas and saints necessary for Utopia, Ernest helped bring its feet to the ground by insisting that these things without power were useless."[1]

You had only to look at Ernie Bevin to see all this expressed in his physical appearance: the heavy build, big head, broad face, the obvious strength of his hands and shoulders, combined with a harsh powerful voice and a rolling walk reminiscent of a battleship in heavy seas. "He would square up to anyone," Attlee wrote, "physically or morally, with relish."

But this appearance was also misleading. Those who drew from it conclusions about the sort of man they were dealing with were often disconcerted to find that at close quarters Ernie Bevin was a more complicated character than Churchill's "working-class John Bull". His unusual confidence, for example, was combined with an equally marked sensitivity to criticism, which he was always inclined to take as a personal attack, and with a strong suspicion which, once it was aroused, put a brick wall of distrust between him and anyone he took against. In one mood he could be warm and expansive, displaying an understanding of human beings and their problems all the more impressive because of the rich experience on which he drew to illustrate it. In another he could be angry and unreasonable, impossible to argue with and full of prejudices. In handling a difficult committee he would show all the patience and skill of a great negotiator; the same day, he would put up everyone's back at a press conference by turning furiously on an unfortunate questioner, and then go on to compel admiration by the grasp he showed of the possible objections to some scheme he was proposing and the care he took to meet them.

These variations of mood—all the way from the sullen to the exuberant, from the truculent to the philosophical—were the expression of a temperament which was anything but phlegmatic, was in fact difficult and passionate. Although at times he made an

[1] The quotation is taken from a review of the first volume of the present biography written by Lord Attlee and spread over two numbers of *The Observer*, 6 and 13 March 1960. This constitutes one of the most valuable portraits of Bevin by a man who had a unique opportunity for knowing him closely.

effort to control it—Lord Bridges, then Secretary of the Cabinet, expressed admiration for Bevin's self-control on one occasion when he got out of the Cabinet Room before exploding—most of the time his expression and manner showed only too plainly what he was feeling.

The author of an *Observer* profile of Bevin was quite right when he said that the Public Schools had shaped the characters of many more than those who attended them. It was from their tradition that the conventional picture of the modern Englishman had been formed, "with his gentleness and self-control, his mild manners and habit of understatement, his well disciplined mind and body, his modesty, decency and restraint". None of this was true of Bevin who

"escaped all public school influences and presents an exactly opposite picture: sprawling in body and untidy in mind, dictatorial in manner and exuberant in utterance, with a streak of ferocity, and a fund of native shrewdness, with every human instinct alive and unblunted, rollicking in his jollity and surly in ill-humour, as cunning as a peasant and yet, withal, as genuine as nature itself."[1]

"Like Churchill," *The Times* said, in a leading article at the time of his death, "he seemed a visitor from the eighteenth century."[2]

2

The *Observer* profile belongs to a period a few years later when, at the Foreign Office, Bevin's personality achieved its final expansive flowering. During the earlier years of the war, he was more concentrated, more restrained, but the process of expansion had already begun and the characteristics of his maturity had established themselves. Apart from those already described, his self-confidence and range of temperament, four others demand separate notice: his imagination, his vanity, his humanity and his loyalty.

"Ernest looked, and indeed was," Attlee wrote, "the embodiment of common sense. Yet I have never met a man in politics with as much imagination as he had, with the exception of Winston."[3] Churchill had an historical, Bevin a human imagination, the one roused by

[1] *The Observer*, 23 May 1948.
[2] *The Times*, 16 April 1951.
[3] Attlee in *The Observer*, March 1960.

great themes and great men, the other by everyday life and ordinary people. Churchill's gift was to see and make others see decisions in the light of history, Bevin's to see and make others see decisions in terms of human beings.

It was this unexpected quality of imagination in Bevin which led Attlee, when comparing Cripps and Bevin, to say that while both men were tremendous egoists, "Cripps had the egoism of the altruist, Bevin the egoism of the artist".

"I like to create, brothers," Bevin had told his union delegates, and now at last he had the chance to give expression to the creative side of his nature. It was not a literary or visionary but a practical imagination, best seen in the concern which he showed, even in the middle of a war, for welfare and rehabilitation and his insistence on humanising administration and the exercise of compulsory powers. It was fed, not from books, but from talking to people and from reflection on his own experience. He relied on his memory to store away all sorts of information, ideas and experiences which he had picked up, and would then produce them, often years later and often in unexpected combination. He had a gift not only of assimilation but of association enabling him not so much to reproduce as to select and combine what he remembered. "Given time," Citrine remarked, "there was hardly a problem for which Bevin could not come up with an answer."

It was in keeping with Ernie Bevin's naturalness of character that he should take an unabashed pride in his achievement. His own career fascinated him and when he relaxed there was nothing he liked better than to go back over its different episodes and marvel at the distance he had travelled from his early days in Winsford and Bristol. "You know, Harry," he said to one of his friends, "I'm a turn-up in a million." He was, and few would have denied it.[1] But for a long time it was a stumbling block to many brought up in the English tradition of good form that he should talk of "my" policy and "my" people, and say "I" where most people would have used "we". Others, particularly in the Labour Party, were irritated by Bevin's claim to have done more for the working class personally, as Minister of Labour, than any other Labour politician and revived the old

[1] Even Herbert Morrison, who had less cause than most people to like or admire him, wrote in his autobiography "Ernest Bevin was in some respects a genius" (p. 201).

complaint of his trade-union days, that he "hogged the credit".

Most people, however, who worked at all closely with Bevin were not much worried by his vanity. They looked on it as a defect which, in a smaller man, would have been intolerable but in a big man with so many other qualities was a surface fault, in some ways an endearing weakness which they were well aware of but did not take seriously. For Bevin's vanity never affected his integrity. No man was less changed by success. He had a proper sense of his own dignity as a minister, but judged—rightly—that it was more important that the Minister of Labour was Ernie Bevin than the other way round, and behaved accordingly. He made no pretence to be anything other than himself, showed not a trace of snobbery and was as impatient of the pomposities of office as of any other form of cant or humbug.

His private life was little changed either, except that he now worked even longer hours than before. He lived modestly, asked no special favours, although he controlled the whole manpower of the country, and when his wife fell ill nursed her himself. His solicitude for his wife, especially during the air raids on London, was one of the things about Bevin which made a favourable impression on Beaverbrook. In 1940 they moved for a time to the Strand Palace Hotel so that Mrs. Bevin should not be alone during raids: later they rented a flat and spent the latter part of the war at 20 Phillimore Court, Kensington. He took pains about his clothes and liked a well-made suit. If good food was put in front of him, he enjoyed it, as he did a drink or a cigar, but he paid little attention to such things any more than he did to what he earned or owned. His consuming interest was his job, and his favourite form of relaxation talking about it.

If anyone tried to invade Bevin's preserves or reduce them, he would fight to defend his authority, and did so with success throughout his ministerial career. The unusual thing about him, however, was to find this instinct for power (which was all of a piece with his appearance and personality) combined, not with personal ambition or a desire for fame, but with a seriousness and disinterestedness of purpose which, in the Labour Movement at least, have more often been associated with a distrust of power. Bevin attached little value to power for its own sake (he took it for granted), nor did he pay much thought, as Churchill did, to his place in history: what he valued power for was the use he could make of it. And the policy he followed in the exercise of it was the very opposite of authoritarian: all his emphasis was on persuading people to co-operate, on the superiority, in the long run, of voluntary over compulsory methods. There is no need to repeat what has already been said about the grounds on which he defended this, but Bevin's attitude was not determined solely by policy, it sprang from the sympathy which he had for ordinary men and women and the desire to help them which was one of the strongest, as it was certainly one of the most attractive elements in his character. He cared about people, and however difficult or obstinate he might be on occasion, it was this fundamental humanity which won the admiration, and in many cases the affection too, of those who worked with him in the Ministry of Labour and the Foreign Office.

This affected not only his policy but his attitude to everyone he met. Whether it was the King, generals, civil servants or dockers, he showed the same gift for cutting through protocol and class distinctions, ignoring rank and titles and talking to people as if they were human beings like himself. It could have unfortunate effects at times, for his comments were often unvarnished and not liked by the self-important or the conventionally-minded; but most people—including the King—responded at once and found themselves talking more naturally to Bevin than they ever had before to a Cabinet Minister. He drove himself hard and demanded a lot of those who worked for him: when he was angry the whole office was aware of the fact and those who encountered him in this mood did so with a good deal of apprehension. But no minister could get more out of his staff.

On one occasion when he was talking in his room at the Ministry to Lord Terrington, he summoned a secretary to take down a note of what had been agreed between them. None of the usual secretaries was available and after some delay a typist from the Ministry pool was sent up instead. This did not put Bevin in the best of tempers and when the girl, unaccustomed to his way of dictating, became nervous and made mistakes, he was sharp with her. She left the room in tears.

Bevin resumed the conversation with his visitor but had clearly got something on his mind. After a few minutes he broke off and sent for the girl again. This time she was more frightened than ever, but Bevin reassured her: "I asked you to come back, my dear," he said, "because I was rude to you in front of this gentleman and I wanted to apologise to you in front of him too. Now run along and type that note."

On other occasions, in the middle of a heated argument with his officials, he would suddenly grin and start to curse them, telling them they were a lot of so-and-so's not fit to be trusted with organising a Sunday-school treat, let alone a nation at war. If they hadn't got him to keep them on the rails, he would add, they would have long since plunged the country into disaster.

Any account of Ernie Bevin which leaves out his humanity and his humour, his warmth and his incongruous charm is incomplete nor can it explain why, in addition to his formidable reputation, he was also, as Attlee said, a much-loved man.

The last of Bevin's outstanding qualities was loyalty (Attlee indeed put it first), loyalty not only to individuals, but to the class from which he sprang.

"He was especially loyal to working people both at home and abroad. Though no violent proponent of the class war, he was conscious in everything he did, whether in home or foreign affairs, of the working people and their interests. There was class consciousness in his nature: no class hatred."[1]

The *Observer* profile made the same point:

"His career . . . has always remained within his class. He never aspired to be a 'bourgeois'; he helped the working class to rise and rose with it. . . . Never, since he became an adult, has he sought to solve his personal problem other than by social action which would benefit his like."[2]

Bevin was slow to give his trust to anyone, but when he did he was like a rock. His dependability was a major asset to both Churchill's and Attlee's Governments. The fact that loyalty was a rallying cry that frequently provoked opposition in the Labour Party did not trouble him at all. Whether in Parliament or at the annual conference, he was never afraid to face an angry party or stand up for unpopular decisions. He was not the man to trim and he would have nothing to do with intrigue. If he was opposed to anything, he said so openly and left no one in doubt where he stood. A great deal of his influence in politics rested on the belief that, whether you agreed or disagreed with him, Ernie Bevin was a man to be relied on. "Once his word was given," Oliver Lyttelton wrote, "nothing would shake it: it had the cachet of a guarantee by the Bank of England."[3]

[1] Attlee in *The Observer*, March 1960.
[2] *The Observer*, 23 May 1948.
[3] *The Memoirs of Lord Chandos* (1962), p. 294.

A comparison with Churchill was one which occurred to many people who worked with both. At first sight this is surprising: there were such obvious differences between them. Not only did they come from two very different Englands, their experience of life, their style and range of interests were totally different. Bevin had none of Churchill's magnetic qualities, his power of captivating men: where Churchill was brilliant, "a glittering bird of paradise" as Beaverbrook called him,[1] Bevin's quality was that of an earthy common sense. Yet in Churchill's Cabinet Bevin was the one man, as Churchill recognised, who could stand up to him on equal terms. However different the expression of their qualities, both were men of determination and temperament, self-educated, pragmatic, proceeding by intuition rather than by logic, with strong pugnacious instincts, strong prejudices and equally strong loyalties.

Like Churchill, Ernie Bevin was a man who made mistakes, and was certainly not free of faults both of temper and judgment; but, like Churchill again, he was a big man, big enough in personality, in spirit and in ideas, to make his faults and mistakes seem blemishes that did not mar or detract from his full stature. Above all, although he came from the opposite end of the social scale, he had the same courage, the same independence, the same confidence to be himself, warts and all, without pretence or inhibition in any company. As Beaverbrook once said of him in a moment of exasperation, he ought to have been born the son of a duke: the fact that he was not, that he owed nothing to birth, education, influence or wealth, makes the man and his career all the more remarkable.

3

Gladstone once remarked that a man at forty-five might as well start training for the ballet as for the Cabinet. Government, however, came more naturally to Bevin than politics and he was much more quickly at home in the Cabinet and Whitehall than he was in Parliament.

Churchill described the members of the War Cabinet as

[1] In a letter to Sir Samuel Hoare quoted in Kenneth Young: *Churchill and Beaverbrook* (1966), pp. 187–8.

"the only ones who had the right to have their heads cut off on Tower Hill if we did not win. . . . The rest could suffer for departmental shortcomings but not on account of the policy of the State. Apart from the War Cabinet, any one could say 'I cannot take the responsibility for this or that'. The burden of policy was borne at a higher level."[1]

In practice, however, the Cabinet system worked well during the war because the War Cabinet devolved so much of its work on to committees. All important decisions were reported to it, but the only questions the War Cabinet itself decided were those which its committees failed to settle or which were brought to it as a final court of appeal.

Strategy and the conduct of the war were accepted by all his colleagues as the special responsibility of Churchill as Minister of Defence as well as Prime Minister. Bevin, although a member of the War Cabinet, had nothing to do with this and was never a member of the Defence Committee in which, during 1940 and 1941, the most important decisions on the war were taken. On one occasion when Churchill tried to mobilise the War Cabinet on his side in one of his disputes with the Chiefs of Staff, Bevin refused to be beguiled by the Prime Minister's appeal to his colleagues for their opinions. He knew nothing about war himself, he declared, and doubted if the opinion of the other members of the Cabinet was worth having either. They had put him in, he told the Prime Minister, to win the war. If they lost confidence in him, they would put him out. But as long as he had their confidence, he should get on with it and not come asking the Cabinet for its opinion on matters about which they knew nothing and which were too serious to be settled by amateur strategists. For once, even Churchill was taken aback and pursued his manœuvre no further.

The procedure for keeping the members of the War Cabinet informed on the military side was known as the "Monday Cabinet Parade". Every Monday, usually between 5.30 and 8.30 in the evening, Ministers (including a number of "constant attenders" outside the War Cabinet) met to hear a report by the Chiefs of Staff with a running commentary by the Prime Minister which on occasion turned into a barbed or angry cross-examination of the Cabinet's military advisers.

Bevin had entered the Cabinet with all the stock Labour prejudices against generals and the military mind. His attitude changed, how-

[1] Churchill, Vol. II, *Their Finest Hour*, p. 12.

ever, as he came to have a better understanding of their problems, and particularly, as the minister responsible for the call-up, of the problem of building up the country's forces after the run-down which had been allowed to take place between the wars. To their surprise, the military chiefs who had started with equally strong prejudices about trade-union leaders, found Bevin to be one of the most helpful of ministers in meeting their needs and a staunch ally when things went badly.[1]

Eden found him equally helpful in talking over his problems as Foreign Secretary. This was the other field, foreign affairs, which was recognised by the Cabinet as being the special province of the Prime Minister. Churchill alone could deal with Roosevelt and Stalin as an equal, and Eden as his lieutenant had some of the same difficulties as the C.I.G.S. in standing up to the impetuous and imperious character of his chief. Eden felt the need of someone to talk to and found Bevin more interested in foreign affairs than any other member of the Cabinet. A friendship sprang up between the two men (they sat side by side in the Cabinet) and at the end of the coalition Eden was eager to see Bevin succeed him as Foreign Secretary if Labour won the election.[2]

Bevin's main concern, however, was with the other side of the war, economic and home affairs. Here there was nothing corresponding to the personal position and driving force of Churchill in the combined roles of Minister of Defence and Prime Minister. From the confusion of committees, however, one had already emerged with the authority to act for the Cabinet over the whole field of government apart from defence and foreign affairs. This was the Lord President's Committee, deliberately kept small (its nucleus was a group of four or five

---

[1] At a Cabinet meeting early in February 1942, when there was nothing but bad news from the Far East and Libya, the C.I.G.S., Sir Alan Brooke, found himself facing a barrage of criticism from ministers. When Bevin asked a question, Brooke remarked later, "I thought at the time that the remark was meant offensively and my blood was up. I therefore turned on him and gave him a short and somewhat rude reply. He said nothing more at the time but came up to me when we were going out and explained in a most charming manner that he had not been trying to get at me but was genuinely asking for information. A typical action on his part and nothing could have been nicer. I apologised for the rudeness of my reply and asked him to dine quietly alone with me when he could ask me any questions he wished. This he did and we had a most pleasant evening together. The more I saw of him in later years, the more I admired him. A very great man." Quoted in Arthur Bryant: *The Turn of the Tide* (1957), pp. 35–6.

[2] *The Eden Memoirs*, Vol. II, *The Reckoning* (1965), p. 550.

ministers) and providing a counterpart to the Defence Committee for home affairs. The rise of the Lord President's Committee was largely due to its first chairman, Sir John Anderson, who proved to have, in addition to great administrative talent and experience, the rare gift of being able to co-ordinate the work of other departments and committees and to adjudicate on their differences dispassionately without arousing jealousy or suspicion. Anderson avoided setting up a department of his own but through Norman Brook (later Lord Normanbrook), who acted as his personal assistant, was able to draw on the resources of the Cabinet secretariat. Thanks to Anderson and the Committee through which he worked, Britain came nearer than at any other time in its history to having an effective instrument for planning.

Bevin was a member of the Lord President's Committee from January 1941 to the end of the coalition in 1945,[1] and missed only 27 of its 299 meetings. Its businesslike atmosphere suited him better than meetings of the Cabinet, which were dominated by the temperamental genius of the Prime Minister, and the contribution he made to its proceedings was second only to that of its chairman. Anderson and Bevin were about as different as two men could be, and Anderson had not much sympathy (as Attlee had) for the more colourful, egotistical and imaginative side of Bevin's character. Bevin, however, won his respect by the care he took to master the Committee's business and read its papers, his shrewdness of judgment and his willingness to turn his mind to problems outside his own immediate sphere of interest.

Bevin and Anderson's ability to work together played a considerable part in the success with which the Lord President's Committee handled one difficult question after another in home affairs. It was important for another reason. It was on Anderson that Churchill relied to frame and present to the Cabinet the recommendations on the allocation of manpower which, beginning with the manpower budget of 1941, became the key instrument of economic planning and control. So successful was Anderson in carrying out this difficult task that he retained personal responsibility for it when he gave up the

---

[1] Only three other ministers remained members of the Committee for the whole of the same period: Anderson, Attlee and Morrison. These three, with Bevin, were, in effect, the War Cabinet for home affairs.

office of Lord President to become Chancellor of the Exchequer (1943).

Far from resenting, Bevin welcomed this arrangement. He was better placed than most ministers to know which demands for manpower were inflated and needed cutting back, but for Bevin to say this aroused suspicion and invited Beaverbrook's retort that, if the Minister of Labour did his job properly, there would be no shortage of manpower. Bevin was shrewd enough to see that if allocations and cuts were recommended by Anderson, whom everyone regarded as impartial, they were much more likely to be accepted. On his side, Anderson was often glad to draw on Bevin's experience and special sources of information when making up his mind. In this way a close partnership was established between the two men which was decisive in determining most questions of manpower policy.

Thanks to the Lord President's and other committees (such as that on Manpower) which were set up for special purposes, a great deal was settled without coming to the Cabinet at all or brought up in the form of agreed recommendations. And it was as a member of these committees, and as chairman of the Production Executive, as much as in his capacity as Minister of Labour, that Bevin acquired his influence in Whitehall, and won the confidence not only of colleagues like Anderson, Andrew Duncan and Leathers, but of the men who ran the organisation, the Cabinet Secretariat and the permanent officials. They had their own standards for judging ministers and rated Bevin as one of the most helpful members of the Government, a man full of resource on whom they could rely to push business forward, to find a way round problems and to get things done.

4

Government, however, cannot be reduced to administration. The big decisions, the decisions with political implications, had to be taken by the Cabinet itself, even if they were pre-digested by a committee. So far as the war or foreign affairs were concerned, the Cabinet was usually content to accept a report by the Prime Minister and endorse what he proposed. But on other matters—the conscription of women, for example, rationing and the manpower allocations—a decision was preceded by considerable and sometimes protracted argument.

With Churchill in the chair, the course of any Cabinet meeting was unpredictable. "I am either sunk in a sullen silence," he said to Oliver Lyttelton, "or else I am shouting the table down." In a talkative mood he would discourse at large and never reach the second item on the agenda. If he was angry or put out, he could be as obstinate, as unreasonable in criticism and violent in reproach as he could be captivating when his mood was sunny. The other members of the Cabinet noticed, however, that he was always careful and took no liberties in his treatment of Bevin—he handled him with kid gloves, Eden said. If Bevin took a long time to come to the point, Churchill made no attempt to cut him short; he paid attention to what he said, particularly if it concerned working-class reaction to any measure that was proposed[1] and on labour questions he relied implicitly on his judgment. "You had to understand," Attlee told Francis Williams, "that Ernie wouldn't stand any nonsense in his own line. Winston understood that."[2] He left him virtually a free hand and intervened less in the business of the Ministry of Labour than in that of almost any other department.

In Churchill's eyes Bevin was more authentic as a working-class leader than any of the other Labour ministers. On his side, although he never belonged to the Prime Minister's immediate circle or stood in the same relationship to him as Beaverbrook or Bracken, Eden or Lyttelton, Bevin shared the admiration of most Englishmen for Churchill as a great war leader and as a man. Neither perhaps was ever really at ease with the other—there was too great a gap between their experience and background for that—and they disagreed about most questions in politics. But each recognised qualities in the other which counted far more than political sympathies and on the question which mattered more than any other, winning the war, there was complete agreement between them.

Oliver Lyttelton believed that this was the great bond between Churchill and Bevin, patriotism, a readiness to put the nation before

[1] On one occasion the Cabinet listened to an enthusiastic report by the Minister of Food on the success of his Ministry's scientists in producing a pill which contained all the ingredients of a balanced diet. The Minister of Food pointed to the great savings this might make possible in food imports. This made an impression on the Cabinet. Bevin, however, said nothing until Churchill, noticing this, asked him why he was so silent. "I was thinking," Bevin replied, "of what a docker would say if he came home from a nine-or-ten-hour shift, and his old woman put a dinner plate in front of him with a pill on it." This ended the discussion.

[2] Francis Williams: *A Prime Minister Remembers* (1960), p. 41.

party which Churchill was naïvely surprised to find in a socialist.

A second, which Attlee as Prime Minister valued as highly as Churchill, was Bevin's loyalty and dependability. In a tight corner he was a reassuring ally, a man who did not panic or run for cover when things went badly. The position of Prime Minister was a lonely one and as the 1940 mood of national unity faded and political differences began to reappear, Churchill at times became depressed and anxious about holding the coalition together. In these circumstances, Bevin's steadfast support (for example over Greece in December 1944), his insistence, however unpopular with some members of his own Party, that the commitment to win the war must take priority over everything else counted far more than anything else in the Prime Minister's eyes.

Attlee was completely overshadowed by Churchill during the war, and most people—including Churchill—failed to detect the qualities he was to show as a Prime Minister. When Attlee first became leader of the Labour Party in 1935 Bevin had regarded him with some suspicion as another middle-class interloper in the Labour Movement, and there could hardly have been a greater contrast in personality, even in appearance, than that between the exuberant Bevin and the modest, dry, laconic Attlee. Once he entered the coalition, however, Bevin felt himself bound by the same loyalty to Attlee as Labour leader as he did to Churchill as Prime Minister. Attlee had the sovereign virtue, in Bevin's eyes, of being straight, you could rely on what he said, and a friendship developed between them which was to prove one of the most successful partnerships in modern British politics.

With the rest of his colleagues Bevin was on friendly enough terms without much regard to politics. Thus he formed a rather surprising friendship with Halifax as well as Eden and in the middle of the war, when there was a renewed outburst of indignation against Baldwin for the lack of military preparedness, he wrote a warm personal letter to the former Prime Minister to express his anger that he should be made a scapegoat for a national failure, for which no one could escape responsibility.[1] The two exceptions, the two men whom he never ceased to distrust, were Beaverbrook and Morrison.

---

[1] Bevin added that if Baldwin ever came up to London and wanted to know what was happening he should come round to the Ministry of Labour, where he would be glad to show him any official papers he wanted to see.

Beaverbrook and Bevin had long disliked each other since the days of the rivalry between the *Daily Herald* and the *Express*. The Beaverbrook papers in the 1930s rarely missed a chance of sniping at Bevin as a trade-union boss; Bevin for his part regarded Beaverbrook himself (a character out of Balzac, as one of his biographers calls him) and the newspaper empire he had created as pernicious influences in English life. Both men therefore entered the Cabinet with a strong predisposition to fight each other.

Trouble began at once over Beaverbrook's claim to take all the labour he needed for the aircraft factories without consulting the Minister of Labour. "We won't be arbitrated on," he declared. Bevin was not the man to let such a challenge pass. There were rows, angry notes, reports of what the Beaver had said in a rage and of what Bevin had threatened to do.

There were, of course, real issues as well as personalities in the dispute, with Beaverbrook as the apostle of improvisation, Bevin of organisation. In 1940 Beaverbrook got a good deal of his own way, and justified it by the number of planes he produced. But he lacked Bevin's staying power. Throughout 1941 he was on the verge of offering or threatening resignation, and his individualism—"I am not a committee man," he wrote to Churchill, "I am the cat that walks alone"—left him isolated apart from the friendship of the Prime Minister. This was sufficient to keep him in the Government and provide him with a succession of offices, but not to overcome—in many ways it increased—the distrust with which he was regarded by the rest of the Cabinet.

Beaverbrook resigned as Minister of Aircraft Production at the end of April 1941. After two months' interlude as Minister of State, Churchill persuaded him to take on the Ministry of Supply. The running fight with Bevin was soon renewed. Beaverbrook objected in particular to Bevin's position as chairman of the Production Executive. Writing to Churchill in October 1941 he complained that, if disputes between the Ministry of Labour and the supply ministries were to be referred to the Executive, as Bevin claimed, "that is to say, any dispute with Mr. Bevin is to be referred to Mr. Bevin". Any complaints about the labour needs of the Ministry of Supply were to be dealt with by Bevin—"who is responsible for the deficiency". Finally, Bevin's proposed "permanent inspection commission in the supply ministries" would deprive the Minister of Supply of any indepen-

One impression of Ernest Bevin in 1943.

Bevin in a different mood.

dence: "he will become a member of One Big Union."[1] He would refuse to attend Cabinet meetings, Beaverbrook declared, when Bevin was due to be present: "As my resistance only makes for unpleasantness at Cabinet, I propose to absent myself altogether."[2]

When Bevin in turn complained to the Prime Minister about Beaverbrook's obstructive attitude, the latter retorted that there would be no trouble at all if Bevin would do his job and supply him with the labour he needed:

"I have done everything possible to persuade you that my conduct in relation to Mr. Bevin is entirely correct. His complaints against me are founded on pressure for labour which I have directed against him. The trouble began on 25 July last. It will be over when I get my necessary supply of labour."

"In the Production Executive (Beaverbrook added) Mr. Bevin is getting so much authority that he overlays the Production Ministries and asserts control. I am not yielding to that movement."[3]

The same month (November 1941) Beaverbrook urged Churchill, as one way out of his difficulties, to create a Ministry of Production and put Bevin at the head of it: if so, he would be willing to serve under him and help to establish his authority.[4] Whether this was a tactical manœuvre or another example of Beaverbrook's volatile temperament, it only increased Bevin's distrust of him. He was no more to be won over by offers of support than intimidated by opposition. Amongst the papers Bevin kept is a letter dated 22 November 1941 (from the handwriting it looks as if it had been written impulsively, on the spur of the moment) in which Beaverbrook wrote:

"Dear Ernie,
I hope very much that you have recovered from the effects of your chill. Visiting factories in winter time is certain catastrophe.
It is in the Council Room that you and I must discharge our duties to the British people.

[1] Beaverbrook to Churchill, 30 October 1941, quoted by Kenneth Young, *Churchill and Beaverbrook*, p. 215.
[2] Ibid.
[3] Quoted in Young, p. 217.
[4] Ibid. p. 216.

The factories will always feel your influence and respond to your leadership.

And here it is that I believe you can get some help from me.

How I would like to give it in complete agreement with you and your policy.

Can we make a platform for you where I can stand at your side? I am sure you can do so if you determine to build it.

With your leadership of men and women in the industrial centres supported by the principal Supply Ministry the war effort can be increased.

Don't bother to answer.

Yours ever,
Max."

Bevin's reply, dictated two days later, was not unfriendly but it did not encourage any further advances. After thanking Beaverbrook for his inquiry and saying that he was back at work again, he went on:

"I entirely agree that it is in the Council Room that our duty to the British people must primarily be discharged and that is where I have always tried to do it, in team work with my colleagues and making my contribution equally to all.

I welcome, therefore, what you say and do not think there is any instrument in the Government machine offering greater scope for this mutual effort between all of us responsible for the different factors of production than does the Production Executive.

Your reference to making a platform for me puzzles me a little and I am not sure that I follow what is in your mind. I have no policy or platform except that of the Government as a whole, arrived at through the War Cabinet, and came into the Government not for any personal position but solely to contribute what I could to our common effort under the leadership of the Prime Minister.

From the men and women in the industrial centres I think I have got good results already and I believe that, handled rightly, there can be still further improvement, but it does lead to the necessity of pooling ideas and putting out the right kind of leadership if we are going to achieve this.

You told me not to bother to answer but I thought it desirable just to let you have my views on the points you put to me."[1]

The next day Beaverbrook wrote back:

"My dear Ernie,

I am so sorry that I failed to make my meaning clear, even if the misunderstanding is not an important one.

It is my hope to persuade you to lead all the people to hard work. It is my belief that you can do more in that direction than anyone else.

[1] Bevin to Beaverbrook, 24 November 1941.

And there is a desire on my part to serve you in such a movement in any capacity you wish. It is my resolve to support and sustain you to the full in that leadership.

Yours ever,
Max

P.S. If you ask me for the Platform—My view is No conscription for women— and Women for Industry only.

M."[1]

Bevin kept his guard up and gave nothing away. A few weeks later the differences between the two men came to a head when Churchill returned from the United States (January 1942) and proposed that Beaverbrook should take over the functions of the Production Executive and become Minister of Production. A fierce battle was at once joined over the powers of the new Minister (see Chapter 6), the upshot of which was that Beaverbrook resigned and Bevin stayed. This settled the matter. Churchill insisted on bringing Beaverbrook back into the Government later and he played an important political role in 1944–5, but he never regained his seat in the War Cabinet or repeated his success of 1940. In the contest between the two men it was Bevin who showed the qualities of toughness Beaverbrook so much admired and Beaverbrook who suffered defeat despite his special relationship with Churchill.

A good many other people, certainly most of the Labour Party, shared Bevin's distrust of Beaverbrook. His dislike of Morrison was a different matter. How much it had to do with the row over the London Transport Bill in the early 1930s is impossible to say. By the 1940s it had hardened into an immovable prejudice. Although they worked together well enough when they had to and addressed each other as Herbert and Ernie, nothing could shake Bevin's conviction that Morrison, unlike Attlee, was not straight. "Don't you believe a word the little b—— says," he would warn his colleagues, and more than once embarrassed his neighbours at a Cabinet meeting by the scornful comments, conveyed in highly audible asides, with which he greeted the proposals of the Home Secretary.

After Bevin's death Morrison maintained that the difficulties between Bevin and himself had been exaggerated and that they got on well enough. This may be true of the last years of Bevin's life, when most people agree that he mellowed and lost much of his old

[1] Beaverbrook to Bevin, 25 November 1941.

suspicion. The evidence for the wartime period, however, points in the opposite direction[1] and the distrust which Bevin showed towards Morrison is of some importance in the history of the Labour Party: it weakened the Labour Party's influence in the coalition, but of course greatly strengthened Attlee's position. If the struggle for leadership had lain between Morrison and Attlee alone, Morrison might in the end have taken Attlee's place. Attlee backed by Bevin was too much for him: Bevin's hostility may have cost Morrison the leadership of the Party and perhaps the Prime Ministership of the country.

Most of the members of the Labour Party, even those who thought Herbert too smart a politician on occasion, regretted Bevin's attitude and thought him unfair to Morrison, too prejudiced to recognise the other man's ability and the difficult job he did during the war. Bevin was unrepentant. To the end of his life he could work up his indignation at the thought of Herbert's untrustworthiness and the skill Morrison showed as the Party's leading tactician did nothing to mollify it: on the contrary, this was exactly what he objected to. Morrison was the politician personified and in Bevin's vocabulary this was sufficient condemnation by itself.

5

At the end of the war in Europe, Bevin told a farewell meeting of his headquarters staff that, when he came to the Ministry of Labour, it had been his ambition to give it the same importance in wartime as the Treasury enjoyed in peacetime. By the time he said this, no one thought it an exaggeration: he had very largely succeeded.

It can be argued, of course, that manpower was bound to be accepted sooner or later as the basis of wartime planning, but it was by no means certain that control of manpower would be left in the hands of a single department or that this would be established as early as the summer of 1940. In previous discussions, before Churchill formed his Government, the Ministry of Labour had shown reluctance to assume any such responsibility. Bevin not only reversed this attitude, but enlarged his claim by adding to the responsibility for the supply of labour, responsibility for the use which industry made of it. Most important of all, he secured from the Cabinet the

---

[1] For two examples, see below, pp. 286–7, 390–2.

powers to put his claim into effect and demonstrated his ability to defend it against all comers. It is only necessary to consider the chances of either his predecessor, Ernest Brown, or his successor, George Isaacs, conceiving so bold a claim and having the strength to make it good, to realise how much the wartime Ministry of Labour owed to Ernest Bevin.

Bevin seems to have begun with considerable doubts about the ability of civil servants to handle questions affecting labour and industry. The Labour Supply Board which he thought of as the most important part of his organisation, and which met daily under his own chairmanship, consisted of two industrialists and two trade unionists brought in from outside. The permanent officials, however, were not so easily beaten: the Civil Service has long experience in dealing with innovations, and in March 1941 the Labour Supply Board was quietly dropped. Shortly afterwards Beveridge left the Ministry and the key post of Director-General of Manpower went to one of the Ministry's own men. To say that Bevin had been captured by his officials sounds clever but misses the point: no minister who fails to come to terms with his civil servants is likely to run his department successfully. If mutual confidence was now established there was accommodation on both sides and no one ever doubted that Bevin was master in his own house.

The change was due to Bevin's discovery that, although the Ministry of Labour had been one of the least important home departments before the war, mainly concerned with unemployment insurance, it had among its senior officials a number of men equal to the demands made on them by its conversion into a key economic ministry thrust into the thick of the problems of manpower and war production. After 1945, in fact, no less than ten of them reached the rank of Permanent Secretary, or its equivalent, in the civil service.[1]

The Permanent Secretary of the Ministry since 1935 had been Sir Tom Phillips. A Welsh country grammar-school boy with a brilliant career as a classic at Oxford,[2] he might as easily have become a don as a civil servant, and still retained the tastes of a scholar. Bevin who had never met anyone like this before was at first mystified by him, then discovered his gifts as a draftsman and from this went on to acquire

[1] Sir Godfrey Ince, *The Ministry of Labour and National Service* (1960), p. 200.
[2] He took a First in Mathematical as well as Classical Moderations, followed by a First in Greats and the Gaisford Prize for Greek Prose.

genuine respect for Phillips's lucidity of mind and disinterested judgment. Phillips did not argue with Bevin, but his warnings could be none the less effective. "Very well, Minister, if you want it, it shall be done, but the consequences will be such and such." Bevin was more impressed by such advice than by any overt opposition to his wishes.

Nor was Phillips at all daunted by the changes which followed Bevin's arrival at the Ministry, including the expansion of its staff from just under 30,000 before the war to 44,500 at its peak in 1943. He saw the opportunities which now lay before the Ministry of Labour and was ready to take them. It was largely thanks to the Permanent Secretary that the changes took place smoothly and that Bevin could count on having a well-run administrative organisation at his disposal.

Bevin and Phillips worked easily together but it remained a working relationship, minister and civil servant, each understanding his role and as far as possible keeping personalities out of it. Bevin was closer personally to the two men who became Deputy Secretaries, Freddie Leggett and Godfrey Ince. Leggett had first met Bevin as long ago as 1916 and struck up a friendship with him on the Government's industrial mission to the U.S.A. in 1926. Leggett's particular interest was industrial relations, which naturally brought him into contact with Bevin as a trade-union leader; so did his service as the Government member of the British delegation to the I.L.O. in the 1930s. More than twenty years' experience had given him a shrewd knowledge of the trade-union world and Bevin was the man whom he most wanted to see as Minister of Labour.

From 1940 to 1942 Leggett was the Ministry's Chief Industrial Commissioner: when Wendell Willkie, on his 1940 visit, asked what this involved, Bevin interrupted Leggett's explanation with the remark, "He doesn't settle industrial disputes, he fondles them." In fact Bevin soon put Leggett in charge of all the Ministry departments which were concerned with matters other than manpower.

Godfrey Ince who became Director-General of Manpower was one of the outstanding civil servants of his generation. Another grammar-school boy, this time from Reigate in Surrey, Ince had graduated with distinction at University College, London, in mathematics and economics at the beginning of the First World War. To the end of his life he retained the physical vigour of an accomplished sportsman, open-faced, direct and warm in manner. After serving in the war, he joined the Ministry of Labour in 1919 and by a curious coincidence served as joint secretary of the Shaw Inquiry at which Bevin made his reputation as the Dockers' K.C. Twenty years later, in 1940, Bevin appointed him as the first head of the Factory and Welfare Department which he set up in the Ministry of Labour, and a year later chose him as Director-General of Manpower.

This was a newly created post which brought together under Ince's control a whole group of departments dealing with national service and military recruitment; labour supply for the whole of industry; training and (one of the most important of the innovations made by Bevin) a Manpower Statistics and Intelligence Department, the source of the manpower budgets on which so much wartime planning came to depend. Ince's job, to direct the mobilisation of the entire manpower resources of the country, was one of the most exacting and responsible jobs in the wartime civil service. Fortunately he had the gifts for which it called: the intellectual power to master the complex and continually changing pattern of manpower distribution, and a temperament which remained imperturbable under the strain of continual crisis. More fortunately still, he was able to establish a relationship of complete mutual confidence with Bevin, and Bevin and Ince together made one of the most effective partnerships of the war.

The division of responsibilities between Ince and Leggett (who were both made Deputy Secretaries in 1942) roughly corresponded to the two major divisions of the Ministry's work, manpower and industrial relations. When Phillips left in 1944 to start the new Ministry of National Insurance and Ince stepped up to become Permanent Secretary, the same pattern was retained: Harold Emmerson took Ince's place as Director-General of Manpower and joined Leggett as Deputy Secretary.[1]

Bevin's relations with his chief officials, indeed with most civil servants, were excellent, far easier than they had been in the union. "He couldn't suspect us," one of them later remarked, "of wanting his job, nor did any of us object if he took the credit for our ideas." Bevin quickly grasped the proper division of functions between a Minister and his officials. They were there to brief him, to advise and

---

[1] H. C. (later Sir Harold) Emmerson, another future Permanent Secretary, had been brought back to the Ministry of Labour in 1942 to replace Leggett as Chief Industrial Commissioner before going on to replace Ince as Director-General of Manpower in 1944.

if necessary warn him before a decision was taken. Once made, he could rely on them to carry it out not only efficiently but loyally. What they wanted from him was the decision itself. This suited him well. A man with none of the subordinate virtues, Bevin worked best at the top, a position for which he was fitted by temperament as well as by ability.

Under the coalition, two junior ministers, one a Conservative, the other a Labour Member of Parliament, were assigned to the Ministry of Labour as joint parliamentary secretaries. Ralph Assheton (later Lord Clitheroe) was already at the Ministry when Bevin succeeded Ernest Brown. Described by one of the officials with a sense of English history as "a leader of the Lancashire squirearchy", he was as far-removed from Bevin in his political views as he was in social background. He went on to become chairman of the Conservative Party organisation at the time of the 1945 election, sufficient indication of where he stood politically; but this did not affect his admiration for Bevin whom he came to regard, equally with Churchill, as the biggest man he had met in the course of his life. This regard was not altered by their sharp disagreement over the conscription of women. Bevin decided on this while Assheton was in the United States for an I.L.O. meeting. It was contrary to the recommendations of the Manpower Committee (of which Assheton was chairman) as well as to his personal convictions, and as soon as he got back Assheton protested to Bevin and asked him, formally, to represent his views to the Cabinet. "No," Bevin replied, "you do that yourself", and raised no objection to the letter which Assheton then drafted. The same evening Ralph Assheton was summoned to a meeting of senior ministers at No. 10 and invited by Bevin to state his objections. A fierce argument developed in front of the other ministers in which Bevin and his parliamentary secretary expressed strongly opposed views and Assheton declared that he could not support him in Parliament. Bevin, however, showed no ill-will, took Assheton home in his car afterwards and, as he left, remarked: "Don't think I shall hold it against you, Ralph."

None the less, not long afterwards, Assheton moved to the Ministry of Supply and his place was taken by George (later Lord) Mc-Corquodale, chairman of a big family printing firm, who had spent the first half of the war serving in the R.A.F. When Churchill offered him the job, the Prime Minister told McCorquodale that he must look after Bevin who was carrying an exceptionally heavy load; he added that since McCorquodale was a large-sized man himself, they might get on well together. This they did, finding common ground in their views on industrial relations, even if they disagreed on politics. Nothing seemed to McCorquodale to show Bevin's earthy common-sense better than his refusal to have anything to do with the move to suppress horse racing in wartime. If men were working hard, Bevin declared, they needed some relaxation to take their minds off their troubles: "When a chap's thinking about who's going to win the 2.30, he isn't thinking about me."

At the time McCorquodale joined the Ministry, Bevin told him with a grin: "Anything you make a mistake about, I will get you out of, and anything you do well I will take the credit for." Bevin was evidently aware of his reputation in such matters: in fact he gave each of his parliamentary secretaries a clearly defined area of responsibility and left them to get on with the job. And in the case of George Tomlinson, his Labour parliamentary secretary, he refused to take credit which no one would have denied him. George, like Ralph Assheton, was a Lancashire man, but of a very different sort, a weaver from Rossendale and one of the most likeable men in the Labour Party. In choosing him, Bevin showed his old preference for lieutenants he could dominate rather than for men who would stand up to him. George's attitude towards his chief came close to hero worship. They shared a common love of ordinary folk, but where Bevin's formidable personality often provoked opposition, George disarmed every prejudice by his unaffected Lancashire speech and humour.

By an inspired choice Bevin made him chairman of the committee set up to report on the rehabilitation of the disabled and, when the report was accepted by the Government, insisted that he should introduce the Bill which won praise from all sides. Nobody would question the missionary zeal George Tomlinson brought to the cause of the disabled, but Bevin was not only responsible for launching the idea in the first place, it was his strong and unwearying support behind the scenes which enabled the Bill to be passed and put into effect before the end of the war. Far from grudging his parliamentary secretary the credit, however, Bevin went out of his way to push him into the forefront, content to see the job done without worrying about his own part in it and delighted at Tomlinson's parliamentary triumph.

CHAPTER 5

The Minister, the Unions—
and the Conscription of Women

I

A few days after taking office, Bevin called his principal officials together and told them: "I shall have a lot of ideas: it's your job to tell me which of them are good and which of them are bad." This was not an easy duty to discharge. Like Churchill, when he was interrupted by disagreement, Bevin would glare fiercely at the speaker in such a way as to disconcert anyone not absolutely sure of his ground. But it was the general experience of his officials that he would listen to criticism, even if it took a day or two for him to concede that they were right and he was wrong. Indeed they were sometimes apt to complain that he picked up their ideas after appearing to pay little attention to them at first, then two or three days later reproduced them as his own, astonishing their authors by the circumstantial account he would give of when and where they had first occurred to him. Whether done consciously or not, this was an irritating habit which had earned Bevin much resentment in his union days, but it does not alter the fact on which, for example, Ince, Assheton and Emmerson are all emphatic, that he had far more ideas of his own than anyone else in the Ministry.

"Few ministers," Ince pointed out, "ever know as much about the business of their ministries as their permanent officials. Bevin did, and could surprise any of us by the information he would produce out of his head about the way in which a particular trade was organised. But Bevin was an exceptional minister in more important ways than that: he knew what he wanted to do, knew where he wanted to go—and how to get there."[1]

[1] Sir Godfrey Ince, in conversation with the author, 19 January 1955.

124

As an example Ince quoted his interest in reconstruction plans from his earliest days in the Ministry, an interest which he never failed to take into account in any of the decisions he took.

Much of what is said about Bevin by the men closest to him in the war repeats what was said about him in his union days: his quickness in grasping a point and reducing a problem to its essentials; his ability to get things moving and his resourcefulness; his imagination in dealing with human and social problems; his tenacity in pursuing a solution; his skill as a negotiator in getting round awkward corners. Bevin had now to deal with highly trained and experienced professionals. They saw the weaknesses which sprang from his lack of the sort of education they had received themselves—his awkwardness in the use of words, for instance—but they were never in any doubt about the quality or originality of his mind. "He could neither read, write, nor speak—and did all three triumphantly." The weaknesses showed up in Parliament where Bevin rarely succeeded in making the most of his case, but what impressed his officials far more was his ability to get things through the Cabinet and its committees, the acid test of a minister in the eyes of the civil service.

The importance Bevin attached to carrying public opinion with him meant a great deal of speaking, yet he rarely delivered a speech that had been written for him. He would use official briefs to provide facts or suggestions but when the time came to dictate he would do no more than glance at the notes prepared for him and set his own argument out for himself. Anyone who examines the notes or typescript from which he spoke will have no difficulty in deciding on the authorship of Bevin's speeches: they bear the unmistakable hall-mark of his own processes of thought. From time to time, when he wanted to clear his mind or stir up someone to action, he would dictate a memorandum or letter of his own: these too are easily recognisable and are among the most valuable of the papers he left behind. Official memoranda and correspondence, for the most part, he was wise enough to leave his Private Secretary, H. G. Gee, or one of the other officials to draft. When he had finished his day he took home a stack of Cabinet and departmental papers; while no one in his department ever saw him open a book, he was an assiduous reader of official documents and owed not a little of his strength as a minister to the fact that he kept himself well-informed. Some of his best ideas were the fruit of reflection on an evening's or week-end's reading: the next morning he would come into the office and with the remark "I've been thinking" (which his officials learned to regard with some alarm), would put up questions he wanted answered, suggestions he wanted considered and points he wanted discussed.

Perhaps the most valuable fruit of his long experience in the trade-union movement was his feeling for what was possible and what impossible at any given time, a power of judgment on which was founded his whole policy of "voluntaryism". Ince illustrated this sense of timing from his handling of the conscription of women.

"Bevin knew quite early in the war that it would be necessary to conscript women, but he impressed it upon us that we had got to do it at the right time—and the right time would be when enough women were in the Forces or in war work to point the finger at those who weren't. Then the time would have come to act."

To treat Bevin's success at the Ministry of Labour as a virtuoso solo performance is to miss the point: the most important element in it was his ability to weld together and inspire a team of men and women who, years later, without exception, still looked back on this as the most exciting period of their official careers. It is hardly surprising that those at the top, in daily contact with the Minister, should feel this: the interesting thing is how far this feeling spread not only among the headquarters staff in London but outside Whitehall altogether, amongst people who only saw Bevin three or four times, perhaps, in the course of the war. There was a very good reason for this: Bevin's conception of the Ministry of Labour and its job produced almost as great changes in the Ministry organisation at local level as it did at the centre.

2

There was no need, of course, to create a local organisation for the Ministry; indeed labour exchanges had been set up, under the Acts of 1905 and 1909, several years before the Ministry itself was created. In 1941 there were eight hundred employment exchanges and another seven hundred branch offices. But their officials in wartime were called on to perform many more and very different functions from those to which they had become accustomed between the wars.

The basis of all manpower policy was registration and, in one form or another, the Ministry's local offices carried out 32 million registrations and filled twenty-two and a half million vacancies. The Ministry was a Ministry of National Service as well as of Labour and the call-up of 7 million men to the Forces was a major business in itself, including all the arrangements for medical examination, deferment and the hearing of objections on grounds of conscience and hardship.[1] An even bigger undertaking was the conduct of eight million individual interviews, mostly of women, one of Bevin's most important innovations which contributed more perhaps than anything else to the successful mobilisation of women for war work. Finally, once mobilisation was complete, the local offices had a scarcely less important part to play in demobilisation and resettlement.

The employment exchanges had acquired considerable experience before the war in handling large numbers of unemployed, but this was a very different business from interviewing hundreds of thousands of men and women, many of whom had never been in a labour exchange before, and persuading or directing them to accept jobs of which they had no previous experience.

New staff had to be recruited and absorbed: at its peak forty thousand out of a total of forty-four thousand men and women employed by the Ministry were working in local and regional offices. Much more important was to instil in all staff, particularly in the existing managers and clerks, a radical change of attitude towards the people with whom they had to deal.

Nowhere was Bevin's personal influence more plain. "I have tried," he said in May 1941, "to humanise the labour exchanges and honestly, I think, with success." Institutions which had been as much associated with unemployment and the dole as the old workhouses with destitution not only undertook their new functions with success but were transformed in the process. Much thought—and imagination—went into producing this change[2]: not content with giving instructions, whenever he went on tour, Bevin arranged to meet and address the local staffs of his Ministry. A description of such a

[1] For these, see Parker, c. 9.
[2] One example quoted to me was Bevin's instruction that the staff in employment exchanges should say "Good morning" to any member of the public coming for advice or interview.

conference is given by Trevor Evans who attended a number with Bevin in his early days at the Ministry.

"Into a regional centre would be invited all the managers of the employment offices in the area. Usually there would be about sixty present. Bevin would open the discussion by outlining his immediate programme. He would end 'I cannot issue regulations which will cover all the contingencies you are likely to meet in the course of your day to day work. You must use your own initiative. If you make a mistake I will stand by you. One thing I will not forgive. That is inaction.'

"This was a clever move on Bevin's part. It gave every local official a sense of playing a part in the national policy of his department. Local men felt at ease with their Minister. Bevin handed his cigarettes round to those on the platform with him. . . . Men from offices in little-known moorland towns or sleepy country market places got up to discuss their problems with their chief. Occasionally he would interrupt to ask them if they had thought of doing this or that, and the local men, who knew their regulations fairly thoroughly, would explain why Bevin's suggestion was not immediately practicable. So Bevin would make a note and see that the regulations were amended to cut through formalities.

"Within six months Bevin not only raised his department to the status of a first-class Ministry . . . but . . . made every official in it conscious of the importance of his own work."[1]

Humanising social administration was a subject on which he talked with conviction. In his first broadcast as Minister of Labour he had spoken of having to deal with "the most difficult material to handle, the human being" and it was a theme which he never tired of reiterating. He was constantly urging industry and the supply departments to take more seriously the need for a properly trained profession of personnel management and he was determined to see that his own department set an example. The results, as in the parallel case of welfare, justified a hundredfold the importance he attached to such matters, for there was no point at which morale was more sensitive than in the contacts between Government and people at local exchanges where, before the war ended, the greater part of the nation had to register, submit to interview and be told what they were required to do.

The Ministry of Labour, of course, like other ministries had its own regional organisation on which a great deal of the detailed work of administration was devolved. Its regional offices were the pivots of

[1] Trevor Evans: *Bevin* (1946), p. 183.

the Ministry's labour supply organisation. Bevin wanted to create, however, between the eleven regional centres and the hundreds of local offices a new level of administration, particularly in the heavily populated industrial districts.

His first attempt to do this by setting up Labour Supply Committees was not very successful. At the end of 1941 he made a second attempt, forming District Manpower Boards in forty-four of the principal towns. By this time the supply of manpower was becoming much more difficult and the methods of dealing with it had to be changed. Block reservation of men in scheduled occupations was to be replaced by a new and more selective system of individual deferment[1] and the District Manpower Boards were created, in the first place, to operate this.

The question whether the Boards should be independent and drawn from outside the official world or a part of the Ministry organisation and staffed by its officers was a matter of sharp political disagreement. Despite the strong prejudice against being "regimented by officials", Bevin came down in favour of an all-official membership for the Boards, arguing that this was the best way to secure uniformity of decisions on deferment between different districts and to guarantee their impartiality. In fact considerable recruitment had to take place to staff the new District Boards (which were made collectively responsible for settling deferment cases) and more than half the chairmen, as well as most of their labour supply and deferment officers, were temporary appointments made from outside the Service. At the end of December 1941 the newly appointed chairmen and the other members of the Boards were summoned to London and addressed by the Minister. Bevin laid much stress on the need to see that their decisions on deferment were firm and equitable, and reminded them that he would have to justify their actions in the House of Commons. Fortunately the experiment worked well: in 1942 alone the Boards settled two million applications for deferment without any criticism being heard in Parliament, and five million in all by the end of the war.

Long before then, they had taken over other functions as well and provided the link between regional and local organisation for which Bevin had been looking. By 1943 there were virtually no untapped resources of manpower left. This meant that if the demands of the

[1] See below, p. 138.

Forces and war industry for more men were to be met, it could only be done by constantly examining the pattern of individual employment in their district, withdrawing younger men wherever possible for posting to the Services or more essential work and finding older men or women to fill their places. In taking over the responsibility for this, the District Manpower Boards were carrying out one of the Ministry's most important tasks and doing so at the right level. They were in fact one of the most successful administrative innovations of the war.[1]

3

A portmanteau phrase like "the mobilisation of the country's resources for war" can be highly misleading if it suggests anything remotely resembling an orderly and uniform operation. In practice, it meant a series of radical reorganisations varying widely from one industry to the next and carried through only by a prolonged and untidy process of argument, improvisation, fresh demands, changes of plan and compromises, which placed a heavy strain on everyone involved.

The Minister of Labour was in the thick of it on a dozen different counts—manpower, dilution, demarcation, training, wages, disputes, the Essential Work Orders—and few days passed in 1940–42 without a group of industrialists or trade unionists coming to see him with a list of objections, complaints or pleas for special treatment. Bevin would listen patiently, at least for a time, but listening was not difficult: what counted was to keep up the pressure all the time to push through changes, however genuine the difficulties or well-founded the complaints. To get the managements and trade unions involved in the engineering industry to accept the provisions of the Essential Work Orders was a major task in itself. It is perfectly true that Bevin could end any discussion by saying "Whether you like it or not, that's what the Government is going to do," and put the authority of his emergency powers behind his wishes. He was not worried about being popular—that was impossible in such a position—but he wanted co-operation, and just when to tell people to stop making difficulties and get on with it, how to strike the balance between consultation and giving orders, was a nice problem of judgment.

[1] For a description of their work, see Parker, c. 18.

Some thought him overbearing, too full of his own ideas to listen to objections and inclined to steam-roller opposition: on occasion this was no doubt true, but others concluded that without a man as positive as Bevin, as determined not to let objections hold up action, the job could never have been done.

After an initial period of suspicion at finding a trade unionist in the Ministry of Labour, most of the employers appear to have accepted Bevin as fair-minded in holding the balance between the two sides of industry. Their quarrels with him were far more over government policy than over any bias in favour of the unions. He spoke the same language as they did and if he was blunt in pressing the Government's needs they preferred this to a smooth evasiveness. Many of them admired him for the job he was doing and for qualities of character which were perhaps more readily appreciated in industry than in politics.

Bevin's relations with the trade unions were more complicated. Churchill had invited him to become Minister of Labour because he believed that a trade unionist could more effectively secure the unions' active co-operation in winning the war than anyone else. This proved to be true but it did not follow automatically from Bevin's appointment or from the thirty years he had spent in the trade-union movement.

Bevin could legitimately claim that no Minister had ever done as much to bring the trade unions into consultation on equal terms, but this did not alter the fact that in practice he was continually pressing the unions to make concessions and accept responsibilities in the national interest which cut across their original function of defending the particular interests of their members. Bevin's policy of working as far as possible through the unions put a further strain on the relations between the officials and the rank-and-file membership, confronting the unions for the first time with the question which has since become so familiar, how a movement built up for defensive purposes can in practice exercise the share of responsibilty which it demands.

Bevin was well aware of the problems this posed, but if he was to carry out his duty as Minister of Labour he could not agree to let the unions' doubts or particular interests impede or divert the Government's policy. While, therefore, he continued to defend the trade-union movement against critics from outside, in private discussions with the T.U.C. he pressed them all the harder. The fact that no one had a better knowledge than Bevin of the weaknesses in the unions' case and that, on the other side, Citrine was all the more determined to assert the T.U.C.'s independence now that Bevin had become Minister of Labour, did not make matters any easier.

There is no need to exaggerate these difficulties. If the discussions were heated at times, disagreement was kept within bounds by the desire of both sides to get on with the war and by recognition of the consequences of any open split. But it is worth making the point that, in sticking to his belief in consulting the trade unions and securing their co-operation, Bevin was not taking the easy way but assuming an additional burden of argument, persuasion and exhortation where others would have been tempted to issue orders and be done with it.

Occasionally Bevin's patience would crack and he would let fly in public. One such occasion occurred in September 1941 over the issue of the Services' demands for skilled manpower. The scale of these demands had already led to criticism in the earlier part of 1941, and in May Bevin asked Beveridge to carry out an inquiry into the use which the different services were making of their skilled men. The committee's first report, published at the end of August, confirmed many of the criticisms and provided further ammunition for those who were already pressing for changes in the Government's conduct of the war. Bevin was in an awkward position. While he had taken the initiative in setting up Beveridge's committee, he was as determined as Churchill to resist the demand which had been voiced in several quarters for a reduction in the size of the Army. In almost every speech he made in 1941 he had gone out of his way to express sympathy with the generals who were expected to build up an army from scratch after war had begun. He declared flatly, not once but several times, that if he had any influence at the end of the war he would use it to see that the Army did not suffer from the same neglect again. "The Labour Movement's attitude towards this whole question of defence," he told his union's delegates, "has got to be re-examined." In the meantime he meant to see that the Services got the men they had been allocated and that British forces were not always sent into battle against superior numbers.

The T.U.C. took a different view and at its congress in early September (1941) passed a resolution calling on the Government "to regard the manning of industry as of equal importance with manpower for the fighting services". In the following weeks a number of speeches by trade-union leaders, supported by the *Daily Herald*, made it clear that the T.U.C. was going to press its views. "We start from the premiss," Citrine announced, "that it is impossible for us to have the biggest Navy, the biggest Air Force and the biggest Army"; and he went on to argue that the Forces could only be adequately equipped if the demands on industrial manpower were reduced. On the 25th the *Herald* published a cartoon showing the War Office and Industry playing cards with "Manpower" as the stake and a mirror so arranged as to show "War Office" "Industry's" cards. The caption underneath read: "Not a Chance." Other newspapers took up the controversy and various groups of employers were reported to be resisting the call-up of skilled men.

To find the *Herald*, of all papers, taking this line made Bevin too angry to contain himself. In a speech at Southampton on the 27th he declared: "My view as Minister of National Services is that the Army has got to be fully manned and fully equipped, with reserves to meet any contingency that may arise." He turned on the *Daily Herald*, "a paper that I helped to build," and accused it (according to newspaper reports of his speech) of "carrying on a nagging, Quisling policy now every day of our lives".

"I say as a Labour man, am I entitled to send a man up in a bomber without providing a journeyman to test the bomber and see it is safe? No, Citrine, nobody can tell me that I shall not call on skilled men. I mean to have sufficient of them for all Services. It shall not be on my conscience that I risked a single airman's life."[1]

4

The press was shocked by Bevin's language and at once took him to task. Citrine issued a statement saying:

"I cannot understand Mr. Bevin's outburst. We are all suffering from some degree of war strain and we must allow for Ministers becoming hyper-sensitive of criticism in such circumstances. We cannot afford the luxury of a quarrel whilst we are fighting Hitler. The Government is fortunate in having the undivided support of the Trade Union movement in its war policy, but that does not mean we have surrendered our independence of judgment or our right to criticise, and we have not the slightest intention of doing so."[2]

[1] Report in the *Daily Herald*, 29 September, and *The Times*, 1 October 1941.
[2] *Daily Herald*, 1 October 1941.

Attlee who had been worried by the growing friction between the two men for some time wrote to both of them and appealed to them to compose a quarrel which could only be "detrimental to the war effort and the Labour Movement".[1] Whether as a result of Attlee's intervention or not, the storm subsided as quickly as it had blown up. Once his temper had cooled, Bevin made no more fuss about accepting the Beveridge committee's report and the Cabinet cut the Service demands for skilled men in the period up to March 1942 from 26,000 to 8,600, half of whom had already been supplied.

But the incident shows the strain under which Bevin was working, combining membership of the War Cabinet and its committees (for example, the chairmanship of the Production Executive) with one of the biggest and most controversial departmental jobs. There was no office in the Government more exposed to criticism and hardly a day passed without a speech or an editorial in which the Minister of Labour was reproved either for being too drastic or not drastic enough in his policy. Bevin held on his way, relying on his own judgment and refusing to be pushed off his course by press campaigns or protests even when they came from the T.U.C. A private exchange of letters with Citrine, however, shows how much each resented the other's attitude.

After issuing his statement to the press, Citrine sat down and wrote Bevin the following note:

Personal

"Dear Bevin,   30 September 1941

"This is a difficult letter to write as I know that I am exposing myself to rebuff.

"I did not see the statement attributed to you about 'quislings' until yesterday morning. I could not understand what had happened to cause you to use this expression, and I did not for a moment assume you applied it to me personally. I thought the reference to the 'Herald' unfortunate, but of course I do not challenge your right to say what you feel.

"With regard to manpower, where you used my name, this represents a genuine difference of opinion, and I do not want it to be distorted into a personal quarrel between us. It is unfortunate that we do not see eye to eye on a number of matters connected with the war. My views, like your own, are sincere and deep, and I must stand by them until I have been shown where I am in error.

[1] Attlee to Bevin, 30 September 1941. He wrote a similar letter to Citrine on the same date.

"I have felt for some time past that you have been resentful of criticism and that you often construed references which I have made, on the express instructions of the General Council, into a personal attack upon you. What I have had to say I have said to your face and not behind your back.

"Whether you believe it or not, no one has paid higher tributes to your ability and capacity than I have.

"I have never regarded you as a superman but definitely as the most outstanding personality in the Labour Movement. You may find it hard to swallow that, but it is true just the same.

"What I have always felt is that you do not give sufficient credit to your colleagues, and that you are inclined to regard those who do not agree with you, and stand up for their opinions, as enemies. I am sorry for this, but I can assure you I am not alone in that opinion.

"I have taken this step of writing you because I do not want any differences of view between us, to be manufactured into a public quarrel which will do no good to the Labour Movement or our national cause. Many attempts have been made during the past few hours to get me to embark on such a controversy, but I have refrained from doing so. It is foolish for people, placed as we are to quarrel, whatever our differences of view may be, and I for my part do not intend to indulge in any vendetta either against you or anyone else.

Yours sincerely,
Walter Citrine"

Bevin's reply written the following day made no concessions at all.

Personal and Confidential

1st October 1941

"Dear Sir Walter,

"I am in receipt of your letter of the 30th September. No one who writes to me, whatever the difficulties may be, gets a rebuff.

"With reference to the Daily Herald, as I have already explained, I did not use the word 'quisling'. What makes the Editor's action so indefensible is that he was told personally that the word was not used. No other newspapers used it except the Evening Standard and the Express.

"For the last four or five months, without naming me, this paper has pursued a very unhelpful course especially in regard to this Ministry and the methods adopted, which have been mainly by suggestion and innuendo and have been particularly insidious.

"In all this I have been extremely patient, out of loyalty and because of my past connection and the Movement's association with it.

"With regard to my reference to you, if the policy of the Government is attacked, Ministers must answer and there is, as you say, room for a genuine difference of opinion, but as you have written to me it may be as well if I express myself clearly. I have tried to establish the closest possible consultation with the Movement through the Consultative Committee, Factory Boards, Panels, etc., and in every way I could. I got established the Production

Advisory Board and through it I thought genuine suggestions would come forward from the Movement to assist the war effort. Time will show whether such constructive proposals will be forthcoming but at the Consultative Committee on various occasions the task of presiding has not been easy.

"There are no grounds for the suggestion that I have been resentful of criticism or construed references made on the express instructions of the General Council into a personal attack.

"Frankly, I have been perturbed at times at what I think, rightly or wrongly, is unwillingness to come to decision and accept responsibility and the desire for reference back, and then, at the same time, to hear criticism of our lack of speed.

"I am glad you do not regard me as a superman, so no one will expect too much, but I am not concerned how people regard me personally. All I have lived for in this world has been for what I could leave behind for the benefit of the people I represent. My record in the Trade-Union Movement is probably the best evidence.

"Your statement about giving sufficient credit to my colleagues and regarding as enemies those who do not agree with me I cannot treat seriously.

"Finally, I do not desire to quarrel with anyone and certainly a vendetta is foreign to my nature. I respect other peoples' positions and opinions but I expect mine to be respected also: and that is all.

For the rest of the war, the two men continued to co-operate, as they had done in the T.U.C. before, without coming any nearer to establishing friendly personal relations. But the friction between them matters little by comparison with what their involuntary partnership achieved. For, however much Bevin and Citrine might rub each other up the wrong way, however sharply they might disagree over particular issues, on broad policy they were in agreement, and this was what counted.

After the General Strike Bevin and Citrine had both worked to convert the T.U.C. from revolutionary dreams of overthrowing capitalism or the piecemeal extraction of concessions to a policy of demanding a voice in the formation of industrial, economic and social policy. The war gave them the chance to achieve their objective, and they made the most of it. Between them they secured an influence for the trade-union movement on the formation of government policy which it has never lost since, whatever the party in power.

This change has had unexpected consequences both for the trade-union movement and far outside it, consequences which have not yet been absorbed either by the unions or their critics and which still remain the subject of political controversy. What can hardly be in doubt is the importance (or, I suspect, the permanence) of the change. A major shift has taken place in the relationship between classes in Britain, as a result of which the working-class majority of the nation has begun to exercise in the political system a power much more commensurate with its numbers. This does not necessarily mean a permanent parliamentary majority for the Labour Party. What it does mean is that since the war both parties, the Conservative as well as the Labour Party, have been compelled to pay much more attention to the demands and grievances, not of the middle class (which dominated politics up to the war), but of the working-class majority of the electorate.

Such a change was no doubt a logical consequence of the extension of the franchise, but it had not taken place before the war. It was the war that made the difference and brought it into effect, and the critical date was 1940, rather than the Labour victory of 1945. For it was in 1940 that the organised working class represented by the trade unions, with Bevin at the Ministry of Labour and Citrine at the T.U.C., was for the first time brought into a position of partnership in the national enterprise of war—a partnership on equal not inferior terms, as in the First World War, and one from which it has never since been dislodged, as it was after 1918.[1]

5

In the course of the summer of 1941 the signs of a general shortage of manpower predicted by the earlier Beveridge Committee began to multiply. More reliable figures were now available than at the time Beveridge had made his inquiry, and in July 1941 the War Cabinet

[1] An American student of British politics, Professor Beer, of Harvard, approaching the question from a very different point of view arrives at the same conclusion:
"The critical moment in the forging of this new social contract," he writes, "was not 1945 but 1940. The major re-adjustment resulted not from a shift in the electoral balance of power, but from a shift in the balance of economic power. In a limited but important sense, the old syndicalist thesis was vindicated. For it was initially not by their votes but by their control over instrumentalities necessary to carrying out vital national purposes that the organised working class raised themselves from their old position of exclusion and inferiority. The Labour victory of 1945 and the consequent adaptation of Conservative policy were later phases of this general process, as was the intense competition between the two parties in their bidding for the votes of a populace conditioned by the Welfare State." (Samuel H. Beer, *Modern British Politics*, 1965, p. 215.)

asked for a fresh survey of manpower resources and demands. The job was given to the Manpower Committee of the Production Executive and the Ministry of Labour, and was completed by October.

This survey put the demand for manpower up to June 1942 at a total of 2 million additional men and women for the Armed Forces and war industries.[1] In fact, even before the report was ready for the Cabinet, the Prime Minister's decision in September to increase the bomber programme raised the figure for the munitions industry by at least another 100,000 and possibly three times that number.

The report then went on to analyse the distribution of manpower between the different groups of industries and to make suggestions where the two million might be found. Two problems stood out sharply: the call-up would only provide three-fifths of the men required for the Services and civil defence, leaving more than 300,000 to be found elsewhere, and close on a million women would have to be recruited for the auxiliary forces and industry.

After prolonged discussion in the Lord President's Committee, Anderson and Bevin agreed to recommend four measures to close the gaps.

First, young men were to be called up at eighteen and a half instead of nineteen.
Second, the system of reservation by occupation was to be changed to one of individual reservation, and made much more selective. (This was the task given to the District Manpower Boards.)
Third, all men and women between the ages of eighteen and sixty were to be placed under statutory obligation to undertake some form of national service.
Fourth, women were to be called up by age groups for compulsory service in the Auxiliary Forces.

Conscription for women was the most controversial proposal of the war and it was at first strongly opposed in the War Cabinet. The Prime Minister expressed his own doubts in a memorandum dated 6 November:

"On the whole I am not yet satisfied, in view of the marked dislike of the process by their Service menfolk, that a case has been established for conscripting women to join the auxiliary services at the present time."[2]

[1] The figures are given in Parker, Table 14, p. 110.
[2] Churchill, Vol. III, *The Grand Alliance*, p. 455.

Churchill's preference was for voluntary recruiting or at the very most for individual selection rather than calling up by age groups.

Bevin, too, had been reluctant to come to a conclusion in favour of conscription, but once he had made up his mind, he stuck to his guns. At the end of an anxious and at times passionate debate among ministers which went on for weeks, he carried his main point: if sufficient recruits were to be found for the Women's Services, there was no alternative to conscription. Its application was to be limited: no married women were to be conscripted and no woman posted to a combatant service except as a volunteer. Appeals were to be allowed on the same grounds as for men, and the only age groups which it was proposed to call up in practice, those between twenty and thirty, were to be given the option of choosing between the auxiliary services, civil defence and certain jobs in industry selected by the Minister of Labour. None the less, the principle was conceded and embodied in the National Service (No. 2) Act which became law on 18 December 1941.

The other measures proposed by Anderson and Bevin were adopted by the Cabinet at the same time, with one change in the obligation of national service which restricted compulsory military service to men under the age of fifty-one. Taken together with the new National Service Act, they completed the statutory powers[1] under which in the next three years the mobilisation of the nation was carried to a higher point than either Hitler's Germany or Mussolini's Italy despite their boasted totalitarian organisation. But of more importance than the legislative instruments was Bevin's greater willingness to use the powers they gave him, a change which can be dated to the same period, the closing months of 1941.

His speeches in the autumn reflected this change. At Stoke in October he said:

"Now things are getting tight. I have to take some more steps in order effectively to distribute this population. . . . It was no good doing it before.

---

[1] The most important of these were:

1. The two National Service Acts, September 1939 and December 1941.
2. Defence Regulation 58A (22 May 1940) which gave the Minister full control over civilian manpower and authorised him to issue orders. It was from this that his power to direct labour derived.
3. The Registration for Employment Order, March 1941.
4. The Essential Work Orders, the first of which was dated March 1941.
5. The Employment of Women (Control of Engagement Order), January 1942.

What was the good of my going to this length when the factories were not built, when I had no place to put the people? It is better to leave as much of the normal life of the nation going on, rather than to call people up and then disappoint them when you have done it.

"Now the time has come when we have to have national service on a more intense scale. I still want to do as much as I can by leadership, but the compulsion will become stricter."[1]

At Swansea in November he told an audience with many trade unionists in it:

"I cannot stop to argue with you why you have been transferred. If we decide you are to be transferred to more effective work, you must go. We cannot tell you every detail and we cannot stop to argue and negotiate about it. While we are doing that we are losing production that we ought to be sending to Russia or somewhere else."[2]

At Middlesbrough, where he called for a million married women to work either full- or part-time in industry, he said:

"I have an inherent faith in our people when they understand. The more willing workers I can get, the better for the country's effort. It is for this reason that, in spite of the criticism showered on me, I have stuck to this question of leadership as long as I can, rather than resort to a drastic compulsion. . . . Well, that may come. Compulsion may have to be applied and is being applied more dramatically than hitherto."[3]

The reasons why Bevin should have begun to talk like this in the autumn of 1941 are not difficult to see. There were at least four.

First, because after the autumn manpower survey the facts were incontrovertible: the general shortage of manpower was no longer predicted, it had arrived. The only way left to fill the gap was to recruit at least a million women either directly for the Auxiliary Forces or to replace men in industry. Bevin's appeals for volunteers had failed to produce anything like that number.

Second, because with the building of new factories and with the expansion of industrial capacity now beginning to show results in increased production, the jobs were there for people to be sent to and they would not be kept hanging about while the other shortages—

[1] Speech at Stoke on Trent, 19 October 1941.
[2] Speech at Swansea, 1 November 1941.
[3] Speech at Middlesbrough, 16 November 1941.

raw materials, equipment, components—had to be overcome before production could really begin.

Third, because the development of the war in 1941, the German attack on Russia and the tremendous battles taking place on the Eastern Front had forced people to realise how much would be demanded and to what pitch hostilities would have to be carried before Germany was defeated. As a result, public opinion generally, including opinion in the unions and the factories, was now in favour of compulsion, which was coming to be recognised as the only way of making sure that people were treated equally and that some did not "dodge the column". To judge by the popular press, Bevin and the Government were well behind public opinion in demanding a greater measure of compulsion and its extension to women as well as men.

Fourth (a point often overlooked), because the Ministry of Labour had now built up the organisation, the staff and the experience required to handle these difficult operations successfully and the welfare services had been expanded to meet the demands that would now be made on them.

In short, a greater degree of compulsion in directing people to the work they should do had become, by the autumn of 1941, necessary, acceptable and practicable.

The result of this hardening of opinion is to be seen, not only in the new legislation and orders (including the conscription of women), but in the number of individual directives issued by the Ministry of Labour. Between July 1941 and June 1942, 32,000 compulsory orders were issued, more than ten times as many as up to the summer of 1941, and in the succeeding twelve months (July 1942–June 1943), 408,000—proof that when the manpower crisis became severe, Bevin was prepared to act quite as drastically as his critics wanted him to. Indeed by the summer of 1943, a total of three-quarters of a million directives had been issued and the number of men and women in the Armed Forces, Civil Defence and the munitions industries of Group I had been raised by 2 million to 46 per cent of the working population, a figure which would have appeared impossible earlier in the war.

Bevin, however, never accepted the argument that what proved possible in the later years of the war showed that he had been wrong in the sparing use of his powers in 1940–41. On the contrary, he regarded the success of the greater degree of compulsion applied after 1941 as the fruit of his earlier reliance upon persuasion, which had given public opinion at the factory and trade-union level time to convince itself that this was necessary. For the striking thing about the period 1942–5 is not that the Government was then prepared to apply sanctions where it had not been before, but that working-class opinion was prepared to accept them, and even demand them, as the only way of making sure that people were treated fairly and equally.

More than that: Bevin could still claim that "voluntaryism" was the foundation of his policy even when he began to issue directions more freely. Up to the end of the war the majority of the British people did the jobs they were called upon to do without the application of sanctions. The number of directions issued, in total, was only a fraction of the number of people brought into industry, moved from one job to another or from one place to another. The greater value of sanctions, as Bevin had always believed, proved to be the knowledge that they were there to be used, if necessary. This supplied for most people a sufficiently powerful incentive to do voluntarily what they would otherwise be compelled to do, and a guarantee that those who tried to evade their duty would not be allowed to get away with it. When this basis of consent was lacking, the attempt to enforce orders by sanctions—as the case of the Betteshanger miners brought before the courts in 1942 showed[1]—could still prove ineffectual and destructive of good industrial relations.

Bevin's policy as Minister of Labour, in short, has to be viewed as a whole, and the achievement of the later years of the war—the high degree of mobilisation, the absence of serious labour unrest, the smoothness with which the registration of women took place—set against whatever was lost in production during the first 12–18 months of his ministry. His policy in both periods was based upon the belief that the strength of a country with the democratic traditions of Britain was only to be drawn out on a basis of consent, and that consent was only to be won by convincing people by experience that there was no alternative to the demands the Government made on them.

[1] See below, pp. 267–8.

6

After the angry debates of July, criticism again died down for a time, only to revive in the autumn of 1941. Frustration was the root of much of it, a frustration springing from the defensive role which was still imposed on Britain and increased by having to watch the German invasion of Russia without any chance of intervention. Although the demand for a Second Front began to be heard, this sense of frustration still found its main expression in criticism of the Government's economic policies, rather than its strategy. *The Economist* noted in November:

"It is a peculiarity of the House—and of the country—that almost complete unanimity about the war is combined with serious doubts about the efficiency of administration, widespread uneasiness about the inequality of war sacrifices and general questioning about the post-war world."[1]

A debate on manpower in the House on 8 October proved inconclusive, although more friendly to the Minister of Labour than in July. This took place shortly after Bevin's row with the T.U.C. over skilled manpower for the Services, and a number of members took up the issue again during the debate. More speakers than usual, however, recognised the difficulties the Minister had to contend with, amongst them (as James Walker, the Labour member for Motherwell, admitted) the vested interests of the trade unions defending the privileges and concessions they had fought for and feared to see lost.

Bevin was closely questioned on the employment of women, and defended the pace at which he was proceeding:

"If I introduced mobilisation of women and made them go to the Employment Exchange, where they were given cards and treated in a hard official style, the scheme would be a complete breakdown. One thing I am very glad about and that is that not only have I got the women to come into industry almost up to the amount required but I have carried the confidence of the parents of this country. I think that is very vital. If you have to build up an enormous amount of welfare work, hostels, and so on—and that is very important when you are taking women of 19–20 and onwards away from their homes . . . you must proceed with tact. . . .

"It was suggested that I should now get to the stage of issuing directives to

1 *The Economist*, 29 November 1941.

women more drastically than I have done. That is the advice from a woman and I am encouraged because it is on lines which we have been wanting to follow and probably now, with the help of the Women's Advisory Committee, it may be that we can tighten up this business more effectively."[1]

When the Debate on the Address opened on 12 November 1941, the Cabinet had not completed its discussion of the new measures it was preparing for a more stringent mobilisation of manpower. The House was in a gloomy mood and its spirits were not raised by Churchill's opening statement that he had no changes to announce in the Government (as *The Times* had urged): "Neither do I consider it necessary to remodel the system of Cabinet Government under which we are now working, or to alter in any fundamental manner the system by which the conduct of the war proceeds nor that by which production of muntions is regulated and maintained."[2] The tone of the debate was more subdued than in July and Shinwell found little support for the rabid attack which he made on the Government; but most speakers referred to a general uneasiness in the country, to the growing demand for a Second Front and to the belief that production was still falling short of what could be achieved.

Ness Edwards, the Labour member for Caerphilly, summed up the general feeling when he said:

"This House, as a rule, wants to travel on the same road. The critics differ from the Government only in their desire to travel along the road with greater speed, greater urgency and greater resolution. . . .

"Now I want to take to task the Minister of Labour. He has shouldered a responsibility which would have destroyed the political life of almost any Member of the House who attempted it and he has acquitted himself in a way that has earned the admiration of the vast majority of our people. His contribution to our ultimate victory will probably be as large as that of any other single person in the Administration. But there are still sections of our resources to which he ought again to direct his attention."[3]

Press as well as Parliament was impatient with the Government's "lethargy". "Still No Plan" was the common complaint and the failure to give a clear lead on what the Government wanted women to do was cited as proof of the hesitation and fumbling which marked the Government's policy. When Bevin appealed at Middlesbrough

[1] House of Commons, 8 October 1941. Hansard, Vol. 374, cols. 1078, 1081–2.
[2] Ibid., 12 November 1941, Vol. 376, col. 33.
[3] Ibid., 25 November 1941, Vol. 376, col. 698–9.

for a million more women to work in industry, half the leader writers in the country proceeded to read him a lesson on the folly of supposing that he could get them without issuing clear directives. "Mr. Bevin prefers 'leadership' to compulsion," was the comment of the *Manchester Guardian.* "So do we all, but do let us have leadership."[1]

As Bevin was at this very time arguing the case in the Cabinet for the conscription of women, he may be supposed to have seen the obvious as clearly as his critics. But the demand for the Government to act had the advantage that, when the Prime Minister introduced the new proposals in the House on 2 December, they met with little opposition, even the conscription of women; the chief complaint was that they had not been made before.

Many of the old criticisms were heard again. Members rose to charge that the country had not yet reached full production; that men and machines were still standing idle; that there was no wages policy; that the Government organisation was inadequate both in Whitehall and the regions; that there should be a Minister of Production. Bevin must have heard with ironical amusement that his was now the favoured name for this elusive post. "Curiously enough," Wardlaw-Milne confessed, "—it is a sign perhaps of the times in which we live—I also came to the conclusion that the right person was the Minister of Labour."[2]

It was a group of Labour members who introduced a new note into the debate, by moving an amendment which assented to the extension of compulsion in mobilising manpower, but demanded, as a corollary,

"that industries vital to the successful prosecution of the war, and especially transport, coalmining and the manufacture of munitions should be brought under public ownership and control, and that the necessary legislation should be brought in as soon as possible."[3]

The case was not well argued: the proposers of the amendment appeared to be unsure whether they were advancing it on the grounds of equity or of efficiency, a confusion which had marked many of the earlier debates on the direction of labour. The interest of the motion lies in its reminder—the first of many in the next three years—of the unsolved issues which underlay the wartime coalition.

1 *Manchester Guardian,* 17 November 1941.
2 House of Commons, 2 December 1941, Hansard, Vol. 376, col. 1086.
3 Ibid., 4 December 1941, Vol. 376, col. 1305.

On this occasion it fell to Bevin, who had himself urged his colleagues in the Cabinet to bring the munitions industry under state control, to re-state the political basis on which the coalition had been formed. He did so without equivocation.

"My entry into this Government and that of every one else concerned, had as its supreme object the winning of the war. That is what the National Government was created for. That is what we were asked to associate together for. . . . . It is a cardinal point of policy for the Government as a whole that neither interest, property, persons, nor prejudice will be allowed to stand in the way of our achieving that great objective. Whatever comes in the way, that is the test that must be applied—none other. If the argument is seriously advanced that there should be further requisitioning of either property, services, or industry, in order to secure a more successful prosecution of the war, the Government will examine any specific claim, and will deal with it on its merits, guided by this one principle."[1]

The question how far Labour as a partner in the coalition Government was justified in pressing for socialist measures was to be a matter of sharp dispute inside the Labour Party later in the war. On this occasion, as many as forty Labour members voted against the Government and a third of the party abstained.

The rest of Bevin's speech was spent in reviewing the stages in mobilising manpower, which, taken together, he claimed, provided evidence of a clear and logical plan. It was in fact (although no one would have supposed so at the time) the last big occasion on which Bevin had to defend his manpower policy in the House of Commons. Two days later the Japanese made their attack on Pearl Harbor. A completely new prospect was opened up by the entry of the United States into the war and the spread of fighting to the Pacific and the Far East. On 12 December 1941 the Prime Minister left for a five-week visit to the United States. When he returned in mid-January it was to face a series of angry debates in which, not the Government's economic policy, but its whole conduct of the war was under fire. When the political dust subsided in the latter part of 1942, mobilisation was no longer an issue nor was the Minister of Labour thought to need advice on how to do his job.

[1] Ibid., cols. 1342–3.

CHAPTER 6

Politics, Manpower and Coal .

I

PEARL HARBOR, by bringing the United States into the war, made sure, as Churchill said, that however long it might take there was no more doubt about the end. But the first results of the Japanese attack were disastrous. Their conquest of South-East Asia, the richest of all colonial areas, proceeded without a check: Hong Kong was lost, Borneo occupied, the *Prince of Wales* and the *Repulse* sunk and the whole of Malaya north of Singapore overrun. Bad news from the Far East was matched by equally bad news from the Mediterranean and the worst shipping losses of the war. When Churchill returned from his visit to the United States in January 1942, he had to face a storm of criticism no longer directed against the defects of economic organisation but against the military direction of the war. Churchill, it appeared, was still irreplaceable but there was a loud demand for changes in his Government, changes of men and changes in the way the war was run. This demand was voiced by newspapers and politicians of almost every shade of opinion from *The Times* to the *New Statesman*.

Churchill, in his most pugnacious mood, insisted on a vote of confidence from the House of Commons, forced the issue to a division and won by 464 to 1, with less than 30 abstentions.[1] But neither the debate nor the vote cleared the air: the criticism continued, and so did the defeats.

However uncompromising he might show himself in debate, Churchill was too experienced a politician not to see the value, and the need, of a conciliatory gesture. His first two moves, however, miscarried.

[1] The debate took place on 27–29 January 1942. The solitary vote against was cast by James Maxton; the tellers were two other I.L.P. members, John McGovern and Campbell Stephen.

The first was an offer of the Ministry of Supply to Sir Stafford Cripps. Newly returned from Russia, Cripps enjoyed an unexpected popularity largely because of a fortuitous association with the resistance of the Russian people. His left-wing ideas won him the reputation of a radical who would bring new blood to a jaded administration and he bore himself, in Churchill's words, "as though he had a message to deliver".

Cripps, however, made conditions, including a seat in the War Cabinet, which the Prime Minister was unwilling to accept, partly because they cut across his second proposal, to create a Minister of Production and appoint Beaverbrook to the post. Cripps therefore turned down the offer and remained outside the Government, a potentially dangerous focus of opposition.

Churchill had little more success with his announcement (on 4 February) of the appointment of a Minister of Production. Beaverbrook's name got a mixed reception and the further announcement on 10 February that manpower and labour questions would be excluded from his authority came in for sharp criticism.

Up to his visit to the United States in December 1941, Churchill had still been opposed to the creation of such a Ministry. Criticism of the organisation at the centre had been revived in January by two powerful articles, "Brakes on Production", which Beveridge contributed to *The Times*.[1] In these he argued that "a Production Executive which does not function as such, with Regional Boards which are almost wholly advisory and have no authority" was "a fundamental hindrance to full production". What was needed was "a supreme informed body to plan and control production to the advantage of the war machine as a whole". Opinion in Whitehall, on the other hand, while admitting plenty of shortcomings when performance was viewed against the idealised image of a "streamlined administration" held "that the Production Executive with its system of committees had not worked badly". And Professor Postan concludes that, "had the issue been decided solely in relation to domestic war production, no major change would have taken place, at any rate not in 1942."[2]

A new situation, however, was created by the entry of the United

[1] *The Times*, 2 and 3 January 1942.
[2] See Postan, pp. 248–74, and also Scott and Hughes, cc. 19–20, for a full discussion of the arguments for and against a Minister of Production.

States into the war, the proposal for a combined Anglo-American programme of production and the appointment of Donald Nelson to be chairman of a War Production Board which would be responsible for the American part in this.[1] Churchill returned from the negotiations in Washington convinced that to represent the British point of view there would have to be a single minister responsible for the whole of British war production and capable of treating with Nelson on equal terms. He was equally convinced that the best man for the job was Beaverbrook, who had made a great impression on the Americans and had also secured a considerable success with Stalin when he negotiated the agreement on British supplies for Russia in Moscow the previous October.[2]

Beaverbrook himself spoke of his new appointment as "a foreign office of supply" which would keep him out of the country much of the time. This may be how the proposal first took shape, but it was soon overlaid by the much more controversial question of the powers the new minister was to be given over war production at home. "Every point of detail in dividing the various responsibilities," Churchill wrote, "had . . . to be fought for as in a battle."[3]

Other ministers besides Bevin were strong in their opposition to giving Beaverbrook overriding powers over production, but Bevin was the heart of the resistance. Whoever had been proposed as Minister of Production, he would have insisted that the appointment should not encroach on his own responsibility for manpower and labour. That this should be in the hands of a trade unionist, at least of a Labour minister, was one of the tacit conditions of the coalition. The fact that Beaverbrook was Churchill's choice for the office made Bevin's opposition doubly sure. Churchill speaks of "the very strong personal antagonisms" which developed between the two men. Bevin is reported to have threatened resignation and insisted that written guarantees should be published safeguarding his own position. Beaverbrook, on the other hand, fluctuated between demands for "ever wider and more untrammelled powers" and a

---

[1] See Churchill's statement in the House of Commons, 10 February 1942, Hansard, Vol. 377, cols. 1402 ff.

[2] Beaverbrook, speaking of the office of Minister of Production, told the House of Lords that "the infant was born in Moscow . . . [and] grew up in Washington" (*The Times*, 12 February 1942).

[3] Churchill Vol. IV, p. 67.

profound distaste for office, a mood of erratic indecision which tried Churchill's patience to the limit.

At the end of a week's heated argument, Churchill produced the draft of a White Paper defining the Minister of Production's powers and providing the guarantees for which Bevin had asked. The duties hitherto exercised by the Production Executive were to be transferred to the Minister of Production with the specific exception of those relating to manpower and labour. These were to be exercised by the Minister of Labour under the direct authority of the War Cabinet. They were to include "the allocation of manpower resources to the armed forces and civil defence, to war production and to civil industry as well as general labour questions in the field of production". To leave no doubt about the extent of the Minister of Labour's authority, the White Paper continued:

"As part of his functions in dealing with demands for and allocating manpower, the Minister of Labour and National Service has the duty of bringing to notice any direction in which he thinks that greater economy in the use of manpower could be effected, and for this purpose his officers will have such facilities as they require for obtaining information about the utilisation of labour."[1]

The constitutional responsibilities of other Ministers concerned with production were safeguarded by allowing them the right of appeal to the Minister of Defence and the War Cabinet. But in the case of the Minister of Labour, it was made clear beyond any doubt that the Minister of Production was dealing with an equal not a subordinate power.

On the morning of 10 February, Churchill sent Beaverbrook the draft of the White Paper and an ultimatum.

"So far as I am concerned," Churchill wrote, "it [the White Paper] is in its final form. I have lavished my time and strength during the last week in trying to make arrangements which are satisfactory to you and to the public interest and to allay the anxieties of the departments with whom you will be brought in contact. I can do no more."

After referring to the Minister of War Transport's claim to an effective say in the types of merchant vessels, Churchill continued:

[1] See Churchill's statement in the House of Commons, 10 February 1942, Hansard, Vol. 377, cols. 1404–5.

"If, after all else has been settled, you break on this point or indeed on any other in connection with the great office I have shaped for you, I feel bound to say that you will be harshly judged by the nation. . . ."[1]

He proposed to lay the White Paper before Parliament that morning. He would only defer making his statement if Beaverbrook had decided "to sever our relations". The alternatives were Beaverbrook's acceptance or his resignation. Beaverbrook accepted.

The White Paper was not a success. No sooner had the Prime Minister finished making his statement in the Commons than Hore-Belisha and Shinwell were on their feet to ask why manpower and labour had been excluded from the Minister of Production's jurisdiction. Churchill's defence of the new arrangements did not satisfy them. Press reaction was equally unfavourable. The most sympathetic comment was that of *The Economist*:

"Plainly the new instrument of decision is appreciably weakened by this reservation. But the reasons for it are equally plain. Mr. Bevin is Mr. Bevin. He has staked out his claim for autonomy vis à vis Lord Beaverbrook, and the Prime Minister has confirmed it. Nor is this just an act of aggrandisement. It has been accepted as a principle of practical policy, that in wartime labour should be controlled by a Labour leader. The only political alternative to Mr. Bevin in allocating manpower and dealing with Labour questions is not the newspaper magnate, Lord Beaverbrook, but the trade-union leader, Sir Walter Citrine."[2]

At this point, however, the particular question of the Minister of Production and his powers became submerged in much bigger issues.

2

Within less than a week of Churchill's announcement in the House of Commons, two further blows to British pride brought indignation with the Government's conduct of the war to the boiling point. On 12 February the German battle-cruisers *Scharnhorst* and *Gneisenau* accompanied by the cruiser *Prinz Eugen* slipped out of Brest, passed through the Straits of Dover and in spite of attacks from British sea and air forces reached German ports apparently unscathed. Three

---

[1] Quoted in Churchill, Vol. IV, pp. 67–8.
[2] *The Economist*, 14 February 1942.

days later, Singapore, the great imperial fortress which had been popularly supposed impregnable, surrendered to the Japanese, and a hundred thousand British and Imperial troops were reported taken prisoner.

The demand that Churchill should reconstruct his Cabinet and divest himself of some of his responsibilities was at once redoubled. *The Times* spoke of "putting the office of Prime Minister into commission", other papers of a "catalogue of catastrophes" and a political crisis for which Churchill must blame his own obstinacy in resisting friendly advice. When the House met on the 17th, the Prime Minister was harried from all sides and there were angry exchanges in which he described the House as in "a state of panic"—a remark greeted with cries of "No," "Withdraw"—and charged his critics with conducting a "rattling process" in the press. A debate hurriedly fixed for the following week gave every prospect of another parliamentary storm and demand for a vote of confidence.

Fortunately Churchill's anger did not impair his judgment. He had, in fact, already started on the reconstruction of his Ministry and on the 19th announced important changes in the War Cabinet. Beaverbrook's advice had been to reduce it to three, besides the Prime Minister, each of whom would be responsible for a group of departments. "The War Cabinet," he wrote, "should consist of Bevin, the strongest man in the present Cabinet; Eden, the most popular...; and Attlee, the leader of the Socialist Party. The other members of the Cabinet should be wiped out. They are valiant men, more honourable than the thirty, but they attain not to the first three."[1]

Churchill did not accept this advice; he preserved the character of the War Cabinet but changed its members. In particular, he brought in Cripps as Lord Privy Seal and handed over to him the Leadership of the House of Commons. Kingsley Wood (Chancellor of the Exchequer) and Greenwood went out; Attlee became Deputy Prime Minister and Dominions Secretary.

The most surprising change of all was Beaverbrook's sudden decision to resign the post over which there had been so much fuss and to give up office altogether. At a Cabinet meeting on the day on which Churchill proposed to make public the changes in the Government (19 February 1942), the old wrangle about the powers of the

[1] Beaverbrook to Churchill, 17 February 1942, quoted in Churchill Vol. IV, pp. 73-4.

Minister of Production broke out again. Beaverbrook launched into a tirade to the effect that his views were being ignored and abruptly left the room.[1] This time his departure was final. Churchill wrote later that "the long and harassing discussions which took place in my presence between him and other principal Ministers convinced me it was better to press him no further."[2] He goes on to speak of Beaverbrook's ill-health and describes him as being in a state of nervous breakdown. It seems more likely, however, that Beaverbrook's ill-health and the attacks of asthma from which he suffered were the product of a conflict in his mind which had been evident throughout than the other way round. In his book *Distinguished for Talent*, Woodrow Wyatt suggests that Beaverbrook's timing of his resignation was connected with the fall of Singapore four days before. "He estimated that Churchill would be driven from power by a dispirited and resentful country. He thought himself the automatic next choice."[3] If this was so, it could well have set up a conflict in Beaverbrook's mind between his loyalty to Churchill and the ambition to replace him as Prime Minister. But, as Lord Moran points out, there is no reason to suppose that this was at the back of Beaverbrook's mind when he was on the point of resigning in the spring of 1941 or again in the autumn. Moran, who was Beaverbrook's as well as Churchill's doctor, suggests another explanation. The determination with which he had fought Bevin and others as Minister of Aircraft Production and Minister of Supply could not conceal the fact that in both cases he had ended by losing heart and seeking to resign. For all his fascination with power and his zest for political battles Beaverbrook lacked the staying power and the confidence in himself which was Bevin's strength. So now, faced with the opposition which his appointment had roused in the Cabinet, he preferred at the last moment to resign rather than to stay and fight it out. "As his doctor," Moran wrote later, "I had been familiar with Lord Beaverbrook's asthma for many years and I did not accept it then as an adequate explanation. . . . I still believe that it was his own profound mistrust of himself that haunted him in office."[4]

Whatever the reason for it, Beaverbrook's departure at once eased

[1] Kenneth Young: *Churchill and Beaverbrook*, p. 229.
[2] Churchill Vol. IV, p. 74.
[3] Woodrow Wyatt, *Distinguished for Talent* (1958), p. 39.
[4] Lord Moran: *Winston Churchill, The Struggle for Survival* (1966), p. 30.

the situation. Oliver Lyttelton was hastily recalled from Cairo to take his place as Minister of Production and when the dust had settled the War Cabinet emerged with two new members, Cripps and Lyttelton, in place of the three who had gone.[1] Eden, Bevin and Anderson, together with Churchill and Attlee, brought its total membership up to seven. The most important change outside the War Cabinet was the appointment of a new Secretary of State for War, Sir James Grigg, in place of Captain Margesson. New ministers were appointed to the heads of five other departments and no fewer than nine junior ministers "placed their offices at the Prime Minister's disposal".

The new appointments were universally welcomed. Relief at Churchill's willingness to yield to the demand for change was fortified by the extravagant interpretation of critics who read more into the changes than time was to justify. Unimpressed by what others made of them, Churchill continued to run the Cabinet in his own way, faithful to his view "that War Cabinet members should also be the holders of responsible offices and not mere advisers at large with nothing to do but think and talk and take decisions by compromise or majority."[2]

There was too much relief, however, at the disappearance of the atmosphere of crisis for analysis to be pressed too far. Churchill was still Minister of Defence and still prepared to defend his responsibility for the conduct of the war when the House met on 24 February, but no one wanted to continue the quarrel and, contrary to the expectations of the week before, the debate was marked by a feeling of anti-climax. The *New Statesman*'s wry comment—"A short period of goodwill in which all the critics will be silenced, a honeymoon period of harmony between Government and public is before us"[3]—was not so very different from Churchill's own—"Thus we gained a breathing space in which to endure the further misfortunes that were coming upon us."[4]

In the sudden calm after the storm, the argument about the Minister of Production's powers was not renewed. The White Paper was quietly withdrawn and when Lyttelton's formal appointment as

---

[1] During his service as Minister of State in the Middle East, Lyttelton had the rank of a member of the War Cabinet but was unable to attend its meetings.
[2] Churchill, Vol. IV, p. 75.
[3] *New Statesman & Nation*, 7 March 1942.
[4] Churchill, ibid., p. 78.

Minister of Production was announced on 12 March, no new division of responsibilities was published in its place. With Beaverbrook gone, Bevin was prepared to settle any difficulties with the new Minister informally, in place of the cut-and-dried definitions of the White Paper, and Churchill announced that "in all matters connected with allocation, distribution and efficient use of labour within the field of war production, the Minister of Production and the Minister of Labour and National Service will work together".[1] Lyttelton's own statement, later in March, suggested that, after all, this might be the best arrangement.

"No system," he told the House, "which seeks to divide the actual allocation of labour judged from a productive point of view, and the provision and welfare of the labour itself, is more than lip service to a formula. Unless the two Ministers charged, on the one hand, with the production of inanimate objects like guns and tanks, and the Minister charged with the lives and hopes and welfare of human beings, on the other, work in close co-operation, nothing satisfactory can be achieved . . . I cannot do more than assure the House that, on these matters, my right hon. friend the Minister of Labour and I feel perfectly convinced that we can work in harmony."[2]

Subsequent events proved these words not to have been too optimistic. In the next two years, the Minister of Production succeeded in providing a stronger link between strategy and production and a better co-ordination of production programmes than had proved possible when these were left to the Production Executive. It took time to accomplish this, but it was done by the methods of co-operation which Lyttelton had foreshadowed and without either destroying the independence of the supply departments or infringing on the jurisdiction over manpower and labour questions which Bevin had successfully defended against Beaverbrook.[3]

The latter question was at least settled once and for all: until the end of the war, Bevin retained the unified control of all manpower problems under a single ministry which he had claimed in 1940. And, although he now ceased to be chairman of the Production Executive, the events of January–February 1942 made clear beyond any doubt

[1] House of Commons, 12 March 1942, Hansard, Vol. 378, col. 1206.
[2] Ibid., 25 March 1942, cols. 1840–1.
[3] Cf. the admirably lucid account in Postan, pp. 248–74.

the strength of his position in the Government. Welcoming the changes, *The Times* commented:

"Mr. Bevin, sitting in the Cabinet as Minister of Labour and National Service, is the one clear exception to the rule that members of the Cabinet should be free from the charge of a domestic department; but his personal position in the country is also exceptional. It would have been difficult to imagine him outside the War Cabinet and difficult to think of a substitute for him in his present post."[1]

After the continuous fire of criticism to which he had been subjected for the past eighteen months, Bevin suddenly found himself accepted as one of the successes of the Government, a man whose personal qualities, the press agreed, quite apart from the importance of his job, made it impossible to leave him out of any war cabinet.

3

The occasional manpower debates in the Commons after the beginning of 1942 now lost most of their former interest.[2] But the problems of manpower policy and the difficulties of translating policy into action remained, and were bound to become more acute as the country approached the limits of its human resources. In determining policy—which demands to accept and which to cut—Bevin worked closely with Anderson and Lyttelton; in carrying out policy the responsibility was his alone.

It would be tedious to recall problems already discussed, but a mistake to suppose that they had ceased to be problems and no longer required much time and effort from the Minister of Labour. Engineering could never find enough skilled men and the drive for greater dilution and the employment of women had to be maintained all the time. For example, between March and December 1942 the number of women employed *part time* in engineering was raised from under 20,000 to 110,000, a development which the engineering employers would have dismissed as out of the question a year before.

[1] *The Times*, 20 February 1942.

[2] The second reading of the Bill to restore pre-war trade-union practices (3 February 1942); a discussion on the mobilisation of women (3 March) and a debate on the operation of Essential Work Orders (21 May) were routine affairs which attracted no attention.

Now that the United States was in the war, production and manpower programmes, as well as strategic plans, had to be revised. Agreement with the Americans on when and where to attack was not easily reached and discussions continued far into 1942. Until these questions were settled, no new manpower budget could be prepared to replace that drawn up before Pearl Harbor and now outdated by events. The new budget was not ready for the Cabinet's approval before December 1942.

In the meantime there were certain decisions which had to be made piecemeal, however uncertain the assumptions on which they were framed. The most urgent of these were concerned with the aircraft programme, building, coalmining and (to be dealt with in the next chapter) shipbuilding and recruitment for the Army.

The labour needs of the aircraft industry had been a matter for dispute from the time that Beaverbrook set up the Ministry of Aircraft Production and Bevin became Minister of Labour in 1940. No other branch of war production enjoyed so continuous an overriding claim on the country's resources—not only in labour, but in raw materials, machine tools, everything that was scarce—from 1940 to 1944. The reasons for this are obvious. The defence of Britain had depended on the R.A.F. Fighter Command in 1940-41, and in 1942-43 the bomber was still the one weapon with which the British could attack the enemy at home.

The priority enjoyed by the aircraft industry, however, reflected not only the strategic importance of the bomber but the failure of the industry to meet the production targets set for it.[1] The M.A.P. argued that a prime reason for short-fall in every programme which it drew up was the failure of the Minister of Labour to supply it with the labour which it required. Bevin retorted with equal vigour that these requirements were grossly exaggerated, that the industry could not absorb the labour which was provided, and that it made inefficient and uneconomical use of the manpower it already employed.

In September 1941, when the Prime Minister called for an unprecedented increase in the output of bombers, the M.A.P. asked for a million additional men and women by the end of 1942. Even when scaled down to 850,000,[2] these figures appeared out of all proportion

[1] See the discussion in Postan, pp. 303-45, and in Inman, pp. 201-7.
[2] This reduced figure for *additional* labour was 150,000 more than the *total* numbers employed in the British coal industry.

to the rest of the munitions programme and, Bevin maintained, out of all proportion to the numbers the aircraft industry could actually use. The deadlock between these two irreconcilable views persisted and in July 1942 the Cabinet asked the Lord President and the Minister of Production to hold an independent inquiry. This confirmed Bevin's scepticism. The investigators sent to examine the methods by which firms forecast their requirements reported that they were largely guesses, "a shot in the dark", with a large margin for reinsurance against contingencies. On 7 October the War Cabinet decided that the only practical course was to concentrate on meeting the urgent needs of particular factories. The controversy over the M.A.P. labour allocation was thus prolonged into 1943.

The evidence in the dispute is complicated by many other factors besides manpower—the fact that aircraft were more subject to modifications and changes of design than any other instrument of war; the unprecedented expansion forced on an industry without the managerial resources or experience to meet it; the legacy of Beaverbrook's method of deliberately fixing targets which could never be achieved in order to stimulate effort. The later history of aircraft production during the war appears, however, to support Bevin's view. Cripps's appointment in November 1942 was followed by a series of drastic reforms, including the first "realistic" programme of aircraft manufacture. It was only when these reforms began to bear fruit in 1943–44 that the aircraft industry was at last able to turn an increase in its labour force into a proportionate increase in production.

The building industry presented problems of a different kind. Its labour force (920,000 at the end of 1941) was too large at a time when manpower was becoming short in every direction. In a directive issued on 27 November 1941, the Prime Minister required the numbers to be reduced to 792,500 during the first three months of 1942 and to 600,000 by the end of 1942.

The difficulty in the way was to persuade the supply and service departments to cut their building programmes and stay within the allocation of labour made to them. Neither the Production Executive nor the Ministry of Works, despite persistent efforts, had been able to secure sufficient information or exercise sufficient authority to do this: new works were still being started at the rate of £15 million and thirty new factories a month.

After the ministerial changes of February 1942 the responsibility for keeping the building programme within bounds and cutting down the labour force rested with three ministers, the Minister of Production (Lyttelton), the Minister of Works (Portal) and the Minister of Labour. They had hardly started to attack the problem, however, when all their calculations were thrown out by the need to provide accommodation for the American forces which were to make Britain an advanced base, first for air attacks, finally for direct assault on Hitler's Europe.

Under the scheme known as "Bolero", the British Government undertook to construct the additional camps and airfields required by the Americans. Over a million men—eventually a million and a half—had to be accommodated in addition to the two million British troops already stationed in the United Kingdom.

If the call-up of building workers was not to stop—and 50,000 were taken in the first six months of 1942—new methods of organising the Government programme had to be tried. Too many jobs were being started without a chance of finishing on time; the transfer of labour from one job to another was too slow and payment by results still inadequately enforced. Bevin suggested creating a single pool of labour for all Government work to be operated by the Ministry of Works. This would enable labour to be switched from one job to another. The rigid divisions between departmental allocations were to be broken down and departments henceforward would be entitled not to a certain amount of labour but to the completion of certain contracts.[1] The Lord President's Committee accepted Bevin's plan—in effect, a return to priorities in place of allocations—and Bevin then set out to get the co-operation of the building industry. On 3 June he called a conference of representatives from both sides and put his proposals to them.

Instead of recommending the reservation of men who might otherwise be called up, he took exactly the opposite course. All those employed in the building industry who were of military age were to be de-reserved at once, but their call-up was to be suspended on condition that they transferred to priority work on government contract.

Equally striking was Bevin's proposal to cut across the traditional lines of demarcation between craftsmen and labourers. It was the

[1] See S. M. Kohan, *Works and Buildings* (1952), c. 5.

latter who were in short supply and Bevin suggested that craftsmen, to be known as "designated craftsmen" and paid, if necessary at craftsmen's rates, should be employed as labourers on work of national importance like "Bolero".

To overcome the resistance of the employers as well as the unions to so abrupt a departure from traditional practice, Bevin called a further conference from all over the kingdom and took the Central Hall, Westminster, to pack them in. It was, Bevin told them, the largest assembly of employers, managers and workmen that he had ever seen gathered in one meeting and, accompanied by the new Minister of Works, Lord Portal, the Minister of Labour proceeded to give his own version of how to conduct industrial relations.

He began by running over what the British people had achieved since he had addressed the conference of trade-union executives in the same hall two years before. Now they were no longer fighting alone, and Britain was to become the great base for allied operations against the Continent. Her ability to do that depended in the first place on the ability of the building industry to build the necessary camps and aerodromes and to build them quickly. This was the reason for asking them to pool all their resources.

"Now I get to a very ticklish point, and that is the crafts. In this job of building up for the great offensive, I want any man to be willing to do anything. If a craftsman is on a job and he is asked to do something in a different trade or to do a bit of labouring, I want him to be ready to do it. I am not troubled about the rate of pay; I do not want to cut down a craftsman's rate to a labourer's because he helps out. If you take a lot of money, we will print it for you and then take it back off you later (*Loud Laughter*). The main thing we are concerned with is speed. If the men on the job will turn their hands to anything just to get the job done, then they can go back to the dignity of their craft and have all the demarcation troubles they like after the war (*Laughter*). But in this one particular effort of great priority, we want to get these things finished, whether it is a factory to speed up our plane production, whether it is the huts or the aerodromes. I appeal to every man to forget his traditions and . . . I believe that all of you, employers, management and men will feel proud, when this thing is over, that you played your part . . . in putting beyond question the claim of the building industry to have done well for a great country."[1]

Inevitably, nothing happened as planned. The "Bolero" plans were continually revised and their execution delayed. The priority

[1] Speech at the Building Trades Conference, Central Hall, Westminster, 9 June 1942.

system soon had to be altered to allow for "super-priorities" and before the end of the year, the Government's building programme was once more hopelessly in excess of the building industry's capacity. In January 1943 a return was made to allocations (now known as "ceilings") in place of priorities. Bevin got fewer men by transfer from less essential work than he had predicted and, although the release of men to munitions work had been suspended, nothing could stop their drifting away from the building sites and transferring themselves to better paid work in munitions factories.

In short, all the familiar features of trying to translate paper plans into the untidy realities of human behaviour reappeared, made worse by the old tradition of casual labour in the building industry. Bevin was not at all surprised: he knew the limitations of planning when it was a matter of dealing with human beings and understood the material with which he had to work. It was his strength as Minister of Labour that he did not let this defeat him. At every stage in manpower planning, estimates had to be scrapped and forecasts revised, an administrator's nightmare. But what mattered was the final result and in the building industry it was not unsatisfactory. In the end the "Bolero" programme was carried out, the departments' needs met, enough men found for such emergency operations as the repair of houses in the V1 and V2 attacks of 1944—and all this without suspending the call-up of men from the building industry. Churchill's directive on reducing the labour force was carried out with no more than a six months' delay for each of the targets—and this despite the fact that the "Bolero" plans were drawn up and had to be fitted in after the directive was issued. Churchill required the number to be cut to 500,000 by the end of 1943: in fact it was down to 496,000 in the middle of 1944.

4

The industry that came nearest to defeating Bevin, and everyone else, was mining.

Coal was the one raw material of which everyone in Britain assumed there would always be enough. Between the wars the nation had become used to an abundance of coal, too many miners and too few markets, as a permanent feature of the economy. It was in 1941 for the first time that an incredulous House of Commons heard that the country was faced with a shortage of coal.[1] In that year total production, the number of men working in the mines and output per man all showed a sharp fall.

First, production. In 1939, this had been 231 million tons; in 1940, 224 million; in the first half of 1941 it looked like being no more than 202 million for the whole year.

Second, manpower. Up to the summer of 1940 the loss of miners to the Forces had been largely balanced (in numbers, though not in quality or fitness) by drawing back unemployed men into the pits: in the second quarter of 1940 the labour force of the industry stood at 764,000 by comparison with 773,000 a year before. The Government had then tried to stop further losses by forbidding recruitment from the mines, but had almost immediately to lift the ban in face of the renewed unemployment which followed the defeat of France and the loss of continental markets. By the spring of 1941, a year later, the labour force had dropped to 690,000, a fall of over 9 per cent in twelve months. In the autumn the age of reservation was raised from 18 to 30, but the industry never recovered during the war from the loss of young able-bodied miners to the Forces which took place in 1940–41.

Finally, productivity. At the beginning of 1939 the annual rate of output had been running at 79 tons for each man employed. A great spurt in the early summer of 1940 raised this to 81, but there had been a marked falling off in the later months of 1940 and in the comparable period of 1941 the figure was down to 74 tons.

This fall in output was matched by an increasing demand for coal as war production began to expand: in the summer of 1941 it was thought that the gap between supply and demand might be as much as 14 million tons in a six-month period.

The first step that needed to be taken was to stop further wastage of the labour force, and in May 1941 Bevin applied the Essential Work Order to the mines. In return for a guaranteed week and a wage increase of a shilling a shift, men were prohibited from leaving and the employers from dismissing them from the industry.

The second step was to recruit more men. In June Bevin broadcast an appeal for 50,000 ex-miners to return voluntarily to the pits— without success. The Cabinet was still reluctant to draft men back from the Army, but in July Bevin ordered the registration of all men with mining experience who had left for other employment. This

[1] See the House of Commons debate, 28 May 1941.

added 30,000 to the labour force, but had to be offset by the wastage from age and injuries: at the end of 1941, it was up to 708,000 but no more.

A third step was to upgrade the labour already in the mines and get a higher proportion of men working at the coal face.

As a result of these measures, production and productivity were both raised in the second half of 1941 and, with consumption lower than had been expected, a coal crisis before the end of the year was avoided.

But the relief was only temporary. In the early months of 1942, although the labour force remained the same size, output per man fell again, to a lower point than ever; and coal production in the first half of 1942 was four and a quarter million tons less than in the latter half of 1941 against a rising curve of war production and demand for fuel.

On 6 April 1942, the Cabinet was presented with the estimates for coal production and consumption in the next twelve months. They showed a gap which could only be closed if the labour force was raised from 705,000 to 720,000. After examining the figures, Sir John Anderson reported that the measures so far adopted would not produce the extra numbers needed and that the Government would have to consider withdrawing ex-miners from the fighting units of the Field Army. Even that, however, would not solve the problem for more than a few months. In an industry which had a higher injury rate and made heavier demands on physical fitness than any other, 40 per cent of the labour force was now over 40 years old.

Despite the protection of the Essential Work Order the labour force was falling at a rate of 28,000 men a year, many of them no longer fit to go on working in the pits. And, most serious of all, the normal recruitment of boys was far too low to make good the losses. As a result whatever steps were taken to bring the numbers up to 720,000 in May, this would not prevent the total dropping back to 705,000 again by the beginning of the winter. However reluctant the Government might be to attack long-term problems in the middle of the war, Anderson concluded, nothing less than the reorganisation of an industry notoriously in need of modernisation would enable it to make an efficient use of its labour force and prevent still larger falls in production.

There were other considerations which pressed the Government to the same unwelcome conclusion. For two years the economic and social conflicts of the years between the wars had been damped down by the sense of a common national danger, but they had been neither resolved nor forgotten. No industry had a worse record of industrial relations than coal, a feud between masters and men which went back two or three generations and in some areas, like South Wales, dominated the lives of whole communities. The bitterness of defeat after the strikes of the 1920s had been kept alive by the miners' resentment at the indifference of the rest of the nation to the conditions of poverty and neglect, the permanent lack of work and lack of hope which hung like a moral blight over the mining districts. These feelings were heightened by the isolation of the mining villages, self-contained communities wholly dependent on a single industry and turned in on themselves. In thousands of miners' families nothing could eradicate the sense of injustice and alienation left by their experience or overcome the cynical suspicion with which they regarded belated efforts to improve the miner's lot now that war had temporarily restored the value of his labour. The sudden and un-expected reappearance of mass unemployment in 1940, following the big effort made to raise output before the fall of France, revived all the old feelings of insecurity. The young men were determined to get out of the pits if they could, fathers and mothers to see no son of theirs go into an industry which had treated them so badly and for which they saw no future.

The miners were not lacking in patriotism, as their record in the fighting services shows, and there had never been any doubt, with their strong political interests, of the miners' hatred of Nazism. But in their attitude to their own industry, there was a "conflict between the miner's undoubted patriotism and his strong sense of wrongs un-remedied".[1]

5

By the spring of 1942, an accumulation of grievances had brought discontent in the coalfields to the danger point. Like everyone else the miners were feeling the accumulated strain of a war now well into its

[1] W. H. B. Court: *Coal* (1951), p. 28. Professor Court's book is an outstanding analysis of the social as well as the economic problems of the mining industry. See, in particular, cc. 1 and 17.

third year, with Britain still on the defensive and still suffering defeats. The extra shifts told on the older men and the miners complained that their rations were not sufficient for heavy physical work. Absenteeism was a particularly sore point. With strong traditions of independence and solidarity, the miners disliked the attempt to enforce discipline under the Essential Work Order and at the same time fiercely resented public criticism which took no account of the greater efforts being made by the majority, or of the difference between shirking and the absenteeism due to injury, sickness and fatigue in an industry with physical conditions far harder and more dangerous than factory work.

The Essential Work Order was unpopular and had only been accepted by the miners' leaders under strong pressure from the Government. The men disliked being bound by the State to an industry in which the mines were still in the possession of the same private owners whom they looked on as their natural enemies. They regarded the Government's intervention, without taking over the mines, or at least forcing the owners to concede a joint national board for the industry, as one-sided, and they blamed Bevin in particular (whose part in ending the General Strike of 1926 they had never forgiven) for allowing the unrestricted call-up of miners after the fall of France and then imposing the Essential Work Order on them. At their Ayr conference in July 1941 the Miners' Federation reiterated their demands for nationalisation and in the following month Nye Bevan expressed their feelings in a sharp attack on Bevin in the House of Commons.

Above all, the miners were angry at the low level of their wages by comparison with those paid for munitions work. In the wage negotiations which accompanied the imposition of the Essential Work Order, Ebby Edwards, one of the Miners' Federation leaders, warned the mine owners (still led by their old enemy, Evan Williams, now 71 and the owners' spokesman since 1919) that they could only expect trouble when "you have men working in this industry where their daughters are working on the other side of the road and taking £2 a week more than their fathers home in wages."[1] The increases granted in 1941 still left the miners low on the list of industrial wages, a comparison which provoked their deep-seated sense of resentment

[1] Quoted, R. Page Arnot, *The Miners in Crisis and War* (1961), p. 320. Dalton describes Evan Williams as "a character worthy of Galsworthy".

towards the rest of the community as well as stirring up domestic strife with sons and daughters in better-paid jobs.

By April 1942, the miners had had enough: if the nation stood in such need of their labour, then let it do something about their pay and conditions. Having made up his mind to protest, the miner proceeded to do so, as Professor Court puts it, "with that marked independence which belongs to his character and to the isolated type of community in which he lives and which had often led him to do it before, not only without asking whether public opinion was on his side, but in actual indifference or hostility to it."[1]

At the same time, therefore, as Ministers were confronted with a production crisis in coal, they were warned by Bevin that they must expect trouble in the coalfields, a warning amply confirmed by the widespread strikes of May and June. On both grounds, something must clearly be done about the coal industry, but, as every man sitting round the Cabinet table well knew, past history made it certain that no issue was more likely than coal to rouse political passions.

There had already been signs that party politics were beginning to revive. The Labour backbenchers' motion of 4 December (1941) calling for the conscription of property had been followed by a renewed demand for the nationalisation of coal in the *Daily Herald*, a demand taken up by Labour members in the debate of 24–25 February. At the request of the Labour Party Executive, the National Council of Labour set to work on plans for reorganising the coal industry.

In March, 1942 for the first time, a Government candidate in a by-election was defeated by an Independent; a month later the Government lost two more seats to Independents (one by a 6,000 majority). Alarmed by the threat to the coalition and to the future of their own Party from such Independent candidates, the Labour Ministers tried to strengthen the electoral truce and commit the Party, not merely to abstain from putting up candidates in by-elections but to giving active support to Government candidates, whatever party they came from. A motion to this effect was passed by the Labour conference in May, but by so narrow a majority (1,275,000 to 1,209,000 votes) that it underlined the dissatisfaction in the Party with official policy. Three of the big unions, the miners, engineers and railwaymen, voted

[1] Court, p. 233.

against the Executive and *The Economist* remarked that "the entire Conference tended to talk as if the war was all over bar the shouting."[1]

In part, of course, the by-election reverses could be explained by the depressed state of the Government's credit. As the third year of war dragged on and there was still so little to show for all the effort that had gone into rearmament, frustration and irritation with the Government were bound to grow. At the same time, with the sense of danger from invasion and the fear of defeat lifted, the political solidarity which had been the great achievement of 1940 and 1941 showed signs of weakening.

One such sign was the agitation for a Second Front which had obvious links with the reviving radicalism of the Left on social issues. Another was the parliamentary row which blew up over coal rationing and showed the strength of Conservative opposition to any proposal which smacked of nationalisation.

The Mines Department had already been preparing plans for rationing domestic fuel and these were taken up with energy by Dalton when he became President of the Board of Trade in the February reconstruction of the Government. In March 1942 he informed the House of Commons that the Cabinet had agreed to rationing in principle and that he proposed to invite Sir William Beveridge to draw up detailed proposals. The scheme was published as a White Paper at the end of April, but by then a powerful opposition lobby had been organised amongst Conservative back-benchers and the Government had to face the threat of an open revolt when the White Paper came up for debate on 7 May.

A variety of motives was involved in this opposition: plain selfishness, resistance from the interests affected, irritation with the miners who would not dig enough coal, irritation with a Government which was failing to win the war quickly enough. But there was no doubt of the political feeling behind the storm with which the proposals had been greeted, or of the part which the Tory 1922 Committee played in defeating them. As one Tory M.P. told Dalton, "they acted as they did because they felt that the Labour Party in the Government was getting too much of its own way."[2] Rationing, especially when

1 *The Economist*, 30 May 1942.
2 Dalton, *The Fateful Years*, p. 400. As President of the Board of Trade Dalton was responsible for the Mines Department. His exuberant manner together with the fact that he was a public-school socialist, and so regarded by many Tories as a

sponsored by a Labour minister, was seen as a first step to government control of the mines and nationalisation. Behind the arguments about the virtues or defects of the scheme, the debate turned into an unacknowledged trial of strength between the parties. The only way out for Dalton was to drop rationing as a separate issue and try to tack it on to the measures under discussion for reorganising the industry. In fact, however, coal rationing had been killed by the Tory opposition, and was only nominally retained as part of the Government's proposals.

6

The revival of party politics, especially on a subject so charged with emotions as coal, bore hard on the Labour members of the Government, particularly on Bevin who had so recently re-asserted his sole responsibility for questions affecting labour.

On the one hand, as a member of the War Cabinet, he knew how far the war still was from being won. On 7 April, the day after the Cabinet's first discussion of coal, Sir Alan Brooke, the Chief of the General Staff, noted in his diary:

"I suppose this Empire has never been in such a precarious position throughout history. I do not like the look of things . . .

"A very gloomy Cabinet meeting. Both Bevin and Alexander reporting Labour discontentment at course of war and difficulty at not being able to give them a full account."[1]

These were the last and heaviest months to get through before the weight of the Anglo-American alliance could be made to tell, and Bevin needed no convincing of the overriding importance of preserving the coalition intact to win the war.

On the other hand, he knew enough of the coal industry and the temper of the miners to be certain that drastic changes would be needed to solve its problems. He was equally well aware that nothing but nationalisation would satisfy the miners and that any proposals

renegade, no doubt helped to stir up partisan feeling. Dalton's chief assistant in handling coal problems was a temporary civil servant who had already been adopted as Labour candidate for South Leeds—Hugh Gaitskell.

[1] Arthur Bryant, *The Turn of the Tide* (1957), p. 350.

which fell short of nationalisation would bring down on his head the accusation that he had once again (as the miners believed he had in 1926) "sold them out".[1]

The Cabinet considered Anderson's recommendations on 10 April. It decided against the recall of miners from the Field Army, but agreed to adopt both his other suggestions. The first (which Bevin had already been urging) was to make an independent inquiry into the obstacles to the recruitment of boys for mining: on the 18th Bevin and Dalton announced the appointment of a committee under Sir John Forster to carry this out. The second, the reorganisation of the coal industry, was put in the hands of a strong Cabinet committee, with Anderson as its chairman and Bevin as one of its members.

After hearing evidence from all the parties concerned, the Cabinet Committee was able to agree on a list of necessary reforms in the operation of the mines: increased mechanisation, the concentration of labour in the most productive mines and seams, the grouping of collieries to improve the technical efficiency of management and changes in the Essential Work Order to deal more effectively with the minority who persistently stayed away from work. The question that remained to be answered was, how far the State should go in taking over the mines in order to control their operation.

The Mineworkers' Federation in its evidence to the Anderson Committee held to the view which it had put to the Sankey Commission in 1920, that the mines should be taken into public ownership. Bevin agreed with them but, in common with the other Labour minister, believed nationalisation to be politically impossible for a coalition Government during a war. His own proposal was for the State to requisition the mines by a compulsory lease for the period of the war and six months afterwards, paying an annual rent equal to the average net profits for the last five years and offering compensation for any capital losses.[2] "Both of us," Dalton recalled later, "were sure that if the owners lost control of the pits now, they would never get it back."[3] To this Bevin added further proposals for six or seven regional directors of production, with a network of consultative committees at pit, regional and national level; a mines

[1] For the events of 1926 and a different view of Bevin's part in them, see Vol. I, cc. 10–12.

[2] His suggestions are contained in a paper, Reorganisation of the Coal Mining Industry, which he drew up in its final form on 27 May 1942.

[3] Dalton, p. 391.

medical service; revision of the Essential Work Order to deal with absenteeism and an independent wages board.

After a great deal of discussion with the miners' leaders and Labour ministers, the National Council of Labour agreed to accept a similar basis—requisitioning instead of nationalisation—for its own plan, and this was adopted unanimously by the Labour Party at its annual conference on 25 May.

Despite Bevin's and Dalton's advocacy, however, the majority of the Cabinet Committee, impressed no doubt by the political temper shown in the Commons debate on coal rationing, preferred to avoid the question of ownership altogether. At the end of May they recommended to the Cabinet that the reorganisation of the industry should be carried out under the existing powers conferred on the Government by the Defence Regulations. A cumbersome system of dual control was established, with the State directing mining operations through regional controls while the owners remained in possession of their property and continued to be responsible for the finance of the mines. This system was to last until a final decision about the future of the industry was taken by Parliament, thereby avoiding the abrupt ending of control which had led to the bitter conflicts after 1918. The discussion of wages and hours was excluded: these were to be settled by national negotiations which had long been sought by the miners in place of the district negotiations imposed by the owners after the defeat of the 1926 strike.

The Anderson Committee's report was adopted by the War Cabinet. A separate Ministry of Fuel and Power was created to take responsibility for the control of the mines and Gwilym Lloyd George appointed as Minister on 4 June. Bevin, like Dalton, believed that these were the best terms they could obtain without threatening the coalition, and accepted them as such. It was harder to convince the Party and the miners' leaders that they could not have got more if they had fought with more persistence. It took three meetings of the Parliamentary Labour Party to get approval of the Government's scheme. The House of Commons debate on 10–11 June showed that many Labour members still remained unconvinced. Greenwood's opening speech was the most critical he had delivered since leaving office, and he explored with sardonic effect the Government's retreat from its own arguments on coal rationing. Other Labour members complained that the Tories and the 1922 Committee had "won". At the close of the debate only eight members went into the lobby against the Government, which had a majority of 321 for its proposals, but the feeling that the episode had ended in a humiliating defeat and that Labour Ministers had let themselves be too easily overborne continued to plague the Party leadership far into 1943.

The miners' dissatisfaction, on the other hand, was tempered by the wage settlement which followed immediately. Bevin and Dalton had not waited for the debate before setting up the independent board of inquiry under Lord Greene to which the Cabinet had agreed. Its terms of reference were to deal with the immediate question of the miners' pay and then investigate the whole question of machinery for the negotiation of wages and conditions. Its recommendations on the first question were clear and prompt. It found that the argument for an unconditional increase of wages was well based and, while not accepting the claim of the Mineworkers' Federation (for 4/– extra), recommended an additional 2/6 per shift. No less important, it adopted the miners' case for a minimum national wage (refused by the owners) and recommended a figure of 83/– a week for underground, and 78/– a week for surface work. This raised the miners at one step from 54th to 23rd in the list of industrial earnings and (partly because of the speed with which the award was published, on 18 June) did more than anything else to halt the spread of strikes in the coalfields.

During the remainder of 1942 Bevin transferred another 13,000 men back to the mines from other industries and another 10,000 from the Forces. By the end of 1942 the mining labour force had been brought back to 711,000 and output raised again, but both began to fall after Christmas and 1943 saw the anxieties of 1942 repeated.

The Forster Committee reported in July 1942, recommending separate provision for the needs of boys entering the mines—in training, wages, medical examination and such amenities as baths and canteens. But Bevin's hopes that these reforms would improve the rate of recruitment were largely disappointed. Mining was accepted as an alternative form of national service to the Forces, but the number of conscripts who chose to go into the pits in the last four months of 1942 was not more than 1,100.

Bevin had no hopes at all of the output bonus which the Greene Board recommended in August. He thought it unworkable in the form suggested and unlikely to raise output, a verdict which was soon proved correct. The main Greene award he believed to be justified: the miners' wages had been too low from the beginning, but he was pessimistic about the chances of the 1942 reorganisation, even with the increase in pay, curing the ills of the industry. These were too deep-seated to be touched by piecemeal reforms. "One thing I would not be moved on," he added in his comments on the Greene Board's August report. "The cost must go on the commodity. We must resist at all costs a Treasury subsidy to this industry either now or in the future. In fact this report emphasises my view (leaving politics out of it) that this industry has got into such a mess that the only thing to do, if we are to be fair to the public, is to take it over and work it on a sound and proper basis."[1]

Bevin's pessimism was to be only too well justified by later events. For more than a century Britain's industrial strength had rested on the foundation of a cheap and abundant supply of coal: in the 1940s the nation was forced belatedly to realise that, if it had been cheap in money costs, it had been purchased at too high a price in human misery and human resentment. Neither Bevin nor anyone else could remove within a year or two the social and psychological consequences of the long, bitter history of the mining industry. The best they could do, within the political limits, was to initiate reforms in the hope that ten or twenty years later these might bear fruit with a new generation of miners. In the remaining years of the war all they proved able to achieve was to prevent the fall in production becoming disastrous: nothing could stop it continuing to fall.

7

Fortunately, few of the problems Bevin had to deal with were as intractable as those of the coal industry. The direction of women, for example, which the Cabinet had approached with such misgivings, proceeded smoothly without attracting any attention from the press or Parliament, apart from a sharp outburst of indignation in Scotland at the transfer of Scots girls to the Midlands. By the end of 1942 he could report that more than eight and a half million women between the ages of 19 and 46 had been registered for national service, a

[1] Notes dictated by the Minister, 20 August 1942.

considerable feat of organisation when one considers the variety of domestic circumstances which had to be taken into account before any direction into war work or the Services could be made.

In May, Bevin sent Winant, the American Ambassador, a private review of the progress made in mobilising British manpower—"I thought perhaps our respected mutual friend on the other side might be interested." Of the 33 million men and women between the ages of 14 and 64, 22 million were occupied in the Armed Forces, Civil Defence, industry and essential services like transport.

"This is the more striking," he added, "when you consider that the 'unoccupied' figure, which is made up largely of married women, includes also schoolchildren over fourteen, persons not capable of work, thousands of women who are taking in lodgers on war work and evacuees, and a large number who are working part time and many full time in the Women's Voluntary Services, nursery schools, canteens and a host of other important fields."

In less than three years, Bevin continued, the numbers in the armed forces and civil defence had risen by more than four million,[1] an expansion only made possible by the numbers of women ready to take the place of men in industry and services. "The net *increase* of women in industry as a whole is 1,132,000. But this does not represent the whole picture, for there has been an enormous transference from the less essential industries and trades into war industry. The total number of women in industry and the services at the present moment is 6,311,000."[2]

Two days later, on 21 May, Bevin had to face complaints from a number of Labour M.P.s on the way the Essential Work Orders were being applied. His reply gave the House some idea of the magnitude of the task he had undertaken. There were six and a half million people under the Essential Work Orders, he told them, a number which would soon rise to eight million. Of course there were going to be complaints of the way in which the Orders operated: he had been criticised for being weak and not ruthless enough, now because he was too ruthless. In fact, legal proceedings had been taken against one in 10,000 of those covered by the Orders, imprisonment imposed on one in 50,000, fifteen of them women.

[1] Mid-1939: 417,000. April 1942: 4,430,000.
[2] Bevin to Winant, 19 May 1942.

"Between the two I think I am about right. I do not take much notice of either, unless there are facts which call for investigation. I worked out a plan when I became a Minister and I have refused to submit to clamour. . . .

"I have got a rotten job but I am not going to refuse to face it. . . . No one in our Movement can ever accuse me of playing to the gallery. I do not care whether I lose a seat in this House or whether I lose my place in the Government. I came into this Government with my eyes open to try to win the war, and when that is done, let others go on and build the peace, if you like, but I knew what was at stake between Fascism, Nazism and ourselves."[1]

Bevin was handicapped by his inability for security reasons to quote in public the figures he had given to Winant, but his statement that two out of three people between the ages of 14 and 64 were now in the Forces or engaged full time in industry came, as the *News Chronicle* admitted, "as something of a shock to most people. No country in the world has ever mobilised its manpower to this extent."[2]

A return, however, for all this concentration of effort had still to appear. In the spring and summer of 1942 it seemed as far off as ever. In the Far East, the Japanese completed their conquest of South-East Asia and turned to attack India. In Russia, the Germans recaptured the initiative in May and drove across the Don towards the Caucasus. In North Africa, Rommel was quicker than the British to mount a fresh offensive. In the Atlantic, the U-boat campaign inflicted heavier losses than ever, a total of 568 Allied ships (more than 3 million gross tons) sunk between January and July, nearly twice the figure for the whole of 1941.

Lacking the stimulus either of danger (as in 1940) or of victory, feeling in the country was edgy and restless, a mood which found expression in the growing agitation for a Second Front.

The reconstructed Government did not long escape criticism. A Gallup poll at the end of March showed only 35 per cent satisfied with conduct of the war, 50 per cent dissatisfied and 15 per cent who had no views. The *News Chronicle*, commenting on the poll, said that it reflected "a widespread sense of frustration. . . . The call is everywhere for *action*. For more vigour; more initiative, a ruthless extermination of waste, muddle and delay."[3] In April *The Times* published an article by Sir Edward Grigg calling for a Combined General Staff

[1] House of Commons, 21 May 1942, Hansard, Vol. 380, cols. 422–6.
[2] *News Chronicle*, 22 May 1942.
[3] *News Chronicle*, 28 March 1942.

under a chairman other than Churchill. This line of criticism found strong support in the press, the House of Lords and in the Commons debate on the war on 19 and 20 May. For the first time critics in the Commons (Clement Davies, Wardlaw-Milne, Stokes) said openly that they had lost confidence in the Prime Minister, while a more moderate group (Grigg, Oliver Stanley, Hore-Belisha, Sir Derek Grimston, Sir Ralph Glyn) pressed for an independent chairman of the Chiefs of Staff Committee to end the Prime Minister's "political interference" with the conduct of the war.

There was little the Government, even Churchill, could do to convince the House of Commons or the nation that they were as eager for action as any of their critics. The long period of preparation and rearmament had been foreseen in the summer of 1940; their plans had been made; all they could do was stick to them and wait for them to bear fruit. But the strain was as heavy as any Government has ever had to bear, a test not only of Churchill's leadership but of the ability of the coalition to hold together under adversity.

The defence of the Government in the House fell to other ministers, but Bevin left no one in doubt where he stood. This was no time for trimming. In Cabinet, in talking to M.P.s and trade unionists, in the speeches which he made up and down the country, his support was unequivocal, his confidence unshakable.

"This Government, like every other Government, gets a lot of criticism, but I have come to Barnsley to assure you that this is the best Government you have ever had. I won't say it is the best you are ever likely to get, but you could not have had a Government more assiduous in its duty than this has been, you could not have had a Government more pertinacious in its efforts than this has been, you could not have had a Government that would have tried harder than this Government has done, with tremendous limitations, to overcome almost insuperable difficulties, and you could not have had a Government that could have been more energetic, more tireless in its efforts and more determined to yield to none until victory is secured. I had been for many years in opposition to the Prime Minister. . . . What I have found in Winston Churchill is that he is a great colleague, he is a great leader, he is tireless in his endeavour, and I have never worked with a man who is more determined to carry on this struggle, however long it lasts, until victory for liberty is achieved."[1]

[1] Speech at Barnsley, 3 May 1942.

It was in the next six months, he told another Yorkshire audience at Shipley, that the tide would turn and Britain pass from defence to attack.

"I cannot tell you where or how. We shall have to take a great deal more punishment in the next few months: we are under no delusion about that. But do not lose heart, do not be too disturbed."[1]

8

On 21 June, however, "the best Government you have ever had" suffered one of the heaviest blows to its credit in the whole war. The garrison of Tobruk, which had been expected by the Government as much as by everyone else to hold out for months, surrendered to a force little more than half its size. First Singapore, now Tobruk. Anger at this fresh disgrace to British arms was sharpened by renewed reports that in the desert fighting British tanks and guns could not stand up to the Germans'. Popular feeling was shown in the by-election at Maldon where the Government candidate was not only defeated by an Independent but secured only 6,000 out of 20,000 votes. On 25 June, Sir John Wardlaw-Milne, with support from members of all three parties, tabled a motion of censure:

"That this House, while paying tribute to the heroism and endurance of the Armed Forces of the Crown . . . has no confidence in the central direction of the war."

It was an ill-judged move, as the event showed: exasperated though they were, few members of the House would have been found to vote against Churchill and the Government if there were any serious danger that they might be defeated. To most people it was inconceivable that there could be any alternative to Churchill as Prime Minister. But inevitably there was speculation: if not Churchill, who?

Since his resignation in February, Beaverbrook, far from retiring from politics, had identified himself with the opposition built up round the issue of the Second Front. He spoke at mass meetings which

[1] Speech at Shipley, 12 April 1942.

the *Daily Express* helped to organise, and in a broadcast to mark the anniversary of Hitler's attack on the U.S.S.R. (on 21 June, the day Tobruk fell), made the charge that "people in high places" had opposed helping Russia in 1941 and were opposing a Second Front now. Coming from a man who had until a few months before been a member of the War Cabinet such an accusation attracted attention. *The Economist* wrote:

"This is plain mischief-making. . . . It is too much Lord Beaverbrook's practice in the speeches which he makes nowadays to puff out a cloud of suspicion against people who are left nameless. It is hard to avoid the view that these attacks have some other and political purpose, however dimly thought out yet."[1]

In the week following Tobruk, Beaverbrook asked Bevin to go and see him. Bevin at first refused, but Beaverbrook was insistent and he finally agreed. No written record of their conversation has come to light and Lord Beaverbrook disclaimed any recollection of it when asked in 1961. The only account is that which Bevin gave to his staff on his return and repeated to others subsequently. According to this, Beaverbrook expressed the opinion that Churchill was on the way out and started to sound Bevin on the formation of an alternative Government. This is not as improbable as it may sound. Beaverbrook had described Bevin to Churchill as the strongest member of his Cabinet and twenty years later still said[2] that, if he had had to form a Government himself, the first man he would have wanted to secure was Bevin, describing him as "a powerful beast", a tribute which he was not prepared to extend to any other member of the wartime Cabinet except Churchill.

Whether Beaverbrook was thinking of himself as Prime Minister or, as Bevin believed, was proposing an alliance in which he would do for Bevin what he had done for Lloyd George in the earlier war, there was no doubt about Bevin's answer. He refused to let Beaverbrook finish and declared that he would at once go to Churchill and tell him the whole story. Beaverbrook's reply was to laugh and tell him that he could, Churchill would never believe him. This proved to be true. Whether in fact Churchill did not believe him or, knowing Beaverbrook, did not take the story too seriously, Bevin was uncertain. But

[1] *The Economist*, 27 June 1942.
[2] In conversation with the author, June 1961.

he seems to have been as shocked by Churchill's attitude as he was by Beaverbrook's. Some time later, when he was recounting the episode, someone asked him how he explained Churchill's relationship with Beaverbrook which intrigued so many people. He thought for a moment and then replied: "He's like a man who's married a whore: he knows she's a whore but he loves her just the same."

At the end of the Tobruk week Bevin was in Liverpool: he was in a grim mood and did not mince his words. Warning employers that they must expect a big new call-up from the munitions industry, he said at a press conference: "We are getting to an intense fighting point in the war, and the delays taking place now over almost every case of de-reservation will have to go by the board."

After opening two hostels for seamen, he went on to address a mass meeting in the Philharmonic Hall, with the Lord Mayor in the chair. There was still trouble with labour in the Liverpool docks (another committee of inquiry had to be appointed in September) but Bevin meant to say what he felt, whether it was popular with his own people or not.

"No other nation," he told the crowded hall, "has ever been mobilised or disciplined to the extent this nation has. We have had a few strikes, not many. I do not want any more. I say to the trade unionists of this country: 'You may get your irritations in the factories, but don't you stop work from now on either in mines, factory, dock or anywhere else for one minute. You don't hurt me, you don't hurt the Government, but that one minute may mean the life of one of your own sons in the fighting field....'

"It is criminal to stop work at this moment. You must not do it. I say to everyone of my own people, whom I have lived and worked for all my life, there will be plenty of time—hundreds of years—to go in for strikes after this, but let us get on with it and get through with it....

"The output of British shipyards is not as good as it ought to be.... You are not doing as well as you ought to be in Liverpool. Employers tell me it cannot be done. The men say it cannot be done. I do not believe it. Apply your minds to it. Get rid of the antiquated ideas and you can turn out more ships and ship repairs than you are doing now. Cammell Laird's [in Birkenhead] have done it. Why cannot others?"

He kept politics out of his speech to the end, but made what he had to say plain enough for anyone to understand:

"I saw in a paper to-day some talk that the better thing would be not to have such a brilliant man at the top, but to have a team of men.... It has been my

Bevin on his way to a Cabinet Meeting in December 1941.

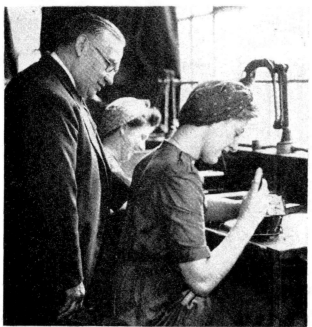

Bevin's interest in training. *Above*. Women trainees at Letchworth. *Below*. Soldiers learning a trade at an L.C.C. Technical Institute.

good fortune to be a first man many times, to lead. But I am willing to sub-ordinate myself in a great cause of this character and I have never served with any body of men in my life, irrespective of party, who have done more to work as a team than the Churchill Cabinet . . .

"This wicked filthy business of trying to break up national unity by playing Winston Churchill off against his colleagues by certain newspaper million-aires is the most diabolical thing I have ever known. He came in, in 1940, when the country was under the weather—and I am prepared as a trade unionist to go on under the banner of Winston Churchill working with him as a colleague, fighting for him to the end."[1]

In fact, when the vote of censure came to be debated at the beginning of July, it turned into a fiasco. A demand for an inquiry into the loss of Tobruk and the deficiencies of the Eighth Army's equipment would have met with wide support: it was only with the greatest difficulty that Attlee and Bevin dissuaded the Parliamentary Labour Party, at its meeting on 30 June, from putting down an amendment to this effect. But the flat statement that the House of Commons had lost confidence in the central direction of the war went further than most members were prepared to go. Moreover, while Wardlaw-Milne made the main charge against Churchill his inter-ference with the professionals, Admiral Keyes, who seconded the motion, argued that the fault was the other way round: his one criticism of the Prime Minister was his failure to override his pro-fessional advisers, especially in the Admiralty. This contradiction between mover and seconder, together with Wardlaw-Milne's extraordinary suggestion that the Duke of Gloucester should be made Commander-in-Chief of the Army, reduced the debate below the level of the occasion. Although Lyttelton had a rough time when he tried to defend the design of British weapons and the tactics pursued in the Libyan battle (Clement Davies demanded that the House should impeach those responsible), the critics failed to impress even themselves. When the House divided after Churchill's closing speech, the abstentions were far fewer than had been expected and the vote was 475–25 in the Government's favour.

Comment in the press afterwards suggests that, while few thought the Government had answered the criticisms made in the debate, nobody had any illusions that an alternative Government could be found to replace it. *The Economist* put the position very well when it wrote:

1 Speech at Liverpool, 28 June 1942.

G

"Apparently it is not yet realised by Mr. Churchill and those beside him that his Government must be its own alternative. . . . Apart from the woe-woe carpers, the critics wish to win the war, not to play politics. They believe that Mr. Churchill, Sir Stafford Cripps, Mr. Eden, Sir John Anderson, Mr. Lyttelton and Mr. Bevin can do the job, but they are by no means sure that the job is yet being fully done."[1]

Having failed either to shake the Government or persuade the Prime Minister to adopt their proposals, there was nothing the critics could do except wait until the Government either began to win the war or plunged the country into some final disaster. When Churchill gave his next report on the war, in September, only three members could be found to speak after him. The long and at times bitter debate on the conduct of the war which had lasted from December 1941 to July 1942 spluttered out inconclusively in the autumn; by the winter it was forgotten in the exhilaration of victory.

[1] *The Economist*, 25 July 1942.

CHAPTER 7

Bevin and Reconstruction

I

ON 25 JULY agreement was at last reached between the British and Americans where to direct their first attack on the enemy before the end of 1942—in North Africa, not later than 30 October. The next day Bevin was speaking at Pontypool when he was interrupted by an attempt to move a resolution calling for a Second Front.

"By creating a division in the country," he told his audience of miners, "our friends on the Left who shout this slogan are creating the very conditions we all want to avoid. . . . Don't talk to me about a Second Front, but help the Government so that there shall be no shortage of coal anywhere. Eighty vital days lie ahead of us. If everybody works with a will, victory is possible much sooner than it ever seemed before."

This, and a speech by Lyttelton giving the same figure of 80 days (which, in any case, proved to be inaccurate), were the only hints that the Allies were within sight of capturing the initiative at the end of the twenty-eight months of almost unbroken defeat which, as Churchill says, had been the record of his ministry so far. "The inner circle who knew were anxious about what would happen. All those who did not know were disquieted that nothing was happening."[1]

They had still to get through the summer and early autumn—the battle of Alamein did not begin until 23 October, the landings at the other end of North Africa until 8 November, and the final stages of the long period of waiting were hard to bear. The Government remained in office, it appeared at times, rather because there was no alternative than because of the confidence it created. If the conduct of the war was exhausted as a subject of criticism, there was plenty of scope for dispute on the social and political issues which had first been raised

[1] Churchill, Vol. IV, p. 493.

in the spring of 1942 and which recaptured public attention from July onwards.

At the end of that month there was a furious row in the Parliamentary Labour Party over increases in old-age pensions. The Government proposed an increase of 2/6 a week and Bevin and Attlee argued the case for accepting this at a meeting of the Parliamentary Party before the matter came to the House of Commons. The Parliamentary Party, however, spurred on by its Administrative Committee, decided to move an amendment rejecting the Government measures as inadequate. The question turned on the assurances Bevin gave to the Parliamentary Labour Party, and repeated in the House, that this was only an interim proposal, that the Government was awaiting the Beveridge report on social security, and that the proposals to be introduced after the report would make proper provision for pensioners.

The debate, on 29 July, followed the familiar lines of division between Conservative and Labour backbenchers with a good deal of party and class feeling. Several of the Labour speakers, however, were more bitter against the Labour ministers in the coalition than against the Tories. Kirkwood declared:

"It was just the same with the coal question. . . . If those who are our representatives in the Cabinet had laid it down, 'Unless you nationalise the mines, we will resign', the Prime Minister, powerful as he is, would have had to concede that to our representatives."[1]

He was followed by Sidney Silverman who asked Bevin whether he dared stand up "and assure this Party and the people that the Government would have collapsed if he had insisted that the increase should be 5s. instead of 2s. 6d."[2]

Attlee would have damped down feeling, but Bevin's temperament would not allow him to:

"It was I," he declared, "in the Trades Union Congress, who wrote the pamphlet asking for a State superannuation scheme. I will not be accused of not trying to do my best. It was I who, on behalf of the Trades Union Congress, as the ex-Minister of Health will remember, raised the question with

---

[1] House of Commons, 29 July 1942, Hansard, Vol. 382, cols. 604–5.
[2] Ibid, col. 613.

the Government before it was debated in this House, and tried to negotiate a new State superannuation scheme. Instead of what I then thought was a proper scheme, this House carried a Bill introducing supplementary pensions. I venture to suggest that the results have been costly to the State, and it is evident that it has not given final satisfaction. That is the position the Government find themselves in now."[1]

The right course, he argued, was to re-cast the whole scheme and that would be done next session after Beveridge's recommendations had been received. Greenwood (now Leader of the Opposition) accepted Bevin's assurance and was prepared to withdraw the amendment, but he was repudiated by those behind him. By now tempers were up. Shinwell interrupted Bevin to ask him how long he thought his Tory friends would continue to cheer him. "I do not mind how long the Tories cheer," Bevin retorted, "and how long the hon. Member hates. I know my friends behind me and I know the hon. Member."

Bevin later apologised but Shinwell refused to be satisfied and got more and more angry.

"Shinwell: Are we to understand that the Minister of Labour is the only person imbued with deep feeling about social question? . . . He is frequently telling the House, much too frequently I think, of his great exploits on social issues in this Government . . . I accept it and I assume that he is obsessed with a deep and passionate sincerity for the well-being of the working classes, but surely he can give other people the same credit.
"Bevin: Sure.
"Shinwell: Then why jeer and sneer—
"Bevin: I only joined in fraternal greetings.
"Shinwell: —unless it is to uphold the venom that has obsessed the right hon. Gentleman against everybody who has dared to criticise his precious Government? . . ."[2]

When Shinwell finished, Buchanan took up the attack. He warned the Minister of Labour that his long connection with the Labour Movement did not entitle him to justify anything he liked in the House. That was to behave like Jimmy Thomas, who had an equally long and honourable record. He must justify measures on their merits and not tell them all the time how much he had done for the working classes.

1 Ibid., col. 631.
2 Ibid., col. 644.

The amendment was pressed to a division and 63 members voted against the Government, the largest adverse vote since Churchill had formed his Government in 1940.

The row over pensions was unimportant in itself, but it was indicative of the strength of political passions and political memories. Many in the Labour Movement harboured the suspicion that at the end of the war the Tories would out-manœuvre them as they had after 1918, that Labour ministers would be hoodwinked or would "betray" the Party as they had in 1931 and that the wartime promises of radical change would be worthless unless the Government could be forced to commit itself now. On the other side, the Tories, who still held the majority in the House of Commons, were restless at the socialistic trend of wartime legislation, particularly the increased intervention of the State in economic affairs, and resented what they regarded as Labour's attempts to introduce controversial changes under cover of the emergency.

Uncertainty about the Government's intentions fed both sides' suspicions. When Bevin criticised the profit motive in a speech to the National Chamber of Trade,[1] Conservatives were up in arms. When two reports commissioned by the Government were published, the Scott Report on the use of land in August and the Uthwatt Report on compensation and betterment in September, everyone from *The Times* leftwards demanded to know what action the Government proposed to take. The Beveridge Report on social security and insurance was impatiently awaited, and questions on the date of its publication reflected a predisposition to believe that the Government was lukewarm towards its recommendations. Two major measures were known to be in preparation, both certain to be controversial, Butler's Education Bill and Bevin's Catering Wages Bill. As early as September *The Times* reported that the catering industry was "agitated" and began publishing a series of letters many of which condemned Bevin's proposals in advance as "socialistic".

India was another subject which stirred political feelings. Civil

---

[1] 15 July 1942. It was in this same speech that Bevin, commenting on the loss of the country's overseas investments, remarked: "I do not know that that is bad. We have spent so much money and time on trying to save them that we might be better off without them. It is certain that, at the end of the war, if we are to live we have got to buy goods for goods and that the rentier, comfortably living on interest alone, will be gone." This produced a flood of angry letters and reproachful editorials in the press.

disobedience and terrorist attacks led the Government to order the arrest of the Congress Party leaders. When Churchill made a statement in the House on 10 September, he was angrily attacked by Bevan and Shinwell, and their challenge to a debate was in turn taken up by the Tories the following day. "For my part," one Labour member declared, "whatever might be the intrinsic merits of the Prime Minister's statement and the actual language used . . . it was the manner of its expression, its truculent, swashbuckling, 'damn your eyes' sort of thing which took us back to the debates on the old India Bill."[1] Another backbencher, remarking on the temper of the House, concluded: "I think I was right in the impression I received yesterday that we were almost back to pre-war party days."[2]

The Government kept Parliament in recess as much as possible during this anxious season: the House of Commons met only fifteen times between August and November. But it was often enough to show the way the wind was blowing. When the Government introduced a Bill to prolong the existing Parliament for a further year, Greenwood supported it but added the warning:

"A year from now I might wish to press rather harder than I am anxious to do to-day that, if the political situation deteriorates further, if the bitterness deepens and if a desire for co-operation does not increase . . . the only solution obviously is a General Election."[3]

The more ardent members were already arguing for an election, on the grounds that the country was undergoing a social revolution which the House did not reflect or express. With a Conservative majority and the support of moderate Labour opinion, the Government could still afford to disregard such a proposal. But the strength of the demand for a commitment to sweeping post-war changes, whatever the political consequences, was shown by the attitude of Stafford Cripps. The duty as Leader of the House of Commons of explaining successive defeats and defending the Government's temporising attitude bore more and more heavily on Cripps. He was critical of the way Churchill ran things and thought the whole of the central direction of the war needed overhauling. These doubts were reinforced by the strong conviction that national morale required the

[1] House of Commons, 11 September 1942, Hansard, Vol. 383, col. 601.
[2] Ibid., col. 569.
[3] House of Commons, 30 September 1942, Hansard, Vol. 383, col. 819.

Government "to give the people some more definite prospect for the future".

In September 1942 Cripps decided that he could not go on and must resign. Churchill needed no prompting to see that for Cripps to resign from the War Cabinet "during this period of oppressive pause" would create a political crisis and provide a focus for radical discontents. He exerted all his influence, backed by strong pressure from the other members of the War Cabinet, to persuade Cripps that, with great military operations impending, duty required him to subordinate his personal views to the national interest. Cripps, to his credit, agreed to withhold his resignation and, once the North Africa campaign was launched, Churchill skilfully found alternative employment for his talents in the Ministry of Aircraft Production. But, before giving up his seat in the War Cabinet, Cripps made clear to the House of Commons his strong belief that great social changes must follow the war and that the Government, simply because it was a coalition, could not avoid the duty of framing a programme of post-war reconstruction in advance.

2

All Churchill's instincts were against doing anything of the sort. He was always, Eden noted, reluctant to turn his mind to anything not directly connected with the war.[1] The greatest of war ministers—and a traditionalist—he had little interest in economics or social reform. Cripps was convinced that a great radical programme would unite and exalt the nation; Churchill, with a good deal more realism, feared that it would rouse all the devils of controversy, divide the coalition and Parliament, divert energies which needed to be concentrated on winning the war and weaken the sense of national unity.

The great complaint of the Left was that the Labour ministers accepted this and did not use their position to force Churchill's hand and commit the country while the radical mood engendered by the war was still strong. Harold Laski, writing at the end of 1941 to his friend Felix Frankfurter about a reconstruction plan which he had drawn up for the Labour Party, said angrily that the critical stage would come when

[1] Cf. *The Eden Memoirs, The Reckoning*, pp. 441–2.

186

"the Labour ministers try to water it down so as to maintain their happy subordination to Winston. That, Bevin always partly excepted, seems to be the only role they really enjoy. The real tragedy is that they are satisfied with their position."[1]

Laski blamed Attlee more than anyone else and a few months later tried to win Bevin over to a move against him. "My dear Ernest," he wrote in a letter of 9 March 1942,

"I shall not easily forget your remarks at the National Executive on Friday—above all your eagerness to represent the inarticulate masses of Europe at the peace. But you heard my plea; you heard the reply Clem Attlee made. I say again that unless you become the first man instead of the second, the confidence of the masses in this movement will rapidly die. The faith and temper victory requires are ebbing away because there are neither great victories nor great measures. It is time for a fighting leader and you are the right person for that place. It isn't your own feeling that should decide but the urgent needs of the common folk who trust you.

Ever yours,
Harold Laski."[2]

Bevin was not to be drawn and made no reply. Not only had he a poor opinion of Laski as a politician—and then, as later, refused to let himself be set up as a rival to Attlee—but he rejected the whole thesis on which Laski based his plea and which, elaborated in *Reflections on*

---

[1] Quoted in Kingsley Martin, *Harold Laski* (1953), p. 152.

[2] R. H. Tawney held much the same opinion. In a letter to Beatrice Webb, dated 6 December 1942, he wrote: "As far as I can see the Labour Party has temporarily ceased not only to count, but to believe in itself. In May 1940 a coalition was inevitable. . . . I am not sure that it will continue to be either inevitable or desirable—that depends on the war—and the price gets heavier and heavier. I am inclined to think that it should come out of the Government as soon as victory is in sight. But if it is to do that, it should be preparing its plans now. . . . The leaders are too immersed in day-to-day duties. . . . The only man who, in my opinion, might stop the rot is Bevin. When I saw him I thought he had greatly grown in breadth of outlook and that he was a possible Prime Minister. . . . If I knew Bevin better I should tell [him] that the moment will soon come when the most useful thing he could do is not to run the Ministry of Labour but come out and make the leadership of the Labour Movement his main task. Whether he has this in mind, I don't know. He must, I'm afraid, carry on where he is until it is clear that the country is almost out of the wood."
I owe this quotation to Mrs. Elizabeth Morgan who is writing a study of R. H. Tawney as a socialist and found this letter among the Webb Papers.

the Revolution of Our Time[1], was common in one form or another to most left-wing thinking during the war.

Laski argued that the war offered a unique opportunity for carrying out a social revolution in Britain by consent; that the overwhelming majority of the people were in favour of such a revolution and only a small group opposed to it; that if the Labour Ministers insisted that the coalition should inaugurate such a revolution, Churchill would be unable to resist; that a programme of radical change, far from being a distraction, was an essential part of the strategy of total war; and that if this were not carried through during the war, the fruits of victory would have been thrown away and an opportunity lost which would never return, leaving the British people to carry out the same revolution after the war by violence and at the cost of much suffering.

Bevin would have nothing to do with such a programme. This was not because he did not care about the sort of world which would emerge after the war or fail to realise the effect wartime developments would have upon it. Speaking in the House of Commons in October 1942, he declared:

"I have never accepted the view that war, post-war and pre-war are separate things. War may be an intensification in the development of our lives, but there is no definite break. Everything you do before a war determines largely what will happen in the war, while everything you do in the war will largely determine what will happen after the war."[2]

But he kept his priorities straight. Overriding everything else was the need to defeat Hitler. Until that was settled—and it was far from being settled in 1942—he had no patience with those who argued that the Labour Party should start to use its position in the coalition in order to impose its views on the post-war settlement. The war was not yet won and to say that a radical programme of reconstruction was "an essential part of the strategy of total war", he believed, was playing with words: the almost certain result of pressing for it in the way Laski and the Left wanted would be to split the Government and the nation and so make it much more difficult to win the war.

[1] Published in 1943. While Laski was writing *Reflections on the Revolution of Our Time* in 1942, Bevin was drafting the clauses of the Catering Wages Bill: there is much Labour history compressed into this contrast.

[2] House of Commons debate on wages policy, 21 October 1942, Hansard, Vol. 383, col. 2060.

In any case, Bevin found the "either . . . or" on which Laski's case depended unconvincing. The iron laws of Marxist analysis (even in Laski's watered down wartime version[1]) no more impressed him than those of classical economics. He did not believe that it was a case of all or nothing, either a social revolution carried beyond the point of no return by the end of the war or total failure.

Bevin had a clearer understanding than Laski of the processes of wartime change which were already changing British society.[2] He thought it perfectly possible to add to these by a number of reforms during the war, without achieving the revolution which the Left wanted, but also without having to abandon the hope of carrying them further after the war. It was not (he believed) what was done in the middle of the war but what was done at its end which would be decisive: that was the time when being in office would really count and give Labour the chance to hold what had been gained during the war and to see that the rest of the reforms and innovations, which it had only been possible to plan while the war was going on, were not left as paper schemes.

Bevin was as determined as Laski or Cripps or anyone else to see that there was no return to the 1930s after the war and no repetition of 1918, but he had very different ideas of how to go about it. The war had to come before anything else: this was the overriding interest for Labour as much as anyone and he shared Churchill's (and Attlee's) view that this placed a duty on the party leaders of both sides to damp down controversy. That meant, obviously, getting fewer reforms than Labour wanted to see introduced, but Bevin took this more philosophically than the Left. He suspected in any case that, on purely practical grounds, even a socialist Government would find it hard to carry out major structural changes in the middle of a war, and that if Laski were to get his way the social revolution would be reduced to a series of splendid declarations, promises and slogans. These might give those who wanted change emotional satisfaction, but they were not hard coin, as the 1914 war had shown. If the worst thing the coalition meant was fewer exciting promises, this mattered less than

---

[1] See the interesting chapter 12 in Herbert A. Deane: *The Political Ideas of Harold J. Laski* (New York, Columbia University Press 1955).

[2] For a description of these, see the classic study by Richard Titmuss: *Problems of Social Policy* (1950).

the fact that even a half-share in the Government gave Labour the chance to do three things which he valued much more highly.

The first was the chance to influence social and economic policy during the war, with virtually a free hand in his own sphere of labour policy.

The second was the chance to take part in and push forward planning for the reconstruction period after the war.

The third was the chance to use their position to shape decisions—if only by blocking what the Tories wanted to do—in the crucial period towards the end of the war when there would be a great cry for freedom from controls and a "return to normality" with the same results as after 1918.

Enough has already been said to show how Bevin made use of the first of these opportunities; how he used the third will appear later on; it is time to turn to the second, the chance to take part in post-war planning, and see what use he made of this.

Before going on to look at his proposals in detail it will be useful to provide ourselves with some sort of map and ask what was the direction in which Bevin wanted to see British society develop after the war.

It is not an easy question to answer. As in the case of his labour-supply policy, Bevin himself was inclined to talk in terms of particular proposals and leave others to infer the general policy from them. None the less, as in the case of labour supply (where nobody in the end doubted that he had a policy), it is worth while to set down the main elements in Bevin's thinking about society and politics, and see what they add up to. If we do this, there are, I suggest, six items to be included.

The first was action by the State to maintain full employment and secure the use of economic resources for social purposes, not simply individual profit. Among the forms which this might take were those recommended by Keynes for keeping up employment, the nationalisation of a number of key industries and services, and control over the location of industry.

The second was action by the State to maintain the wartime policy of "fair shares" wherever anything was in short supply—including physical controls (e.g. over building), rationing, food and housing subsidies.

The third was a comprehensive social security service to provide all citizens, as a right, with assistance in cases of need, poverty and sickness.

The fourth was an extension of the State's provision of education to create greater equality of opportunity—for instance, by raising the school leaving age to sixteen, and providing better technical education and industrial training.

To these four items, all involving positive action by the State and increased social expenditure paid for out of increased taxation, must be added two more.

One was a new conception of industrial relations beginning with better wages and conditions (to be secured by joint negotiation) and extending to something approaching a partnership on equal terms between management and workers.

The other was the extension of joint consultation with the unions and the employers to the whole range of government economic and social policy.

What sort of a programme do these six items add up to? Not a socialist programme in the sense that, if carried out as it very largely was by the Labour Government after 1945, it would produce a socialist society. If there were items in it of a socialist character (such as nationalisation), there were others (such as full employment and the welfare state) which owed as much to Keynes and Beveridge as to socialism, while Nye Bevan was so much impressed by Bevin's insistence on Government consultation with the trade unions and employers' associations as to declare that this was incompatible with socialism and amounted to the replacement of the parliamentary by the corporate state. No doubt if the last two items had stood alone and had been pressed to the exclusion of the other four, there would have been substance in this claim.[1] But the corporate state, based on vocational rather than territorial representation, fits such a programme as a whole even less than the description "socialist", in the sense in which the Left used it of a decisive change to the collective ownership of the means of production, distribution and exchange.

What it does fit, surprisingly well, is the sort of society which has developed in Britain in the twenty years since the war, a hybrid society to which neither of the terms "capitalism" nor "socialism", as these have been understood historically, can be applied with much success. The main features of this post-war society are:

[1] See below, pp. 275-6.

First, a mixed economy, partly in public, partly in private ownership, with both sectors subject to constant intervention by Government, a managed as well as a mixed economy.

Second, a commitment by all parties to the maintenance of full
employment.

Third, the welfare state.

Fourth, extension of the State's provision of education in order to
provide greater equality of opportunity.

Fifth, great improvements in the position of the industrial worker
and increased recognition of the role of the unions in industry, even
if this still falls far short (on both sides) of Bevin's conception of a
partnership between management and men.

Sixth, a great increase in Government consultation of interest
groups (chief among them, the T.U.C. and the employers'
associations) and the attempt to build up a tripartite pattern of
co-operation (Government—unions—employers) in carrying out
economic policy.

Substantially, these correspond to the main features which I have
listed as characteristic of Bevin's thinking about society and politics.
That still leaves open, of course, the extent to which this was the sort of
society Bevin wanted to see. Even if he had foreseen the direction in
which the policies he advocated would lead, he was unlikely to have
put down on paper anything like a general statement of it. The papers
he left behind, as one might expect, contain hints but no more. All
one can say with any certainty is that he pursued such policies with
great consistency and that they correspond, to a surprising extent, to
the sort of society which has developed in Britain since the war.

The process by which this came about, of course, is a complex one
and many other people besides Ernest Bevin contributed to it. None
the less as the pages which follow will show, few members of the wartime coalition were as active in thinking about the post-war settlement. The fact that some of his proposals came to nothing and that
few of them could be put into effect before the war was over did not
worry Bevin: he grasped the importance of getting ideas into circulation inside Whitehall and realised that if there was a chance to
take action when the war came to an end, much would depend on
how far detailed plans had been prepared in advance.[1]

[1] Emanuel Shinwell, one of the most vocal of wartime critics, was astonished
when he became Minister of Fuel and Power in 1945 to find no plans prepared for

3

The evidence for Bevin's part in post-war planning is to be found in a number of memoranda and letters left behind among his papers. These provide only a fragmentary record of his activities but they are sufficient—in default of access to the records of the Cabinet and its committees—to show the range of his interests and the persistency with which he pursued them.

Three general points are worth making by way of introduction.

Until 1944–5 Bevin made little, if any, attempt to relate his proposals to a party programme, but left on one side the question what Government would be there to put them into effect—or reject them— at the end of the war. This is particularly noticeable in the case of the White Paper on Employment Policy where, in the face of scepticism and opposition from his own party, he persisted in his efforts to take the commitment to maintain full employment out of party politics and make it binding on all parties.

Next, he was as much interested in the development of international affairs after the war as in domestic policy, and in particular showed an unusual grasp of the part economic and social questions were to play in international relations.

Finally, there is evident in his post-war plans as in his wartime policy the same distinctive gift for seeing what others saw in terms of economics, or class conflict, or political power as human problems. This led him to take an interest in matters, like rehabilitation or the neglect of adolescents, that had nothing to do with politics; and it was this concern for ordinary men and women, running like a scarlet thread through everything he touched, which, more than any economic or political theory, inspired his ideas about the sort of world he wanted to see after the war.

His first interest, naturally enough, was to see that the reforms he had introduced as Minister of Labour should be consolidated and

---

nationalisation of the coal mines: "The miners expected it almost at a wave of the ministerial wand . . . and I had myself spoken of it as a primary task once the Labour Party was in power. I had believed . . . that in the Party archives a blueprint was ready. Now as Minister of Fuel and Power I found that nothing practical and tangible existed. There were some pamphlets, some memoranda produced for private circulation, and nothing else. I had to start on a clear desk." (Emanuel Shinwell: *Conflict without Malice* (1955), pp. 172–3.)

extended after the war. In April 1942 he asked his officials to prepare plans for building up the Ministry's Factory Department, and gave them a directive on which to work.

"I would like to see one comprehensive inspectorate covering the protection and health of the industrial community in whatever sphere of employment they may find themselves. At present there is too much dispersal. . . ."

He wanted to see the Factory Medical Service, the rehabilitation service which he was beginning to develop,[1] and the Industrial Health Research Board all continue after the war. They would need to be linked more closely with the Factory Department and this to become a permanent part of the Ministry of Labour, instead of returning to the Home Office. So would personnel management, the much expanded welfare services and post-war training and apprenticeship schemes which could be expected to throw up problems affecting the health and working conditions of young people. To be effective, all these reforms needed to be backed by efficient inspection.[2]

An address which he gave at the first meeting of the Joint Industrial Council for the Ophthalmic Optical Industry[3] provides evidence of his plans for developing the self-government of industry

"in the hope that by the time hostilities cease, there will be not a single industry of any kind in the country that has not wage-regulating machinery of some kind or another. . . . It is no use talking about 'homes for heroes', the thing to be done is to agree and create the machinery for a decent standard of living."

The J.I.C.s, however, should not be limited to wage questions:

"Employers and unions will have to be alive to the fact that they will have a different people to deal with from those who went away into the Forces . . . I do not think that the future generation will quite accept the old conception of employers and work-people that existed before this war. There will be a different kind of equality. There will be managements and operatives, but it will be functionary not dominating, and in that sense a Joint Industrial Council gives the right opportunity and place to meet on level terms, each contributing from its particular function to the success of the industry as a whole."

[1] See below, pp. 217–9 & 288–90.
[2] Note by the Minister, 21 April 1942.
[3] 1 July 1942.

He underlined a point which he was to make again to the National Chamber of Trade later in the month:

"People have been taken out of their normal run of life, they will come back with great expectations. At the same time they will come back to an impoverished country.... We must remember that ... our foreign investments have gone. This time we have to maintain this country not by living on investments abroad but by selling goods abroad and taking in return their products."

To do this Britain would need to call on all the technical skill she could find and industry would no longer be able to afford the old lack of co-operation between managements and men.

Another matter on which Bevin felt British industry would have to change its ideas drastically was training. This was one of the questions he seized on as soon as he was shown the draft proposals for the Education Bill, in March 1942.

"How are we going to meet the position, unless everybody has an element of training either in industry or agriculture, that will arise from the fact that foreign investments will be gone and in all probability the structure of the Empire will be changed? To maintain our standards of living we shall in all probability find ourselves very much in the position, even when we have won the war, that Germany was in during the last twenty years."

He was strongly in favour, as he had always been, of raising the school-leaving age to sixteen, but he thought that the whole system of technical education needed changing. Could apprenticeships be shortened? Could they break down the limitations on training created by fear of unemployment? Who were the right people to develop technical education, the Local Authorities or industry?[1]

He asked his parliamentary secretaries to meet the parliamentary secretary of the Board of Education in order to investigate these questions, and the next month he took them up with his Joint Consultative Committee. Industrial training and technical education in Britain, he suggested, were out of date and had fallen behind the best practice abroad: he persuaded the Committee to start an inquiry covering all the main industries of the country.

One of the first industries in which Bevin was able to use his

[1] Bevin's Notes on the Education Document, 19 March 1942.

position as Minister of Labour to get employers and unions to agree on a new scheme for apprenticeships was building. But this was only a beginning of what he believed would need to be done if the building industry was to meet the enormous demands certain to be made on it at the end of the war. In a memorandum which he dictated in 1942 he argued that the industry would have to accept state control of building for at least a decade after the armistice. It would be better to declare this at once and take the reorganisation of the industry resolutely in hand during the war, so that a long-term plan could be worked out for twenty to twenty-five years ahead. "I take the view," Bevin wrote, "that the rebuilding of Britain may need to absorb, directly or indirectly, as much as a quarter of the insurable population."[1]

One proposal he made, to help the small builder, was a holding company.

"By this means, the smaller builders could be brought under control, without losing their individuality: pooling of equipment, co-operation in training schemes could be facilitated and the way paved for the regionalisation of the industry."

The demand for houses at the end of the war, Bevin continued, would raise social questions of the first order and some indication as soon as possible of how the Government proposed to meet it would have a great effect on morale. He was critical of housing estates with their segregated populations and standardised small houses: he wanted Local Authorities to offer a much wider variety of housing, if necessary with power to sell. But his most interesting proposals were a National Housing Corporation, to provide mortgages at cheaper rates and safeguard the purchaser against the speculative builder, and a scheme for a New Britain Loan to link national savings certificates with housing so that people could start saving for their future homes in wartime.

He suggested a National Hire Purchase Corporation to provide the same sort of facilities for the purchase of furniture and household equipment. He wrote scathingly of the way in which hire purchase had been exploited, the exorbitant prices charged for trash, and of his anxiety lest this should repeat itself at the end of the war when the

---

[1] The Post-War Building and Housing Programme, a memorandum, which Bevin sent to Sir William Jowitt, the Paymaster-General who was in charge of reconstruction plans, in April 1942.

pent-up demand for supplies of all sorts was released. The Corporation he proposed would not only prevent this but would also be able to insist on higher standards in furniture making.[1]

Many of Bevin's ideas for post-war development, of course, can be traced back to his experience as a trade-union leader. A good illustration of this is a letter which he wrote to Arthur Greenwood in November 1941,[2] while Greenwood was still in charge of post-war planning, urging him to prepare plans for the reorganisation of British ports in order to cut the transport costs of British trade. Both on Merseyside and Clydeside, Bevin wrote, the Port Authorities were moribund as far as any new ideas were concerned and dominated by private shipping interests which gave no thought to the national interest. When the war broke out the equipment of both ports was out of date and their methods of handling cargoes inefficient. Middlesbrough was another port with big scope for development. Twelve years before, Bevin had tried to interest Ramsay MacDonald in a scheme for developing Tees-side along the lines of the Port of London Authority. He urged Greenwood to look at this scheme again, bearing in mind the possibilities of Middlesbrough as a port when trade with Germany started up again after the war.

More was required, however, than a form of public management for individual ports; there would have to be a national body capable of looking at the facilities provided by the ports as a whole. "Owing to the haphazard method of development," Bevin calculated, "at no time have the ports been used to more than 30 per cent capacity, which means that transport has to carry a capital cost out of all proportion to what is reasonable." Anyone who thinks he was exaggerating can find melancholy confirmation for the picture he drew in the Rochdale Committee's Report on the ports—twenty years after the date of Bevin's letter.

At the end of September 1942, Bevin made an effort to put together in a single scheme the elements of a post-war industrial policy, and to help him focus his ideas he dictated a list of the main questions he needed to answer first. He put the list in his bag and took it away to think about over the week-end: it consisted of eleven items.

---

[1] Ibid.
[2] Bevin to Greenwood, 28 November 1941.

1. Demobilisation Plans
    (i) Fighting Services
    (ii) Munition Industries
2. Public works—organisation, control and finance.
3. How far should preference be given to (1) disabled men (2) ex-servicemen, in submission for employment?
4. Should women be replaced by men to the maximum?
5. Lowering of pension age? Compulsory retirement.
6. Raising of school-leaving age.
7. Policy on hours.
8. Policy on wages—should compulsory arbitration remain?
9. Should functions of government departments be changed?
10. Future of International Labour Organisation.
11. Should compulsory military service be retained?

One result of Bevin's week-end reflections was an examination of all the Orders and Regulations issued by the Ministry of Labour since he took office—there were ninety-four altogether—to see which ought to be embodied in post-war labour policy. Without waiting for this to be completed, Bevin put down a number of points to provide a guide for his officials.

First, compulsory arbitration: this ought, he suggested, to be continued for six years after the war to avoid either inflation or deflation. At the same time, he threw out the suggestion of a court of investigation, perhaps a regional court, to which disputes ought to go in the first place.

Second, the Essential Work Orders: Bevin was eager to keep the weekly engagement and guaranteed wages, if only to prevent industry slipping back into the pre-war habit of part-time work, "when the Unemployment Act was merely a subsidy to meet what was growing into a conspiracy between the trade unions and the employers".

Third, welfare: he wanted new legislation on the lines of the Factory Acts to enforce welfare inside the factory. Should the Factory and Welfare Board, he asked, become a permanent part of the machinery of State? Outside the factory, he made the suggestion that the Unemployment Assistance Board might be converted into a National Welfare Board with more positive duties in addition to the relief of unemployment.

Among the other matters on which he touched were the employment of women, canteens, juvenile employment and technical education:

"In other words," he concluded his paper, "can I now, out of the experience of war produce a new industrial code of conduct, inspection, enforcement and welfare?"

If so, he would call a full conference of the T.U.C. and the employers and put his proposals to them.[1]

4

Although it never crossed Bevin's mind that he would ever be Foreign Secretary, he did not fall into the mistake of supposing that Britain's economic problems after the war could be solved in isolation. His experience as a trade-union leader had made him acutely aware of the impact of international conditions on British industry and employment. As early as the autumn of 1940, he had a talk with R. A. Butler and Sir Alexander Cadogan about liaison between the Foreign Office and the Ministry of Labour. He followed this up by producing a long memorandum (drafted by Leggett) on foreign policy and the reform of the diplomatic service and sending this to the Foreign Secretary, Lord Halifax.[2] His starting point was the neglect of social and labour questions by the Foreign Office, the Dominions Office and British representatives abroad. No British Prime Minister or Foreign Minister had ever attended the I.L.O. or given it any encouragement. No British Embassy ever thought of sending a report on labour developments, working-class movements or standards of living: such questions belonged to a separate world, of no concern to diplomacy.

"Since I have been in office and had an opportunity of reading the telegrams, there does not appear to me to be one I have yet come across which gives us any line what the working people are thinking and feeling in the various countries. ... Only extreme left movements become of interest because they menace the existing order of society."

[1] Note from the Minister to the Secretary: Post-War Policy on the Industrial Side. 3 October 1942.
[2] There is no precise date in the copy among Bevin's papers. It was written at some time between the beginning of October 1940 and Halifax's departure from the Foreign Office (December 1940).

One reason for this was that Britain chose her diplomatic representatives from a narrow circle of people, and their contacts abroad were limited to the same groups. The fundamental reason, however, was the traditional conception of foreign policy which was limited to national security, the protection of overseas investments and the increase of facilities for trade, all matters for decision with men at the top in politics, finance and business. "The fact that a country can be wealthy at the same time as the large majority of the people may be poor, and that this has become a dominant factor in national and world politics, has not entered far into the field of diplomacy."

No Government in the twentieth century, Bevin pointed out, would think of conducting its domestic policy without paying attention to economic and social questions, the level of wages and employment, the attitude of the employers' and workers' organisations and the effect of such matters on politics. It was time the British Government broadened its conception of foreign policy to take account of similar questions abroad and their impact on the policies of other states.

Apart from the obvious criticism that recruitment and training needed to be widened, Bevin made four other proposals. The first was to form the Foreign Office, Diplomatic and Consular staffs into a single service. The second was to give those selected for it and for the Dominions, Indian and Colonial Services a probationary period in other government departments and some experience of commercial and industrial practice; the third, to organise a regular interchange of staff between the overseas services and home departments like the Board of Trade and the Ministry of Labour. Finally, he suggested that in every important embassy abroad there should be an officer, recruited from outside the foreign service if necessary, specially appointed to watch and report on industrial and social developments.[1]

---

[1] As an experiment Bevin persuaded R. H. Tawney to go to America in the autumn of 1941 and spend some months attached to the Washington Embassy. The experiment, as Tawney ruefully admitted, was not a success. Bevin, however, was not deterred and tried again with Archie (later Sir Archibald) Gordon. Gordon more than justified his hopes and by the end of the 1950s Labour Attachés had been appointed in eighteen countries. Sir Archibald Gordon recalls that when he had been briefed on his appointment and was leaving the room, Bevin called him back and said: "Here. If you make mistakes yourself, you must stand on

When Eden became Foreign Secretary he consulted Bevin freely about the reform of the Foreign Service. He was not always prepared to go as far as Bevin wanted him to: for example in a letter of 5 June 1942, Bevin wrote to him:

"One thing I forgot yesterday in our hurry—I do not like your paragraph regarding women. If I were you I should take the bold line of admitting women to the Service. . . . Go right out and accept terms of equality, purely on the basis of ability, it would be far better."

This had to wait until 1945 when Bevin himself was Foreign Secretary. None the less, anyone who reads the White Paper of January 1943[1], the foundation of the post-war foreign service, will see the influence of Bevin's ideas on Eden's proposals, and by an ironical twist which neither man foresaw in 1942–43 it fell to Bevin as Foreign Secretary to put the Eden reforms into effect.

Bevin wanted to do more than introduce reforms in the foreign service: he wanted to persuade the Foreign Office to look at international relations from a new angle.

"In any country," he wrote in his 1940 memorandum, "men and women tend to be divided by class, by politics and by religion. They are brought into the closest human contact and sympathy with each other . . . as human beings subject to common dangers, poverty, illness and death. . . . Nationalism is likely to become stronger and not weaker, and countries will tend more and more to develop industries which will be bound to be the cause of a certain amount of competition and division. It seems necessary to look for a binding form for peace not in the Customs Unions or economic groups, although these will emerge, but in those matters in which all human beings, irrespective of nationality, have a common interest. These are security against poverty, care in sickness and trouble, protection against injury, provision for old age, all of which tend to assist in the great impelling human desire to have a home, to rear a family in decent and independent circumstances and to have a life in which work and leisure are properly adjusted. In short, international policy should be based not on the increase and safeguarding of the total trade and income of individual countries, but on the provision by international co-operation of the needs of human individuals. . . . Britishers, Poles, Chileans,

---

your own feet, but if it's somebody else, remember I'm here." In fact, he adds, Bevin took great interest in what he did and gave him strong support all the time.

[1] *Proposals for the Reform of the Foreign Service*, Cmd. 6420. The text is conveniently reproduced in Lord Strang's *The Foreign Office* (1955).

etc. being looked at together as human beings whose well-being is inter-
dependent."

"The above conception would lead to directed planning of the use of international resources and capital instead of national or international financier
investments for profit, with human betterment a major objective and not
merely an incidental result. The movement of peoples with the necessary
capital resources, the planning and undertaking of works of public utility,
transport development, etc. would be undertaken on an international and
not a national basis. . . . If instead of stopping at reports on unemployment
and emigration, the nations had joined together in financing development
and movement, the course of events would probably have been entirely
changed."

The whole passage is characteristic of the way Bevin's mind
worked, starting from his own experience and trying to work his way
towards a general principle. His experience of international
relations was obviously very much more limited than his experience
of industrial relations, but he had thought about it, and if he expressed himself in a vague and naïve way, he had equally clearly got
hold of an idea. Since the war international action to help meet
economic and social needs has become so much of a commonplace
that it is easy to forget how radical a departure this has been from the
traditional agenda of international relations; as radical as the change
in domestic politics which followed industrialisation. Bevin was not of
course by any means the only man to grasp this, but his continual
hammering away during the war at this sort of approach to the post-
war settlement had its effect on official as well as public opinion, and
it was during his period as Foreign Secretary that this enlarged
concept of foreign policy first took effect.

5

An important source of Bevin's ideas about foreign policy was his
experience in the International Transport Workers' Federation and
the International Labour Organisation.[1] In April 1942 the emergency
committee of the I.L.O. met in London, and Bevin was the obvious
choice to welcome the delegates on behalf of the British Government.
His phrase, "This is a people's war: it must lead to a people's peace"
caught popular attention and the *Manchester Guardian*, congratulating

[1] See Vol. I, pp. 407–10, 577–8.

him on his speech, wrote that "more directly than any other British minister he faced the fact that the peace will be a revolutionary peace".[1]

In his opening address Bevin told the delegates:

"In war, out of the desire for self-preservation we are ready to undergo control, regulations and discipline beyond the belief of what most of us would have thought possible. As soon as the 'cease fire' sounds, there may be a danger of a tremendous reaction. It is then, I suggest, that the statesmen of the world and all those responsible for the leadership of mankind must stand together resolutely and hold on to some form of controls while the foundations of peace, stability and orderly development are being worked out."

This meant two things: suppressing the desire for immediate gain, and recognising that no nation could think only of its own limited interest.

"The less you discuss things as countries, and the more you can face them as problems affecting all countries, the more likely are you to find a correct solution. . . . Immediately you get away from the purely nationalistic outlook or from the limited vision that arises from your own interests, and proceed to grapple with problems as such, then the national barriers that nationalism or narrow interest creates are broken down."

Bevin urged the conference to pay special attention to the primary producer and not concentrate on the industrialised areas of the world:

"You cannot have a decent civilisation if you leave the peasants of the world under-paid and, in spite of the fact that he grows the food of the world, under-fed and in poverty. You cannot afford to allow the industrialist to have his cartel, his price-fixing arrangements and all the other devices while leaving the primary producer unprotected."

He ended his speech by quoting Clause 5 of the Atlantic Charter with its promise to secure for all improved standards of living and security. That promise was meant to apply universally, irrespective of colour or race.

"It really means the end of exploitation as we knew it in the nineteenth century. . . . To achieve this end, mere revolutionary upheaval will not do. It would probably have the effect of setting us further back. It takes so long to rebuild."[2]

[1] *Manchester Guardian*, 21 April 1942.
[2] Speech at the meeting of the I.L.O. Emergency Committee in London, 20 April 1942.

What was needed was to pool experience and knowledge through bodies like the I.L.O.

The Times was as impressed as the *Manchester Guardian* by Bevin's speech and after reporting prominently his views on reconstruction, added: "Mr. Bevin's personality and office fit him to give a lead to this country and to the Government in this immense task."[1]

Bevin kept up his advocacy of support for the I.L.O. in and out of season. In June 1942, for instance, he took up with Eden the familiar objection that, while the British could be trusted to carry out punctiliously any engagement they accepted under an I.L.O. Convention, other countries could not. This argument, Bevin pointed out, never seemed to apply to political and commercial agreements, only where labour questions were involved: was it, in effect, simply a convenient way of avoiding obligations the British did not want to accept?[2] When the time came to set up the new world organisation in the United Nations, he was able to put the full weight of the British Government's support behind the I.L.O.'s claim to be given a proper place among the other specialised agencies, and the success of the I.L.O. in surviving the war and maintaining its identity owed more to Bevin's efforts on its behalf than to those of any other single person, in Britain or any other country.

His interests, however, were not confined to the I.L.O. He studied with care the Treasury's proposals on the future of British monetary policy and wrote to Eden and the Chancellor to express strong disagreement. As they stood, he declared, the proposals were a return to the automatic gold standard "with just a little more rope before the poor unfortunate debtor is hanged". Bevin reminded his colleagues that he had already been forced to take part in one general strike by a return to the gold standard and he had no intention of seeing the country involved in a second. He went to Keynes and, with his help, drafted a clause to make clear that no country should be required by the proposed international bank to adopt a deflationary policy which would have the effect of creating unemployment.[3]

At the end of the year Bevin drew together his ideas on the post-war settlement and set them down in the light of the Cabinet discussions

---

[1] *The Times*, 22 April 1942.
[2] Bevin to Eden, 18 June 1942.
[3] Bevin to Eden, 24 April 1942.

in which he had taken part. He sent a copy of his memorandum to Eden with the remark:

"I cannot help feeling that so far we have been thinking too much in terms of political groupings, derived from the old balance of power. . . . I am deeply concerned to get a different approach to post-war organisation—an approach which recognises that while man as a political animal tends to look backwards, as an economic animal he is forced, whether he likes it or not, to look forward."[1]

In support of this view, Bevin wrote:

"The achievement of collective security is not solely, or even primarily, a political problem, and any plan which sets itself merely to secure a balance of political forces will not last. We have to find an economic basis for collective security if individual nations and peoples are to recognise that they have a stake in maintaining it.

"We must aim at an organisation which will develop to the full such international services as transport and communications; which will encourage customs unions and develop international banking and lending facilities. It is such means as this that will guarantee the necessary cohesive force between nations while leaving political independence and theories of sovereignty untouched."

As a starting point Bevin suggested that they should look at three possible frameworks in which to develop such links: the United Nations as the nucleus of a world organisation; a European Commonwealth, beginning with the exiled European Governments and including Britain; the British Commonwealth, including India and the colonies. They should begin, he thought, by studying the common needs of the nations in these groups and the organisation of the common services necessary to deal with them; "foremost amongst these are feeding, transport, currency, economic rehabilitation and customs."[2]

On the political settlement at the end of the war, Bevin had not yet formed particular views of his own. There was one political issue, however, on which he already held strong views, self-government for India. In September 1941 he wrote to Amery, the Secretary of State for India:

[1] Bevin to Eden, 8 December 1942.

[2] Memorandum entitled: The Four-Point Plan and Post-War Settlement: The economic basis of international organisation. No date, but attached to a letter of 8 December 1942.

"I have given a lot of thought to the talk we had some days ago on India. I have gone again through the papers and I have also been thinking about it in relation to possible developments as a result of the war in the East. I must confess that leaving the settlement of the Indian problem until after the war fills me with alarm. . . .

"It seems to me that the time to take action to establish Dominion status is now—to develop or improvise the form of Government to carry on through the war but to remove from all doubt the question of Indian freedom at the end of the war."[1]

Holding these views, he was a strong supporter of the offer which Cripps took to India in March 1942, and was disappointed with the refusal of the Congress Party to consider it seriously, the more so since he believed that self-government for India was the key to the future of the Commonwealth.

Bevin had only paid one brief visit to India, on the way back from Australia in 1938. But the impression then made on him by the conditions of life of the Indian people was one of the sources of his conviction that no international settlement would last unless it did something to relieve the mass poverty of the over-populated, under-developed countries of the East. He told an American correspondent how he stood and watched the construction of a house in an Indian town.

"Stone, mortar, all the building material passed hand to hand up human chains strung from the bottoms and tops of long ladders. The sight sickened me and I never forgot it. I made up my mind that some day I would do whatever I can to raise the standard of living in India."[2]

There was one thing which was within his power to do as Minister of Labour even in wartime. In June 1940 he put up a proposal to bring young Indians over to Britain for six months' industrial training. The proposal was turned down by the India Office and the Viceroy, but Bevin was persistent, reopened the matter as soon as he became a member of the War Cabinet and overcame not only the official opposition but the wartime difficulties of transport. The only conditions he made were that the men chosen should come from all religions and castes, that they should have had three years' experience in industry and that they should be selected by an independent board

---

[1] Bevin to Amery, 24 September 1941.
[2] Broadcast by Fred Bate, N.B.C. correspondent in London, 14 September 1941.

on which Labour was represented. At the same time he helped to find technicians and instructors who could be sent out to India to start training schemes there.

On arrival the trainees were sent to the Letchworth Training Centre, where they lived in a hostel with an Indian manager and were given instruction in fitting and machine operations. For the next two months they moved to a centre near Manchester, lodging with working-class families, and then went on to complete their course with private employers. Bevin took a close personal interest in the progress of his experiment. He visited Letchworth several times to see how it was going and used his union contacts to make sure that the young men were given an insight into British methods of trade-union organisation and industrial negotiation. And he never failed to entertain each group of trainees at his Ministry before they left and to speak to them of his hopes for India. Inevitably, in wartime, such a scheme could have no more than a limited scope. But it is a very good example of the practical imagination Bevin could bring to bear on a problem when his interest was roused: by the end of 1945, 700 Indians had been through the course and their subsequent careers justified his faith in what could be made of such an experiment.

6

Another I.L.O. body met in London that summer, the Joint Maritime Commission on which Bevin had served in the 1930s. His object now, he told the Commission, was to secure an international charter which would establish standards of social legislation applicable to seamen of all nationalities, Indians and Chinese no less than Europeans. On the last day of 1942 he dictated a memorandum, nine pages long, reviewing what still needed to be done to bring British practice into line with the I.L.O. recommendations on seamen's protection and welfare which he had persuaded the Government to adopt in 1940.

Shipping, indeed, all the way from building the vessels to unloading their cargoes at the docks, was a major problem throughout 1942: on top of the heavy losses inflicted by the U-boats, there was now the need to assemble sufficient ships to carry the invading force to one end of North Africa and build up the striking power of the Eighth Army at the other.

At the end of February 1942 Bevin tried the experiment which he was later to repeat with the builders: summoning a conference representative of all the interests in the docks and putting the position to them himself. This was Bevin's home ground and he trod it with assurance: there was no trick on either side of the industry with which he was not familiar. He warned his own union officers that they could throw away the advantages of the guaranteed week by letting the men abuse it:

"Over one third of the men in the North West, particularly in Liverpool, are breaking their guarantee: that is, they do not get it because they do not turn up every day. . . . We cannot retain a man in a port in a time like this who is not going to play the game."

Neither Government nor nation would stand for such malpractices: the alternative was military control of the ports. If the unions did not like the prospect of this, then they must discipline their own members.

It would be nine months, he told the dockers, before they could look for relief of the shipping shortage. During this period the Government asked for a day, or a day and a half, to be saved on the turn round of every ship. That could be done if the restrictions on overtime and double shift working were abandoned.

"You have all got your books of rules, you have all got your port customs and practices, and I propose to register them en bloc. This will save everybody trouble; and when the time comes for us to restore them they can be put back without question. I said in the House of Commons the other day that those things are property rights. It has taken years to get them—I have spent a few years getting them myself—and I should be the last in the world to come to the men and ask them to give them up if it were not absolutely necessary."

When he had finished, another hour was spent in questions and criticism from the floor. But Bevin would not let the meeting break up without bringing them back to face the main problem of greater output. Whatever the shortcomings in welfare, whatever the grievances on this or that particular issue, the Government had gone further in giving the dockers a square deal than ever before and they had a right to expect them to respond.

"Transport is one of the vital keys to success and we cannot allow it to be monkeyed about with by anybody. It should not be said that this war was lost through any lack of will or decision. . . . If employers or foremen or men, by manœuvres of any kind, delay equipment getting to the troops, then prosecution must follow."[1]

What mattered more, in fact, than any threat of prosecution, as Bevin well knew, was to impress the unions and the employers with the urgency of the situation. Following the conference, the T.G.W.U. issued instructions to all its members that the Working Rule Book was withdrawn—"Rules and regulations must go overboard—All rules both sides."[2] That the biggest union in the country could be brought to do this openly was proof of the authority Bevin still commanded in the trade-union movement and of his willingness to use it.

Both in the docks and in shipbuilding one of the great obstacles to the efficient employment of the labour force was particularism. The men worked in gangs and were accustomed to work for a particular employer; they resisted the traditional pattern of employment being broken up in order to move men from a yard where they were temporarily unemployed to others alongside, where there was a shortage of labour. In the docks, as in the Merchant Navy, Bevin succeeded in introducing a system of pooling labour under the Essential Work Orders, but in the shipyards, where mobility of labour was already hampered by the strict lines of demarcation between the different crafts and unions, he had no success. In August 1942 he renewed his earlier proposal to transfer the control of all shipbuilding labour to the Admiralty in the hope that the men would then feel that they were being allocated to a group of shipyards rather than to a particular employer, and that this would allow greater freedom of movement between jobs. Not only the Admiralty but his own local officials were opposed to his scheme: apart from the difficulty of wage differentials, which was formidable in its complexity, it ran counter to all the traditions of the industry, and it was argued that more would be lost than gained by disturbing them. Whatever could be done to increase output must be done within the existing framework of employment.

Bevin had no choice but to abandon his proposal: the only course was to keep on trying to break down in detail the resistance to the

[1] *Quicker Turn Round of Ships.* Confidential Report of Speeches at the Conference of Representatives of the Port Transport Industry, 27 February 1942.

[2] *To All Dock Workers.* Leaflet issued by Area No. 6 of the T.G.W.U. and signed by the Area Secretary and the Dock Group Secretary.

expedients already described.[1] It was a slow, irritating and untidy process, the very opposite of the clear-cut rational plan which the critics of labour policy demanded. It met only one test, but in the end it was the only test that mattered: it produced sufficient vessels to prevent the Government's plans being hamstrung by lack of ships. Repairs to the two-and-a-half million tons of damaged shipping which had been the great problem in the spring of 1941 were cleared very much more rapidly than had been expected and the production of new merchant ships for the year exceeded by 50,000 tons the target figure of 1.1 million. Encouraged by these figures, the Prime Minister revised the figure for 1942, putting it back to the original target of a million and a quarter tons. There was a similar improvement in naval production. In the first half of 1941 this had been running at no more than 68 per cent of the tonnage which the yards were expected to complete: the comparable figure for the first half of 1942 was over 82 per cent.

In September 1942 Bevin brought 1,500 representatives of the shipbuilding industry together in London for the third of his industrial conferences. This time he took the chair and left the main speech to be made by A. V. Alexander, the First Lord of the Admiralty. But from the moment the questioning and criticism from the floor began, Bevin was in the thick of it, and it was he who in the closing speech tackled the most difficult question of all, what had happened to the shipbuilding industry in the twenty years' depression between the wars. He refused to offer any easy promises for the future: there was bound to be dislocation at the end of the war and almost certainly a contraction of the industry in face of the development of air transport.

"But this I will say: if an industry is contracted through economic or international reasons over which you cannot exercise control, we are not going to leave the men sticking about an industry that has gone; we have got to find alternative employment, alternative trades, alternative opportunities. . . . That does not apply only to shipbuilding but to cotton and many other industries, and the Government in its reconstruction effort is approaching all these problems recognising that the thing we must preserve, and not allow to rot again, is the skill of our people."[2]

---

[1] See pp. 26–29, 59–63 above.
[2] Verbatim report of the Conference of the Shipbuilding and Ship-repairing Industry with the Minister of Labour and the First Lord of the Admiralty. Central Hall, Westminster, 23 September 1942.

Above. Bevin opening Merchant Navy House at Newport. *Below.* The Minister of Labour and his wartime Director-General of Manpower, Godfrey Ince.

With Indian trainees outside the Ministry of Labour.

A Tyneside shipyard worker told a reporter as he came out: "Both sides got a load off their chests; now we feel more like doing what has to be done." The figures bear out his remark. Production for the year 1942 finally topped the target of a million and a quarter tons of merchant shipping by 50,000 tons, and in the first half of 1943 naval production reached 92 per cent of the Admiralty's estimate.

The foundation of this increased output was the success of the Ministry of Labour in raising the labour force engaged in shipbuilding and repairing from 203,000 in June 1940 to 272,000 by June 1943, an increase of a third.[1] This solid achievement was obscured, in other industries as well as shipbuilding, by the inflated demands which all departments made as a reinsurance, by the habit of blaming inadequate supplies of labour for every other deficiency in production and by the constant propaganda for higher output and the recruitment of women. The figures which would have shown the Ministry's success in meeting demands were kept secret to avoid helping the enemy.

In the autumn of 1941, when the manpower budget for the period up to June 1942 had been revised, gloomy views had been expressed in Whitehall about the possibility of realising the targets set. When the count was made in the summer of 1942 it took the Prime Minister and everybody else by surprise. If the number of women recruited for the services and industry had fallen below, the number of men was above the original estimate in both cases.[2]

In a personal minute to the Minister of Labour, the Prime Minister wrote:

"I see that you drafted nearly a million men and women into the Services, thereby fulfilling the great bulk of their requirements, and at the same time added 800,000 to the labour force on munitions.

---

[1] In marine engineering there was an even larger increase, over half, from 59 to 89 thousand, in the same period.

[2] The figures which Bevin gave to the National Production Advisory Council for the twelve months ending 30 June 1942 were:

Intake of men into the Forces = over three-quarters of a million, 40,000 over the original estimate.

Intake of women into the Forces = nearly one-quarter of a million, 100,000 below,
Net increase in men engaged on Government Munitions work = 284,000 men, 7,000 above original estimate.

Net increase in women = 522,000, 58,000 below.

"I congratulate you on this great performance.

W.S C.

24.9"[1]

Although the figures were not made public, enough was known of them in Parliament and Fleet Street to still any criticism of manpower policy.

Towards the end of July Bevin gave the House of Commons a report on the work which the Factory Department of his Ministry had done. His insistence on the importance of welfare services which had originally been criticised as "pampering the workers" was now greeted on all hands as a far-sighted recognition of how to get the best out of them.

"Mr. Bevin," the *Manchester Guardian* wrote, in a friendly, if patronising, leader, "may be criticised for some things he says and does, but no one can doubt his knowledge of industry and of the men and women in it and his appreciation of them as human beings first and instruments of war second. And that counts for a great deal."[2]

Even on wages, which were the most controversial part of the Government's economic policy, Bevin had the better of the Commons debate in October. Sir Alfred Beit expressed the feelings of many Tory backbenchers when he criticised the Minister of Labour for being "more concerned in creating a strong and even unassailable position for labour after the war than fitting wages into the general framework of the national war economy."[3] But when this had been said, when the anomalies between pay in the Services and in munitions had been drawn out and the risk of leaving wages to be settled by collective bargaining re-stated, the fact was that none of the critics had a pratical alternative to suggest.

[1] Churchill, Vol. IV, p. 797.

[2] *Manchester Guardian*, 23 July 1942. That Bevin was far from satisfied with what had been achieved so far is shown by a personal letter of 13 November 1942 to the Minister of Food, Lord Woolton. "On Saturday last I went into a dock canteen at Hull. It was very poor, the place was filthy and the standard of service left much to be desired." There were far too many complaints of similar character, especially of the poor quality of food served. Bevin asked Woolton if he would not consider setting up a non-profit-making association to run canteens on the same lines as the National Service Hostels Association.

[3] House of Commons, 21 October 1942, Hansard, Vol. 383, col. 2039.

Bevin's defence of his policy followed the same lines as in the Cabinet the year before. The labour force of the country had been re-deployed without the industrial disputes which occurred in 1914–18; an average increase of earnings of 47.5 per cent had not led to dangerous inflation and had been matched by an increase in production per man hour of over 44 per cent. Did Sir Alfred Beit and his friends really want state regulation of the whole of the economic life of the nation, with profits and salaries as well as wages brought into the debate? The one thing they could not do, Bevin insisted, was to single out wages and treat them in isolation. Collective bargaining backed by arbitration was part of a consistent labour policy, including the Essential Work Orders and the direction of labour, which the House must either accept or reject as a whole. When he sat down, it was clear that the House would settle to accept, some because they felt Bevin had succeeded better than in any of his earlier speeches in explaining his policy, the rest because they judged it best to leave well alone.

7

By October 1942 the Ministry of Labour had prepared its autumn survey of the manpower position. This had been awaited with impatience in Whitehall. The manpower budget had now become the principal instrument of government planning and many decisions were held up until the allocation of manpower had been settled. The responsibility for recommending this to the Cabinet rested on Anderson, but he could not frame his manpower budget until the Ministry of Labour provided the figures in its survey.

In the summer of 1942, for example, the War Office put up a case for raising Army recruitment beyond the limit fixed before Japan and America entered the war. Churchill handed the matter over to a committee consisting of Anderson, Bevin and Grigg to investigate. They decided to recommend an increase of 100,000 and Bevin and his officials had just finished earmarking the sources from which this could be found when the War Office put in a further memorandum arguing the case for an additional quarter of a million men; the Air Force followed this by asking for a further 120,000 and the Admiralty for another 22,000.

These claims illustrate well the central role of the manpower discussions. With each Service and department thinking simply in terms of its own needs, the total requirements were likely to add up to a figure out of all relation to the resources available: it was the responsibility of the Minister of Labour to bring them down to earth and confront them with the hard facts of the situation.

In 1943, the Ministry succeeded in doing this with a greater wealth of detail than before. Its autumn survey was divided into three parts. The first provided a comparison of the distribution of the population in 1939 and 1942; the second set out the current distribution of the labour force between different groups of industries, while the third summarised the requirements of the Forces and supply departments up to the end of June 1943. In the note with which he presented his Ministry's conclusions Bevin underlined the fact that there were no longer sufficient men and women to meet these requirements without a further drastic cut in civilian standards of living. They had at last come within sight of the limits of mobilisation.

The obvious consequence of this was a reversal of the procedure hitherto followed. Instead of agreeing on the allocations of manpower required and then asking the Minister of Labour to find the bodies, the Cabinet had to start from the other end, deciding how many men and women could be made available and then settling how demand could be accommodated to supply. In order to focus the issues more clearly, Anderson extended the period under review to the end of 1943. When this had been done, the gap between supply (950,000 men and 650,000 women, a total of 1.6 million) and the combined demand of the Services and war industry (over 2.6 million) was revealed as more than a million. Even this was based upon a number of optimistic calculations: withdrawing half a million more from the less essential industries, drawing nearly a million more women into part-time work, calling up every fit man of military age from the building industry and over 100,000 from munitions work.

A gap of a million, perhaps more, was too large to be closed by juggling with the figures or devising a compromise. A major cut would have to be made. Of the three directions in which this might be looked for—the Services, munitions or civilian standards—Churchill preferred to press hardest on the first. Bevin was critical of his assumptions: he thought the plans for the expansion of aircraft production and shipbuilding unrealistic in view of their past performance

and, left to himself, would have cut back their labour force in favour
of the Services. But he did not press his view, and with a few modifications in detail the War Cabinet finally approved Churchill's proposals on 11 December.[1] Apart from a reduction in the age of call-up
from eighteen and a half to eighteen, which Bevin announced to the
House of Commons the same day, none of these decisions required
legislative action: nor, for obvious reasons, was any hint of them given
in public. The press, the morning after Bevin announced the new call-
up, was full of leading articles on manpower, almost without
exception friendly to the Minister of Labour; but not a word was said
about the decisions the Cabinet had just taken or the task which the
Ministry of Labour had been set of finding another one million six
hundred thousand men and women when the country was already
more fully mobilised than in 1918.[2]

8

By the time the allocation and cuts for 1943 had been settled, October
(1942) had turned into December and the war had passed its
climacteric. With Alamein and the landings in North Africa,
Stalingrad and the failure of the German drive to the Caucasus, the

---

[1] The allocations for the eighteen months July 1942–December 1943 were fixed
as follows:

(Figures in thousands of men and women)

	Original Demands	Cut Imposed	Addition to Manpower Strength on 30 June 1942
Navy	323	} —75	434
Shipbuilding	186		
Army	809	—380	} 351
Ministry of Supply	148	—226	
R.A.F.	472	—225	} 750
M.A.P.	603	—100	
Civil Defence	—	—75	—75
Miscellaneous	135	—19	116
	2,676	—1,100	1,576

[2] In November Bevin gave these figures in a lecture at the Staff College (Cam-
berley): In 1918, 27.8 per cent of the adult population (men and women) were in
the Forces, Civil Defence and Munitions; in mid-1942, over a year before the peak
of mobilisation in the Second World War, the corresponding percentage was 30.1
per cent.

swing in the balance of advantage awaited since America's entry into the war at last began to show itself. The effect on morale was immediate. The doubts about the conduct of the war were swept away, and overnight Churchill's critics found themselves isolated and on the defensive.

The victories in Africa and the feeling that the Allies were now beginning to win the war were bound to open a new chapter politically. Just as the economic organisation of the war had given way to its military conduct as the main topic of controversy at the beginning of 1942, at the end of the year this in turn was replaced by the debate on post-war reconstruction.

While the outcome of the operations planned in Africa was still in doubt the Government had done its best to damp this down. The opening of the new parliamentary session in November, however, already made plain that the plea of national unity would no longer be accepted as an excuse for postponing discussion of future plans.

Churchill himself was plainly uneasy about the consequences of the change. "The most painful experience would lie before us," he said in a broadcast at the end of November, "if we fell to quarrelling about what we should do with our victory before that victory had been won."[1] The King's Speech at the opening of Parliament (11 November) contained a good many references to post-war questions, but Cripps, in announcing the Government's programme for the session, declared:

"The times are clearly inappropriate to bring forward legislation of a character which is likely to arouse serious controversy between the political parties."[2]

The debate which followed the King's Speech showed the difficulties which the Government could expect to encounter. A strong body of Conservative members was determined to block any attempt by Labour to introduce measures to which they objected. The majority of the Labour Party was equally determined to prevent a return to the Britain of the 1930s at the end of the war. Each side attacked the other for political dishonesty and the subordination of national to class interests. The Conservatives accused the Socialists of exploiting the wartime emergency for partisan purposes; Labour and

[1] *The War Speeches of Winston Churchill* (1951), Vol. II, p. 370.
[2] House of Commons, 11 November 1942, Hansard, Vol. 385, col. 40.

Liberal members denounced the Tories for attempting, under cover of national unity, to preserve the status quo and cheat the nation of the new deal which had been promised at the end of the war. Both sides combined to attack the Government for the vagueness of its intentions which each claimed was being used by the other partner in the coalition as a cloak for furthering its own interests. Tories complained that there was not a true Conservative in the War Cabinet, that the Labour ministers got their own way by threats of breaking up the coalition; the Left on the contrary represented Attlee and Bevin as "the dupes if not the allies of reaction", cast for the role of Mac-Donald and J. H. Thomas in another "National" Government.

Until Christmas 1942 Parliament's time was largely spent in preliminary exchanges and the taking up of positions. The Government had not yet put forward any proposals on which battle could be joined. "We shall see," one Labour member remarked, "how each party behaves when the real fight starts in February."

Before embarking on this, however, we may end this chapter with a brief account of one reform sponsored by Bevin which all parties were prepared to accept as non-controversial.

Much experience in workmen's compensation cases had convinced him that a scheme ought to be developed for the rehabilitation of the disabled, whether civilians injured in an air raid or men wounded in combat. It was not at all clear why such a scheme should be organised by the Ministry of Labour rather than the Ministry of Health or Pensions, but Bevin did not worry about that. He argued that as Minister of Labour he could not leave the disabled unemployed in wartime, persuaded the other ministries to co-operate and succeeded in starting an Interim Scheme in the autumn of 1941. It was arranged for officers from the Ministry of Labour to visit injured men or women before they left hospital and discuss with the doctors as well as the patients what work they might be able to take up on discharge. Where necessary, training was provided in government training centres and the Ministry of Labour undertook the responsibility of finding jobs.

Bevin's grounds for taking an interest in the matter were the contribution the disabled could make to the war effort, but his real aim was to develop a comprehensive rehabilitation scheme, including those injured in industry, as a permanent social service. In December 1941 he succeeded in setting up an inter-departmental committee with wide terms of reference under the chairmanship of George Tomlinson. While George disarmed all suspicions, Bevin stayed in the background but left no doubt that he would provide powerful reinforcement if there was any delay or obstruction.

In October 1942 he secured the support of the Lord Mayor of London for a demonstration at the Mansion House of how far injured people could be restored to normal activities. Pleading for willingness to offer the disabled jobs, he announced the agreement of the Minister of Pensions to no reduction in pensions for those who took up work and urged that they should be paid standard wages. He laid great stress on the psychological effects of allowing an injured man to feel that he was still a full man and outlined his plan of a permanent National Rehabilitation Service to provide training for anyone injured during or after the war. "We do not want, when persons have been injured for any cause, to leave them as a burden on their family and friends, or to cause them to be demoralised by a feeling of dependence on public charity."[1]

A note of 28 November 1942 shows the way in which he saw the application of rehabilitation to industry. He had been looking into the facilities for helping men injured in the heavy industries of South Wales, not only miners but steel, copper and tin-plate workers. He was appalled by what he found and set to work to bring the Mines Department, the Ministry of Health and the Welsh Board of Health together to take effective action, pointing out that considerable sums of money were being paid out in health and unemployment insurance as well as compensation for injury, with nothing to show for it. "Where I think the mistake will be made," he wrote, "is if the miners attempt to do this alone. What is needed in South Wales is a chain of rehabilitation centres that will treat the whole community including miners."

When the Tomlinson Committee reported at the end of 1942, it endorsed Bevin's suggestions and proposed a national scheme, to be administered by the Ministry of Labour, under which all disabled persons, whatever the cause of their injuries, should be eligible for help. With the report published as a White Paper,[2] Bevin set another committee to work on the problems of organisation and before the

---

[1] Speech at the Mansion House, 14 October 1942.
[2] January 1943, Cmd. 6415.

year was out had a Bill ready to lay before Parliament.[1] This was Bevin at his best, grasping a need, having the imagination to see what was involved in human and social loss and how this could be reduced, then using patience, skill and persuasion to draft and pilot a scheme through until it became law.

[1] See below, pp. 288–90.

CHAPTER 8

The Catering Wages Bill,
the Beveridge Report, and Coal Again

I

BY THE time Parliament met in mid-January 1943, the impression given by the victories in North Africa had been strengthened by the capitulation of the German Sixth Army at Stalingrad where the German commander, von Paulus, finally surrendered on 30 January. Neither party in the House of Commons now felt any inhibition about putting pressure on the Coalition Government over post-war reconstruction. The first challenge came from the Right and was aimed directly at Bevin whom the Tory backbenchers regarded as the most dangerous of Labour ministers. This explains the otherwise surprising choice of issue for a trial of strength: the Catering Wages Bill.

In the previous summer Bevin and his officials at the Ministry had decided to attack the problem of the catering industry and the wages and conditions of the half million people employed in hotels, restaurants, cafés, bars, boarding-houses, the whole gamut from the West End clubs and luxury hotels to industrial canteens. Bevin had by now succeeded in setting up six joint industrial councils for the different branches of retail distribution, an industry which, like catering, displayed great variety in conditions of employment all the way from Harrods or Woolworths to the village store.

Catering was left as the largest industry in the country in which wages, hours and conditions of work were uncontrolled by collective agreement or statutory regulation, and the Ministry had good reason to believe that, in some cases at least, the arbitrary way in which employers treated their staff and the lack of redress both needed investigation.

For this reason, despite the obvious importance of the industry in a time of food rationing, Bevin had refused to give it the protection of the Essential Work Order. He now made up his mind, if the employers would not set their own house in order, to force them to do so. Having organised road haulage before the war, when he was no more than a trade-union leader,[1] and mastered the problems of retail distribution, he was not to be put off by the argument, employed in both these cases, that the nature and variety of the jobs involved made regulation impossible. He took great pains to prepare his case, studying earlier attempts which had failed and devising a form of procedure which he believed would get round the difficulties in the way. In November (1942) he put his proposals to the Lord President's Committee and the War Cabinet and secured their agreement. On the 27th of the same month he informed the House of Commons, in answer to a question, that a Bill was being drafted and would be introduced in the New Year.

As soon as he had decided to take the matter up, Bevin saw representatives of the catering employers and discussed his ideas with them. He met with an unfavourable reception and as early as July 1942 *The Times* reported that the powerful Hotels and Restaurants Association was unalterably opposed to his proposals. Other sections of the trade were canvassed and at the end of July a Catering Trades Joint Committee was set up to fight any form of regulation of the industry.

Two months before, the Conservative 1922 Committee had defeated the plan for fuel rationing, and the catering employers were hopeful that the Committee would be prepared to intervene a second time against a scheme which they described as socialistic and a threat to free enterprise. The Committee as a whole was unwilling to act again so soon after its earlier success, but a Parliamentary Committee on Catering was set up with Sir Douglas Hacking, a former chairman of the Conservative Party Organisation, at its head, and by November this had secured the support of some 200 M.P.s, nearly all of them Conservatives.

The object of Bevin's opponents was to prevent his Bill ever seeing the light of day. They kept up a barrage of questions in Parliament and of unfavourable reports and letters in the press, organised meetings and, not without success, spread the impression that Bevin was trying to establish himself as a dictator of the catering industry. When Bevin made it clear that the Bill would none the less be intro-

[1] See Vol. I, pp. 544–6; 618–19.

duced, the Parliamentary Committee took up Cripps's statement[1] that the times were inappropriate for legislation likely to arouse serious controversy, and claimed that this was a "pledge", which the Government was now breaking, to abstain from introducing any measure to which there was opposition. Their other line of attack was to call for a preliminary investigation, the practical result of which would have been to delay the Bill indefinitely and, with luck, kill it.

When the Government gave no signs of retreating before the opposition's show of parliamentary strength, Hacking and his committee saw Bevin on 14 January (1943) and presented a formal request for an inquiry. Bevin, equally formally, rejected their request and a fortnight later the text of the Bill was laid before Parliament with the carefully selected signatures, in addition to Bevin's own, of Sir Kingsley Wood, Sir Donald Somervell and Sir John Anderson—two Conservatives and the most respected Independent member of the Government.

Bevin had won the first round in what had clearly become a trial of strength between him and the Conservative backbenchers. Although the open opponents of the Bill alone outnumbered the total parliamentary strength of the Labour Party, he had succeeded in holding the support of his colleagues in the Cabinet through several months of skilfully organised propaganda which played on every prejudice against socialism, bureaucracy, trade unionism and the dictatorial ambitions of the Minister of Labour. He could count on a parliamentary majority for a measure backed by the official leadership of all three parties in the coalition and, once the text of the Bill had been made public, moderate opinion swung round to Bevin's side. Not only the Labour and Liberal press, but both *The Times* and *The Financial Times* were favourable to his proposals: "It would be too much to expect," the latter remarked, "that an industry of such a size should remain untouched by the trend towards collective agreements which the war has fostered."[2]

His opponents, however, were far from giving up the fight. The secretary of the Catering Trades Joint Committee (Mr Roy Snell) issued a statement to the Press charging that "under the pretext of improving conditions and helping these trades, Mr Bevin is really seeking to obtain absolute control of them in his own hands," and

[1] See above, p. 216.
[2] *The Financial Times*, 30 January 1943.

calling the Bill "the biggest bluff ever attempted by a Cabinet Minister".[1] A heavy adverse vote in the Commons might still lead the Conservative members of the Government to change their minds and press for the withdrawal of the Bill. More than once Bevin had lost support in the House by the clumsiness of his parliamentary performance and, if he could be goaded into losing his temper, might do the same again and lend support to the picture of him as a trade-union boss trying to ride rough-shod over opposition. Bevin was as well aware of the dangers as his opponents and he approached the debate on 9 February knowing that not only his scheme for the catering industry but his reputation as a Minister and his standing in the Government were at stake.

The most unusual feature of the Bill itself was the appointment of a permanent Catering Commission to examine and keep under review wages and conditions (including health and welfare) in both the main branches of the industry, the supply of food and drink and the provision of accommodation. Where the Commission found that such matters were satisfactorily settled by collective bargaining they need do no more than suggest any improvements that occurred to them. But where there was no satisfactory provision, or no provision at all, for negotiation, the Commission could recommend the Minister of Labour to appoint a wages board. Once such a board was set up it would have statutory power to fix wages and the time allowed for rest periods and holidays. Employers were required to keep proper records of wages and hours and to submit these, when asked, to inspection by the Ministry's officer; where an employer was found ignoring a board regulation he could be fined.

The publication of the Bill had already proved false some of the more exaggerated reports which had been spread about Bevin's intentions. It had for instance been widely rumoured that he meant to pack the Catering Commission with trade-union nominees. The schedule attached to the Bill made it clear that, of the seven members of the Commission, only two were to represent the workers in the industry, two were to represent the employers, while three (including the chairman) were to be independent persons. It was still true, however, that apart from a press conference at the end of January this was the first chance Bevin had had, after months of tendentious

---

[1] *Manchester Guardian*, 1 February 1943.

propaganda, to put his case publicly. The manner of its presentation could not have been better adapted to the mood of the House: he stated his arguments calmly, only once allowing himself to be goaded into a show of feeling. When he was describing the method of appointing the Commission, the word "independent" brought a cry of "Oh!" from the Conservative benches behind him. Wheeling round, Bevin retorted angrily:

"It comes ill from hon. members in this House to sneer at that. You have trusted me since 1940 with powers and have never questioned my exercise of them. If you have never questioned me in ordering millions of people about the country, why do you question my integrity in appointing a Commission now for this purpose? Am I good only for one purpose and not for another?"[1]

When the result was announced, 283 votes in favour, 116 against, there were cheers and counter cheers. Sir Douglas Hacking was on his feet at once to ask the Leader of the House (Eden) whether "he is not now satisfied that there is controversy". The press the following morning was favourable. Under the heading "The Tory Growl", the *Manchester Guardian* wrote:

"The Catering Wages Bill has much importance in the history of the Government. It is the first time the Government has stood firm against a Tory cave."[2]

The *Guardian* commented on the personal animus which the opposition had shown against Bevin and, like *The Times*, urged the Cabinet to stand firm. But it was equally clear that the fight was not yet over and Sir Douglas Hacking and his friends prepared to contest the Bill line by line when it came to the committee stage.

2

The Labour Party rank and file had been delighted with Bevin's performance and Transport House reprinted his speech as a party pamphlet,[3] with a "black list" of those who had voted against the Bill.

---

[1] House of Commons, 9 February 1943, Hansard, Vol. 380, cols. 1201-2.
[2] *Manchester Guardian*, 10 February 1943.
[3] *Square Meals and Square Deals*.

But before the final stages were reached and the Bill made law, another issue arose which revived all the latent dissatisfaction with the coalition's policy and involved Bevin in another angry row with his own Party.

The publication of the Beveridge Report on Social Insurance early in December 1942 is one of the key dates in the wartime history of Britain. Beveridge set out in detail a comprehensive scheme for social insurance for all citizens against sickness, poverty and unemployment together with proposals for a national health service, family allowances and the maintenance of full employment. No official report has ever aroused greater popular interest or enthusiasm. Beveridge's proposals, presented with the passion of a reformer as well as the authority of an expert, crystallised the ill-defined but widely felt desire to make a radical break with the past and create a society freed from its social evils and inequalities. Here at last was a programme, more than that, a manifesto, on which people could fasten. For or against "Beveridge" became the test of allegiance to the future or the past, and those who were "for" were in no mood to listen to qualifications or doubts.

The Government was far from sharing the popular view. The responsibility for Beveridge's appointment in the first place had been Bevin's. In June 1940 Bevin had asked Beveridge to take charge of the new department for welfare which he proposed to set up in his Ministry, and later invited him to carry out the first of the manpower surveys. Beveridge's ability was unquestioned, but he was a difficult colleague, and after trying to fit him into the Ministry of Labour, where he held the rank of Under Secretary, Bevin had to give up the attempt in face of the opposition of his permanent officials who declared that he was impossible to work with. As a way out of the difficulty, Bevin hit on the idea of recommending Beveridge to Greenwood for an inquiry into social insurance. Even in 1941 Bevin was far from regarding this as unimportant, but Beveridge himself was indignant at being removed from the wartime business of the Ministry of Labour and relegated to what he described as a backwater.

The last thing Bevin or anyone else expected was that Beveridge's investigation would produce not just a technical report on social insurance, but a new declaration of human rights brought up to date for an industrial society and dealing in plain and vigorous language with some of the most controversial issues in British politics. Still less had they foreseen that by the time it came to be published, within a few days of the news of the victories in Africa, the public would be avid for just such a blueprint, at once bold and detailed, of the future they wanted to see and Beveridge himself only too willing to play the part of a prophet pointing the way to the Promised Land.

In short, it would have been hard to think of any document more calculated to embarrass a coalition Government which had managed to hold together for two years by a tacit agreement to avoid such controversial issues and now found them thrust upon it, ironically enough, by a Report issued under its own authority.

Faced with an awkward situation, the Government made matters a great deal more difficult for itself by committing every conceivable mistake in its handling of the Report. First of all, its publication was delayed, the only result of which was to attract attention and arouse suspicion. Then, after an initial outburst of enthusiasm by the Ministry of Information, the B.B.C. and the Army Bureau for Current Affairs, extraordinary measures were taken to clamp down on all official publicity, a futile prohibition in view of the un-precedented interest shown in the Report by the press of the whole world. Finally, in face of the mounting demand from all sides that the Government should declare its intentions, ministers maintained a stubborn silence for three months, the only result of which was to involve them in an angry collision with Parliament, the press and public opinion.

While nothing has so far been published about the Cabinet's reception of the Beveridge Report, there is little doubt that the Government's attitude was determined by that of the Prime Minister. Immersed in all the difficulties involved in the conduct of the war and Britain's relations with her American and Russian allies, Churchill is reported to have taken strong exception to the Report, to have refused to see its author and forbidden any government department to allow him inside its doors.

Social security was hardly a subject to kindle the Churchillian imagination but the real reason for his attitude was less the proposals in the Report than the instinctive hostility he showed towards any-thing which threatened to distract attention from the war. If he could have had his way, Churchill would have consigned Sir William Beveridge and his Report to oblivion.[1]

The position of the Labour ministers[2] was an uncomfortable one. Here was a plan which had an obvious attraction for anyone in the Labour Movement and was certain to arouse enthusiasm in the Party. On the other hand, there were all the arguments in favour of not pursuing controversial proposals to the point where they might weaken the coalition, particularly on an issue on which the Prime Minister himself was strongly opposed to any commitment to put the recommendations of the Report into effect and was certain to be supported by his Conservative colleagues. Towards Beveridge himself the Labour ministers were hardly in a more friendly mood than the Prime Minister himself. At a time when they had imposed a self-denying ordinance on themselves in the interests of national unity, he had stolen the Party's thunder and allowed himself to be built up into a popular hero who was prepared to give the common man what the politicians denied him.[3]

---

[1] He did the next best thing by omitting any mention of either from his Memoirs. In Appendix F to Vol. IV, however, he reprinted a paper circulated to the Cabinet on the day he left for the Casablanca Conference (12 January 1943), several weeks after the Beveridge Report was published. In this the Prime Minister wrote: "A dangerous optimism is growing up about the conditions it will be possible to establish here after the war." He went on to list the plans for social improvement which were being pressed—the abolition of unemployment and low wages, better and longer education, great schemes for housing and health, the abolition of want—all without raising the cost of living or reducing the value of money. "Our foreign investments have almost disappeared. The United States will be a strong competitor with British shipping. We shall have great difficulties in planning our exports profitably." To this had to be added other burdens which could not be avoided: help in the reconstruction of Europe, greater aid to the colonies, large military forces to occupy the enemy countries. "The question steals across the mind whether we are not committing our 45 million people to tasks beyond their compass. . . . "It is because I do not wish to deceive the people by false hopes and airy visions of Utopia and Eldorado that I have refrained so far from making promises about the future. We shall do much better if we are not hampered by a cloud of pledges and promises which arise out of the hopeful and genial side of man's nature and are not brought into relation with the hard facts of life" (Vol. IV, pp. 861–2).

[2] Morrison had joined Attlee and Bevin in the War Cabinet in November 1942.

[3] Attlee remarked that "Beveridge seemed to think the war ought to stop while his plan was put into effect". When Francis Williams asked him if Beveridge was right in claiming that it was because his plan got such a response that the War Cabinet turned against it, Attlee replied: "He is a little bit elevated there, I think. He seemed to imagine he was going to be a leader of the nation or of the House of Commons. Always a mistake to think yourself larger than you are." Francis Williams: *A Prime Minister Remembers*, p. 57.

Whatever Ministers' feelings about its author, however, there was no doubt, in view of the public's reception of the Report, that something would have to be done about it; it could not simply be pigeon-holed. As a way out of their difficulties, Bevin proposed that the Report should be handed over to a committee for detailed examination. This was a not unreasonable course and Bevin later maintained, against the critics, that no Government could have been expected to swallow so far-reaching a plan without looking carefully at its recommendations and their cost. In fact the committee reached a very considerable measure of agreement. Its report recommended that all three of the Assumptions and five of the six Principles on which the Report was written should be accepted, together with sixteen of Beveridge's twenty-three Recommendations. Only one was rejected (the conversion of industrial assurance into a public utility) and six more (including those on workmen's compensation and widows' pensions) reserved for further discussion.

There does not appear to have been any disagreement in the Cabinet when the Committee's report was presented in February (1943): Churchill's attitude towards the proposals (if not towards their author) seems to have become more favourable. "I think," Attlee told Francis Williams, "that Winston planned to come in as the first post-war Prime Minister and thought it would be a nice thing to have the Beveridge Plan to put through as an act of his Government. He didn't want it done by the wartime coalition."[1] In a note dated 14 February, the eve of the debate which the Government had arranged in the House of Commons, the Prime Minister underlined the importance of seeing that the scheme to be produced "should be an integral conception and not merely what is left after the critics have pulled out certain weak points". He proposed that a Commission should be set up to "work from now until the end of the war, polishing, reshaping and preparing for the necessary legislation". But he would not agree to initiate the legislation or commit the Government to the expenditures involved before the war was over.

Churchill gave two reasons for his attitude. The first was uncertainty about the country's economic position at the end of the war, and the choice which might have to be made between social insurance and other urgent claims on limited resources.

His second reason was that a positive commitment to carry out the

[1] Ibid., p. 57.

Beveridge plan could only be undertaken by a Government and House of Commons "refreshed by contact with the people".

"We must not forget that we are a Parliament in the eighth year and we have been justified in prolonging our existence only by the physical fact of the war situation and for the purposes of the war. We have no right whatever to tie the hands of future Parliments in regard to social matters which are their proper province. I could not," he added, "as Prime Minister be responsible at this stage for binding my successor, whoever he may be, without knowledge of the conditions under which he will undertake his responsibilities."[1]

There is no evidence that Bevin or any other of the Labour ministers disagreed with these arguments or thought it possible to go further as long as the war lasted.

3

This then was the compromise with which the Government met the House of Commons: acceptance of three-quarters of the Beveridge plan in principle, a start on the work of preparing legislation, postponement of a decision on implementing it until the end of the war. To the enthusiasts who wanted to see a Ministry of Social Security created at once, this was bound to appear inadequate, but it could well have been presented to the House—as it appeared to the Labour ministers in the Cabinet—as the best that could be done by a coalition Government in the middle of a war. The Government, however, having already made a series of blunders in its initial handling of the Report, now capped them all by the maladroit way in which it presented its case to Parliament.

The long time, three months, which had elapsed between publication of the Report and a government announcement made it essential that, when a statement was made, it should be forthright, not evasive, in character. Churchill, however, was ill and the two Government spokesmen, Anderson and the Chancellor of the Exchequer, Kingsley Wood, were the worst possible choices in his place. Instead of announcing right at the outset of the debate[2] that the Government accepted the greater part of the Report and then and

---

[1] Churchill, Vol. IV, Appendix F., p. 862.
[2] The debate took place on 16–18 February 1943.

there stating frankly the limits which it felt necessary to impose on its own power of action as a wartime coalition, both the Government spokesmen left the impression that ministers were lukewarm, if not actually opposed to Beveridge's proposals. After a conventional tribute, Anderson plunged at once into detail, ignoring the broad principles of the Report (which the Government in fact accepted), underlining the tentative nature of the Cabinet's conclusions and enlarging on the difficulties in the way of implementing them. It was a speech devoid of imagination, reducing the bold outlines of Beveridge's scheme to the level of administrative detail and failing to make plain the reasons for the Government's refusal to put it into effect at once. The result was to exasperate those who wanted a firm commitment to action then and there without rallying those who would have recognised the force of Churchill's argument in his note to the Cabinet.

After the first day's debate, it was only with difficulty that Attlee, Bevin and Morrison dissuaded the Parliamentary Labour Party from supporting an amendment rejecting Anderson's statement as inadequate, and after Kingsley Wood's speech dealing with the financial implications of the Report on the second day of the debate there was open revolt. No doubt the Chancellor was right to point out the uncertainties of the country's economic position after the war, but the satisfaction with which he appeared to do so sounded suspiciously like the sort of reasoning which, before the war, had dismissed attempts to deal with unemployment as financially unsound.

Some forty of the younger members of the Conservative Party led by Quintin Hogg and Lord Hinchingbrooke threatened to go into the opposition lobby, and with the Labour Party up in arms, the Government faced an adverse vote as high as 150–180. Angered by the criticism they had encountered, the members of the Cabinet decided to stand firm and assert their authority. A three-line whip was hurriedly issued and, while Anderson talked to the rebellious Tories (Churchill was still absent ill), Attlee and Bevin tackled the Parliamentary Labour Party.

Both sides were exasperated and Bevin's vehemence in expressing himself did more harm than good. Accusing the Parliamentary Party of disloyalty, he stood by the statements made in the House and declared that, if the Party pursued the amendment which had now been tabled, it would amount to a vote of censure on the Labour ministers, a warning which was generally taken as a threat to resign. The meeting showed its resentment at Bevin's outburst by voting, with only three dissentients, to support the amendment.

When the debate was resumed on the third day, Morrison was put up to make the speech which, if it had been made at the beginning, might have carried the House with him. In a brilliant performance, reversing the emphasis of the previous speakers, he laid stress on how far the Government had gone towards accepting the Report. But the Labour Party had gone too far to draw back and even Morrison's skill as a parliamentarian could not remove the impression left by Anderson and Kingsley Wood. The Conservative revolt had been halted and the Government was assured of a comfortable majority. But many who voted for it did so reluctantly and 119 votes were recorded on the other side. The veteran Lloyd George, arriving specially for the occasion, led a small group of Liberals into the opposition lobby. Of the twenty-three Labour members who voted for the Government, twenty-two were ministers; ninety-seven voted against and some thirty more abstained.

The next day the press took the Government to task for its mishandling of a crisis which should never, *The Times* declared, have arisen. "The effect on the Government's reputation at home and abroad," the *Guardian* wrote, "has been highly damaging. Ministers have received virtually no credit for their proposals and the debate was turned into a melancholy series of exhortations and confessions of disappointment."[1] At the end of the three-day debate, as *The Economist* tartly remarked, nobody knew any better than at the beginning what the Government really meant to do about the Beveridge Report: "for all its fair words the Government lacks the courage either to accept or refuse the Beveridge plans."[2] The result of the debate was the opposite of that which Churchill had sought: to strengthen the demand for a clear statement of the Government's intentions not only on social insurance but on the whole question of post-war planning.

While the Government recovered from the hammering it had been given in Parliament and the press, Labour ministers had their own house to put in order. The Labour vote against the Government had not been the spontaneous revolt of a minority but a formal decision of

[1] *Manchester Guardian*, 19 February 1943.
[2] *The Economist*, 20 February 1943.

the Parliamentary Party, and could not be passed over in silence. On the Monday after the Beveridge debate (22 February 1943), Attlee called a meeting of all the Labour ministers. Everybody expected that there would have to be some straight speaking when they met the Parliamentary Party, but Bevin's view of the matter surprised his colleagues.

According to the Standing Orders of the Parliamentary Party, he declared, he had broken the rules by voting for the Government and against the Party's decision. Since this was the case, he demanded to be expelled from the Party or, if not expelled, then publicly acquitted.

After having his say, Bevin left abruptly, not attempting to conceal his anger. Nobody seems to have known what to make of his statement. Obviously one minister could not be expelled without all the other ministers who voted with the Government being expelled at the same time; equally clearly, public acquittal of Bevin would have to be accompanied by acquittal of all the Labour ministers. Either course would make the Parliamentary Party look ridiculous.

The next day the Administrative Committee of the Parliamentary Party met to prepare its defence. It claimed that it had done no more than Sir Douglas Hacking and his Conservative group when they voted against the Catering Wages Bill less than a fortnight before. If Tories were to be allowed this freedom, why should Labour members be denied it? The amendment had neither been framed nor intended as a vote of censure. The Party knew perfectly well that Labour ministers were pressing as hard as they could inside the Government; the action of the Parliamentary Party was intended to apply extra pressure from outside.

This defence was sufficiently ingenious to get over an awkward patch and to prevent any disciplinary action being taken. The National Council of Labour, the Parliamentary Labour Party and the National Executive of the Labour Party all met on the 23rd and 24th, listened to what was said on both sides and did nothing. The National Council and the National Executive re-affirmed their earlier support of the Beveridge proposals but left it at that. Bevin was a member of neither body and stayed away from the meeting of the Parliamentary Party, but by the time Labour ministers met again on the 25th, he was in a calmer frame of mind and agreed to let the matter drop.

The dispute over the Beveridge debate, however, had precipitated a quarrel between Bevin and a group of Labour M.P.s which had been brewing for some time and led to an estrangement between him and the Parliamentary Labour Party lasting until the summer of 1944. The quarrel, on Bevin's side at least, sprang less from disagreement over policy than from the claim by this unofficial opposition to criticise or vote against the coalition, despite the commitment the Party had accepted to support it. Bevin was not alone in resenting this. As early as February 1941, for instance, Chuter Ede, then Parliamentary Secretary to the President of the Board of Education, wrote to Attlee as leader of the Party protesting at a speech by Nye Bevan.

"It is quite clear that a minority of the Party are trying to recreate the position of 1929 and 1931 when those members of the Party who were in the Government were treated as if they had no real connexion with the main body. . . . Unless some steps are taken by the competent authority to enforce loyalty to Party meeting decisions we must face a steady disintegration of Party discipline."[1]

Anybody who reads the wartime debates in the House of Commons must be struck by the frequency (and sometimes the venom) with which a group among the members of the Labour Party repudiated their representatives in the coalition and accused them of abandoning their socialist principles and preparing to repeat MacDonald's "betrayal" of the Party in 1931. Attlee and the other Labour ministers who had grown up in the Labour Party and were accustomed to its looseness of discipline bore with these attacks as best they could, but Bevin, coming for the first time into the world of parliamentary politics, found it hard to suffer in silence.

He objected to them not only for personal reasons but because they offended against the principle to which he attached more importance than any other, that once a party or a trade union or a Government had accepted an obligation by proper majority decision its individual members were bound by it. In his trade-union days Bevin had fought battle after battle for this principle, and he could not see why it should not apply in politics as well: respect for collective decisions seemed to him the foundation of democracy. Once the Labour Party, he argued, had made an agreement to join a coalition Government and put aside party politics for the duration of the war, the commitment ought to be

[1] Chuter Ede to Attlee, 14 February 1941.

observed by all its members and not simply by those who had accepted office.

This was not a view which any party, least of all one with the traditions of the Labour Party, was ever likely to accept and Bevin's attempt to press it at the time of the Beveridge debate got little support. This disagreement on an issue to which he attached great importance affected Bevin's relations with the Party for over a year.[1] While he remained on friendly terms with Attlee and other individual ministers and M.P.s, he withdrew himself from contact with the Parliamentary Party, declaring angrily that he would have nothing to do with "playing the party game" in the coalition and going his own independent way. In the course of the dispute over the Beveridge debate he is reported to have declared that, if he were expelled from the Party, he would continue to act as Minister of Labour, having entered the coalition in the first place as a representative of the trade unions, not of the Labour Party. This mood lasted until 1944. He took no part in the 1943 Party Conference: and there is no record of his having attended any of the meetings of the parliamentary group between the row over Beveridge and the renewed row over Regulation 1AA in May 1944.[2] In February of that year he was conspicuously absent from a three-day conference of Labour ministers, the Administrative Committee of the Parliamentary Party and the National Executive called to discuss party policy.[3] It was only in the last twelve months of the war that the breach was healed and Bevin was willing to take his proper place as one of the leaders of the Labour Party as well as a member of the coalition Government.

[1] Amongst Bevin's papers is a letter from the Labour Chief Whip, William Whiteley, dated 20 December 1943, in which Whiteley attempts to overcome Bevin's objections by a number of suggestions "for regulating party conduct during the coalition".

[2] See below, pp. 303–9.

[3] 25–27 February 1944. 'Critic' wrote in the *New Statesman* on 9 March 1944: "The most disquieting thing about a conference which seems to have been generally very much on the right lines was the absence of Ernest Bevin who sent a letter just saying that for various reasons he could not attend. He is said still to be disgusted with the Party about Beveridge and other matters on which he has differed with his colleagues. It remains one of the chief problems of the Labour Movement that its three most influential figures—Bevin, Morrison and Citrine—so seldom contrive to see eye to eye about anything."

4

The uproar over the Beveridge debate convinced Churchill that he could not avoid saying something about post-war plans if the Government was not to suffer a serious loss in its authority.

His broadcast to the nation on 21 March 1943 was a skilful performance. He began with the frank admission that his purpose was to damp down party controversy in order to concentrate on winning the war, but he went on to outline a "four-year plan" for economic recovery and social reform which made a strong appeal to moderate opinion. Preparations were to be put in hand and preliminary legislation, if necessary, introduced before the end of the war: the plans, however, would only be put into effect and the four-year period begin when the war was over.

The Economist wrote of the relief with which the Prime Minister's "conversion" was heard. "His reluctance to discuss the problems of peace remains, but it is no longer a refusal. On Sunday, he stated the agenda for peacetime planning."[1] Others were more critical, especially at the postponement of action until after the war. But Churchill had successfully redressed his line without abandoning it. He met the criticism that the Government showed too little concern for post-war problems and was unwilling to give a lead, without burdening the coalition with the task of undertaking a major programme of domestic reforms while still engaged in defeating the enemy.

For the rest of the war Churchill's view determined the coalition's policy. There was much discussion of post-war problems, much writing of reports and preparatory planning, but few specific measures of reform requiring legislation were carried (the most important was the Education Act of 1944) and fewer still put into effect before the end of the war. The first and perhaps the most controversial exception remains the Catering Wages Act.

The week after Churchill's broadcast, the Bill reached the committee stage. Sir Douglas Hacking and his friends fought it line by line. Though their numbers were reduced to seventy in a division, after two full days of debate only four clauses had been passed, with fourteen more and two schedules still to be taken. No other Govern-

[1] *The Economist*, 27 March 1943.

ment measure during the war had been subjected to such treatment; but if his opponents hoped to provoke Bevin into another outburst they failed completely. Never before had he shown such complete mastery of his case and of his temper.

When the opposition tried to revive the charge that he was treating catering as a sweated industry, Bevin turned their attack to his advantage. So far from this being the case, he replied:

"I rest my case on the entitlement of every worker to have his basic conditions protected either by collective agreement or by State regulation . . . I stand on this, that where ordinary persons have to apply for employment in the labour market, they are entitled to a foundation in their wages system. May I put it another way—that what is good wages in one generation is bad wages in the next decade. The great thing, if evolution and progress are to be made, is for the workpeople to have that sort of organisation which will provide adaptation and development with the progress of our civilisation and our industrial development."[1]

On the third day Sir Douglas Hacking unexpectedly threw up the fight, which was clearly not going to defeat the Bill, and asked leave to withdraw the amendments which he and his committee had tabled. Other Conservatives, however, were critical of Hacking's action (which suggested that he was interested only in stopping the Bill, not improving it) and maintained the debate for two more days.

Now that he was within sight of victory, Bevin remained as patient and reasonable as he had shown himself from the beginning. The committee procedure allowed him to display more fully than ordinary sessions of the House the skill which had won him fame as a negotiator and he was adroit in making concessions—such as his suggestion that the Commission should have the power to call in expert assessors—which met objections without impairing the effectiveness of the Bill.

By the time the Third Reading was reached on 20 April, the atmosphere of the House had changed completely. Many who had opposed or been suspicious of Bevin's proposals were won over and he received congratulations from all quarters.

"Political opinion," *The Times* wrote, "underwent a remarkable change during the passage of the Catering Wages Bill. . . . Threatened at the outset with uncompromising opposition, it won first the acceptance, then the open

[1] House of Commons, 25 March 1943, Hansard, Vol. 387, cols. 1857-8.

approval of early critics. The original alarm over apprehended encroachments on private rights yielded to general recognition that the proposals would further the interests not only of the men and women employed, but of the industry as a whole and of the public. . . ."[1]

Several other projects in which Bevin was interested reached the table of the House of Commons in 1943. One was the Foreign Service Reform Bill, in the preparation of which Eden had frequently consulted him; another, the White Paper on Training for the Building Industry, was the product of discussions between Government, employers and unions, which the Minister of Labour and the Minister of Works jointly laid before the House in May.[2]

But the Bill which most excited Bevin's interest was the one to become known as the Education Act of 1944. According to Chuter Ede, the Parliamentary Secretary who shared the responsibility for the Bill with the President of the Board of Education, R. A. Butler, Bevin contributed more to it than anyone else outside the Department. Chuter Ede found him ready to talk over the detailed provisions and give advice at any time; he was full of suggestions of his own and when the Bill came up for discussion in the Cabinet gave powerful support, without which, Ede believed, it was unlikely to have been passed during the war.

The idea of widening the opportunities which had been open to his generation, so that secondary as well as primary education should be accessible to all, strongly attracted Bevin. He wanted to see as many as possible continue their schooling until eighteen and then go on to a university. The fact that by calling up young people he had forced many to interrupt, or lose their chance of, a university education was very much in his mind, and he helped to devise the Further Education and Training Scheme (jointly administered by the Ministry of Labour and the Board of Education) which provided grants after the war for those who had reached the required standard to go to university or be trained for a professional career.

He was just as interested in those who left school at the minimum

---

[1] *The Times*, 9 July 1943.

[2] Cmd. (1943) 6428. Its main proposal was to set up a Building Apprenticeship Training Council which would represent all the interests concerned and make a serious attempt to train the future craftsmen the industry would need to meet the post-war demands on it. The White Paper won approval from everyone (including the Prime Minister), and the Council met for the first time in June under the chairmanship of Sir Malcolm Trustram Eve.

age and pressed Butler to include in the Act provision for continuation courses up to eighteen.

"I regard this," he told the T.U.C., "as absolutely vital . . . I want to see the conception of parental responsibility go up to eighteen and not stop at fourteen. The fourteener goes into life with a terrible handicap against the secondary and public-school boy . . . and we must save that situation."[1]

This proposal was linked, on the one side, with his campaign to secure better training and technical education for boys entering industry (the first fruit of which was the building apprenticeship scheme), and on the other with the measures to be described in the next chapter for creating a service which could give boys and girls proper advice about jobs.

Bevin's attention was attracted to another social problem, domestic help, by the evidence which the Ministry accumulated in the course of interviews with several million women. In a note of 2 January 1943 he wrote:

"I am constantly receiving reports that, owing to the intensity of the call-up, great difficulties are arising, both where there is illness and where children have to be cared for while their mothers are at work. Also the morale of our people is being affected by the many cases of women who work long hours and who have to do their own housework in addition.
"I have been turning over in my mind whether at this stage of the war we could organise a collective service which would ease the problem."

Would something on the lines of district nursing be the right direction in which to look? There must be some way, Bevin reasoned, of providing domestic help, freed of the stigma of servility, in homes where it was most needed, not simply where people had the means to pay for it. The other demands on womanpower made it difficult to organise a workable scheme in wartime, but Bevin did not give up. In March 1944 he asked Violet Markham and the national woman officer of the T.G.W.U., Florence Hancock, to look at the problem, giving them a lot of his own ideas to start with and asking them to look at the future of domestic service in institutions as well as private households. The best plan he could devise for the moment was to give Local Authorities power to supply "home-helps" in maternity cases or where a mother of children under five was ill, and in November 1944

[1] Speech to the T.U.C. at Southport, 6 September 1943.

this was extended to provide domestic help to other households where there was hardship caused by sickness and emergency.

He made more progress with the problem of domestic staff in hospitals, where at one time beds had to be closed because of the difficulty of finding kitchen workers and cleaners. The hospital world had old-fashioned ideas about the pay, hours and conditions of staff, and Bevin was averse to using his powers of direction until something was done to improve them. In 1943 he set up a committee under Sir Hector Hetherington to inquire into the wages and conditions of domestic workers in institutions, and when the committee reported, directed women to work in those hospitals which were prepared to adopt its recommendations. A Standing Advisory Committee, Bevin's familiar device, was created to keep the position under review.

The same year, 1943, also brought the Minister of Labour in to find an answer to the shortage of nurses. The opposition to proper pay and conditions had broken down in the course of the war. Two committees under Lord Rushcliffe recommended new salary scales for midwives and nurses, and the Government provided a subsidy to implement them; the Nurses Act settled the status of assistant nurses and between June 1941 and May 1943 the nursing staff in hospitals had been increased from 89,000 to 93,000. But there were still 12,000 vacancies and early in 1943 the responsibility for filling them was transferred from the Ministry of Health to the Ministry of Labour.

Nursing was a difficult profession in which to apply compulsion, if only because some sense of vocation was essential, and at no time were women directed into nursing. The problem was how to make the best use of those who had already taken it up as a career. Bevin took care to set up a National Advisory Council to advise him on how best to set about it. He began by registering every woman with nursing experience and establishing clear priorities for the kinds of nursing—in T.B. wards and mental hospitals, for infectious cases and the chronic sick—in which the shortage was most acute. In September 1943 he applied the Control of Engagement Order so that appointments could only be made through Ministry of Labour offices, placed thirty-one nursing appointment officers in the principal towns and gave each of them a local advisory committee. The following spring he began to direct newly qualified nurses to one of the priority fields, in face of vigorous opposition from the teaching hospitals. These steps were accompanied by a big campaign to attract volunteers and, while they did not end the shortage of nurses, were sufficiently successful to raise the numbers from 93 to 98 thousand between May 1942 and May 1944.

Bevin still felt that the importance of health in industry and the effect of illness on production were not properly recognised. Even before the war, the British Medical Association estimated that 31 million working *weeks* were lost to industry each year through sickness, and the war had raised the figure sharply. The *Daily Express*, backing Bevin's efforts to draw attention to the scale of the problem, calculated that for every day's output lost by strikes about four *months'* working time was lost through illness.[1]

In March 1943 Bevin appointed a strong Industrial Health Advisory Committee with twenty doctors, industrialists and trade unionists as members and himself as chairman. In April he convened a three-day conference on industrial health at the Caxton Hall, inviting everyone concerned with the subject to attend and backing it with the support of a powerful government representation. He presided over the opening session at which the Minister of Health also spoke; his two parliamentary secretaries and the Permanent Secretary of the Ministry of Labour were enlisted as chairmen of the other meetings and Bevin turned up again to speak at the end of the conference. No less important was the impressive list of experts he assembled to speak and the care he took to see that the press was briefed to report the whole proceedings prominently. His object was to put industrial health on the map and to give it as wide an interpretation as possible, covering not only factory medical and nursing services, but medical research, the design of buildings and machinery, communal feeding and personnel management. This was a logical extension, Bevin argued, of the protective legislation which had begun with the Factory Acts and he did everything he could think of to give this side of his Ministry's work a decisive push forward.

5

Bevin meant what he said when he told the Labour Party that the Government would use the time before the end of the war to work out the details of a comprehensive plan for social insurance along the lines recommended by Beveridge and fit this into the other measures for

[1] *Daily Express*, 13 March 1943.

social security which were being planned. In April 1943 he put up a number of ideas to his staff on unemployment relief.[1] Every discussion of this question, he pointed out, recognised that there was a limit to the weight of unemployment which the Insurance Fund could carry, and safeguards had to be provided to prevent the Fund becoming over-strained. That was the reason for the Means Test.

"There you had a clear recognition that if unemployment reached, either in numbers or duration, a certain point, then the problem had to be dealt with distinctly, and it was met by the introduction of the Household Means Test as if the workman was to blame for unemployment having reached that point."

This was a mistaken way of tackling the problem.

"The unemployment period,[2] with a means test and all the other checks against so-called abuse and maintaining the finances of the Fund, created in the minds of our people a kind of 'cleverness'. They felt they had to 'beat the State' and hence the whole spirit of the administration with all its conflict grew up."

He suggested a new approach. So long as unemployment did not exceed 8 per cent (a figure proposed by Beveridge), it could be taken as largely due to the normal turnover of labour and provided for by insurance without special safeguards. If it rose above 8 per cent, Bevin suggested, the way to prevent it crippling the Insurance Fund was, not by imposing checks such as a limited period of relief or the Household Means Test, but by recognising that a new situation of mass un-employment had developed, calling for emergency action by the State. At that point no further attempt should be made to meet the cost of unemployment relief by insurance; instead, the State should use other means to provide work and stimulate employment.

Bevin asked his Permanent Secretary to get out a draft scheme for unemployment relief along these lines, so that they could look at it more closely. Apart from its intrinsic interest, Bevin's proposal illustrates his attitude towards the Beveridge Report. While not questioning the principle of a comprehensive scheme such as Beveridge had suggested, this was the sort of improvement which he believed could be made on a number of points if they were subjected to a careful examination.

[1] They are contained in two notes dictated by the Minister on 3 and 12 April 1943.
[2] The limitation of benefit to twenty-six weeks.

It was in these terms that Bevin defended the Government's treatment of the Report when he went north to address the Scottish T.U.C. in Aberdeen at the end of April. Describing it as "the culmination of ideas on social services over the last forty years", he warned his audience not to think that the Report by itself represented social security. That was a wider thing, involving continuity of employment, wage standards, housing and much else.

"What it really is is a co-ordination of the whole of the nation's ambulance services on a more scientific and proper footing.
"We not only did not reject it," he declared, "we examined it with speed. I have had a good deal of experience of Government Departments and I assert that at no time has a report received more urgent consideration and examination than this one has. But it has to fall into its proper place with all the other things that I have enumerated . . . It is easy to adopt a thing in principle and leave somebody else to draft the Bill, but what we are doing is to bring the whole of this thing together and try to fit it into one blue-print or plan which can give it legislative acceptance."[1]

For ninety minutes the crowded audience of trade unionists listened intently as Bevin surveyed not only the task of mobilisation and the changes which he had made as Minister of Labour, but the steps the Government was taking to prepare for the post-war period. Reminding them that he had carried through a Bill to restore everything given up by the trade unions when the war was over, he asked them to consider carefully whether some of the changes introduced during the war were not of more value to the workpeople they represented than those which they had given up. If so, they might be well advised not to press for an indiscriminate restoration of pre-war practices.

Even more important was the need to think about the difficult period of transition from war to peace:

"It is clear that there will have to be at the end of the war the continuance of control, and it will depend to a very large extent on what part the unions are prepared to play whether or not the nation will get a breathing space in order to lay the foundations of a sound economic and political policy.
"For example, there will be a great shortage of materials and goods at home. If rationing in some form is not continued, prices will rise, those with most wealth will be able to buy up goods and the stabilisation we have striven so hard to maintain during the war will be lost. . . .

[1] Address to the 46th Annual Scottish T.U.C. at Aberdeen, 30 April 1943.

"It must not be forgotten that if we are to rebuild our export trade—and it is absolutely vital to us . . . we cannot allow our goods to be absorbed in the home market at high prices, leading possibly to an orgy of speculation at home such as followed on the last war. A proportion must go for export, with a rationing system which will help to rehabilitate the rest of the world as well as secure fair distribution at home and restore international trade."

This was a necessary reminder of the difficulties the post-war period would bring and Bevin laid stress on the responsibility of the unions, if they were to avoid a repetition of what happened in 1918–26.

"As regards wages, if we can pursue the stabilisation policy into the post-war period with as much success as during the war, we can avoid inflation and, what is more important, deflation. This is not a question of political tactics and we must see that irresponsible people do not make it one. What we have to decide is whether we can accept such political responsibility for that period as will prevent a collapse and a return to the misery we had between the two wars, with the possible danger of being led into a third."

Bevin was always at his best when he was addressing an audience of trade unionists. He had nothing like the same success when he had to deal with the Parliamentary Labour Party. In May there was another row over pensions. The Bill which Bevin had promised the previous summer to clear up a number of anomalies again angered those who had hoped for more generous treatment, especially of old-age pensioners. Ness Edwards declared that the Labour Party had been cheated and fobbed off, placing the blame for this on Bevin. Nye Bevan told him that if the Labour members of the Government could not bring the Government's policy more into line with Labour's wishes, it would be better to break up the coalition. "A Labour Opposition would serve the interests of the nation far better than for Labour ministers to be hostages" to a stubborn and reactionary majority.[1]

Answering for the Government, Bevin was clearly not happy about the case he had to make. This time he tried to placate the critics, with not much more success than when he had been fierce with them the year before: sixty-one of them, refusing to listen to Bevin's plea, voted against the Government.

Only two days before the debate Bevin had written a strong letter to the Chancellor of the Exechequer, saying that it was "nauseating

[1] House of Commons, 20 May 1943, Hansard, Vol. 389, col. 1341.

I

to have these debates over the question of our old people" and urging him to get rid of the controversy by a generous gesture over supplementary pensions, the cost of which was out of all proportion to the political controversy which the question roused. But the Treasury stuck its heels in and Bevin was left to put up the best case he could in the Commons debate and carry the odium for failing to secure more.

Even on an issue about which he felt as strongly as the repeal of the 1927 Trade Disputes and Trade Unions Act,[1] Bevin and the other Labour ministers had to accept as final the Prime Minister's refusal to consider it during the war. The question of repeal "as a gesture of national unity" had first been raised in September 1940. It was turned down by Churchill in April 1941 on the grounds that any proposal to repeal or even alter the Act during the war would stir up controversy. Citrine raised the matter again in April 1943 but with no more success than before. Neither the Labour ministers from inside the Cabinet nor the T.U.C. from outside could persuade the Prime Minister to change his mind. The only result of talks between Labour ministers and the General Council was to leave both sides angry, Citrine accusing Bevin and the others of failing to press hard enough and of failing to give the T.U.C. as much information as they got from Conservative ministers. The Act was not repealed until after the war.

When the Labour Party conference was held at Whitsuntide, 1943, Attlee succeeded in getting a vote for a continuation of the electoral truce, on the grounds that to break it would mean the end of the coalition and deprive Labour of any chance to influence post-war planning. The majority was a convincing one, 2,243,000 to 374,000. But the debate made clear the frustration and anxiety in the Party: frustration at their inability to take advantage of the swing of public opinion towards change, at least as long as the coalition lasted; anxiety lest the Labour ministers should agree to a continuation of the coalition after the war, to a National Government, a Khaki Election or some other Tory trick which would again rob Labour of victory.

Bevin took no part in the Conference, which was perhaps as well in view of the feeling between him and some of the members of the Parliamentary Party. He limited himself to using his influence with the union leaders to get a solid vote in support of Attlee and to making

---

[1] This was the Act passed by the Baldwin Government (in which Churchill was Chancellor of the Exchequer) after the General Strike.

sure that the T.G.W.U. vote was swung against Morrison in the contest for the Party Treasurership. But when the conference was over he wrote an interesting private report on it at the request of the American Ambassador, John Winant.[1]

Characteristically, Bevin began by expounding the role of the trade unions as the trustees of the Labour Movement.

"When agitation by certain sectional interests is going on, [the unions are] particularly tolerant and apparently quiescent until they become conscious that the fabric or the constitution of the Party is being seriously challenged."

When that happened, the unions acted more by instinct than by premeditation, but their intervention was decisive. An example of this had been the motion to allow the Communists to affiliate to the Labour Party, a motion defeated (despite the support of the miners) by a large majority with all the unions, except the miners', throwing their weight behind the platform.

A still more important issue on which the unions had intervened decisively was the continuation of the coalition.

"Again there was a sectional group of the Party who had carried on for some months wrecking tactics with the object of breaking up national unity at this critical moment. . . . The unions' common sense was brought to bear and their action undoubtedly interpreted correctly the feeling of their members all over the country that it would be just suicide to do anything of the kind."

"Now comes the next point," Bevin continued. "Will those in authority in the Party misinterpret this decision?" The Party was naturally anxious that no one, out of love of office or because of his influence in the Party, should be able to commit them at the end of the war without a special conference being summoned to decide the question,

"Whether they would continue in some form of coalition, possibly not this one, to see us through the transition period, or alternatively whether they would restore the position and fight the election purely as a party."

Bevin left no doubt that he thought Attlee had been right to give a pledge that it would not be committed to coalition after the war without consultation. Nor did he conceal the real anxiety which had

[1] Bevin to Winant, 23 June 1943.

been expressed again and again during the conference over post-war measures.

"Did the people who were talking reconstruction really mean what they said? ... Were the discussions with the U.S.A. and Russia proceeding fast enough? Were we, in addition to the fighting, really grappling with these problems or were we going to be carried back into the old pre-war unemployment? ..."

So far as domestic affairs were concerned, there was insistence on social security, not only in implementing the Beveridge Report but providing guarantees against unemployment.

"I think it can be taken," Bevin added, "that the expressions on these matters at the Conference were merely a re-echoing of what we may now regard as the social consciousness of the nation."

And he ended his letter by referring once more to

"the keen and awakened mentality which is watching every step the Government takes in order to secure for the common man a better deal than he has hitherto enjoyed."

6

As soon as the House of Commons had reassembled after the Christmas recess, a two-day debate on manpower had been held in secret session (20–21 January 1943). This gave Bevin greater freedom than he had enjoyed in previous debates to set out the facts which must henceforth govern manpower policy. Since the shortage was due to the high degree of mobilisation already achieved, there was little ground for criticising the Minister of Labour's performance of his duties and the House seems to have accepted as logical Bevin's argument that, now the untapped reserves of manpower were drying up, the needs of new production programmes must largely be met by transferring men and women already employed on war industry.

There was a further debate on 23 February 1943 in which the Labour Party pressed for welfare measures to ease the hardship of transfer. Particular criticisms, however, were combined with warm praise for Bevin's early recognition of the importance of welfare. The opposition came from Scots and Ulster members who disliked any policy which meant transferring workers from their areas and monopolised most of the debate with their complaints.[1]

No serious criticism of Bevin's handling of manpower and labour problems in fact was heard before July 1943 and in his broadcast of 21 March Churchill singled out the Minister of Labour for special recognition.

"Mr. Bevin," he said, "is attacked from time to time, now from one side, now another. When I think of the tremendous changes which have been effected under the strain of war in the lives of the whole people, of both sexes and of every class, with so little friction, and when I consider the practical absence of strikes in the war compared to what happened in the last, I think he will be able to take it all right."[2]

The press reported with pride Bevin's claim that "Britain's manpower is the best organised in the world"[3], and made a pointed comparison with the United States where a much lower degree of mobilisation had been accompanied by much more widespread labour troubles.

"Part of the difficulty," *The Economist* commented, "can be briefly expressed by saying that America has no Ernest Bevin. A labour policy as drastic as that in force in Great Britain can be carried through only by a man whom the rank and file of Labour—all of them—trust."[4]

. If Bevin ever bothered to look back at the earlier criticisms with which the press (not least *The Economist*) and his critics in Parliament had berated him for his lack of a manpower policy, he must have regarded the praise they now accorded rather wryly. Fortunately, even when he resented criticism as unfair, Bevin had never been much moved by what either press or Parliament thought of his policy: he was the better prepared, therefore, to discover before the end of 1943

[1] There was a particular concern felt in Scotland about the moral dangers to which Scots girls might be exposed in the less strict atmosphere of the Midlands and the South. To remove these anxieties Bevin invited a commission, on which the Churches as well as industrialists and trade unions were represented, to come and see for themselves the arrangements which had been made to look after the girls. The commission interviewed nearly 3,000 of them and was impressed by what had been done.

[2] *The War Speeches of Winston Churchill*, Vol. II, p. 434.

[3] Speech at Bristol, 27 March 1943.

[4] *The Economist*, 17 April 1943.

that he was no longer the paragon he appeared to be for a time in the spring of that year.

With the disappearance of the Production Executive at the end of 1941, the functions of its Manpower Committee had been taken over, not by the new Minister of Production (Oliver Lyttelton, later Lord Chandos) but by the Minister of Labour. Whatever ambitions Lyttelton or his staff may have had for extending their authority, any suggestion that they might enter the field of manpower ran up against Bevin's determination to keep control of this in his own hands. No such proposal was in fact ever seriously entertained,[1] and the procedure for determining labour policy and handling its administration remained unaffected by the creation of the Ministry of Production.[2] Substantially, this still rested on the work of three committees. At the top (until September 1943) was the Lord President's Committee which was responsible for making recommendations on the allocation of manpower to the Cabinet. After Anderson became Chancellor of the Exchequer he continued to handle these questions as chairman of a new Cabinet committee on manpower of which Bevin and Lyttelton were the other members. At the official level, the Labour Co-ordinating Committee provided a means of reaching agreement between the Ministry of Labour and the supply ministries over the whole range of labour-supply problems. The third committee, the Headquarters Preference Committee, again composed of representatives of the supply ministries meeting under the chairmanship of the Ministry of Labour, operated the system of preferences for industries or firms whose labour force it was decided to strengthen. It was through the last of these committees that the Ministry of Production was eventually brought into the allocation of labour in the later stages of the war.

As the list of preferences grew, the old difficulties over priority re-

---

[1] See Postan, p. 254: "Nobody could seriously argue that the control of manpower had been unduly dispersed and needed greater concentration than the Minister of Labour had been able to give it. In any case so weighty was Mr. Bevin's personality and so great was his authority in the War Cabinet as to place outside the realm of practical politics any project for taking away from the Ministry of Labour control of the allocation and distribution of manpower. The proposal was not in fact seriously pursued and dropped out of discussion almost at once."

[2] For a discussion of the relations between the Ministry of Production, the Ministry of Labour and the supply ministries, 1942–5, see Postan, pp. 252–69 and Scott and Hughes, cc. 20–22.

appeared and were accentuated by the super-priority given to air-craft production in 1943 (see below, pp. 250–2). In the autumn of that year the War Cabinet gave the Minister of Production the job of determining which other products should be "designated" as of equal importance with aircraft in receiving a labour preference. When the special priority for aircraft was withdrawn at the beginning of 1944, the Ministry of Production continued to designate the products which were entitled to a super preference. Instead of opposing this development, Bevin welcomed it: his Ministry worked closely with the Ministry of Production in controlling the list of designated work and his own relations with Lyttelton were friendly. With this addition, the machinery for manpower control which had been worked out in 1941 lasted until the end of the war.

The main source from which it was still possible to bring fresh supplies of labour into the war effort was the large number of women who could not be transferred because of domestic responsibilities as wives and mothers. By finding them employment within reach of their homes (often a difficult requirement to meet) it was possible to release other women who could be moved. This was a policy easy to formulate but requiring much patience and skill on the part of the Ministry's staff to translate into the industrial arrangements which were involved. And the number of immobile women who were free to take full-time jobs, even in their own districts, was not unlimited. The final resource was to make better use of those who could only take on part-time work. Hitherto the women in this category, who were already carrying the burden of running a home in wartime and looking after children, sick or elderly people, had been left to find their own part-time employment. In April 1943, however, Bevin issued an order requiring them to undertake work to which they were directed.

This was going to the limit in demands on the civilian population. The Cabinet could only rely on Bevin's judgment that it would not promote trouble and he had lengthy discussions with both the Women's Consultative Committee and the trade unions before issuing his order. No complaint was made, however, either in Parliament or the press and by midsummer 1944 the number of women in part-time employment had risen to 900,000. In the face of much scepticism Bevin was delighted to have proved that the part-time woman worker could play a role of quite unsuspected importance in maintaining war production and civilian services.

By May 1943 the Ministry of Labour was able to report on the progress made in carrying out the manpower allocations of December 1942. The figures covered the first half of the eighteen-month period July 1942–December 1943. Of the three Services, the Army and Air Force had already received over 80 per cent of their entitlement for the whole period; the Navy was behind but had so far asked for less than half of its approved intake. The industrial picture was less satisfactory. The Ministry of Supply had failed to reduce its labour force, which had expanded instead by 20,000; industries and services for which no provision had been made at all had increased their manpower by 130,000 and the Ministry of Aircraft Production was 360,000 short of its half million allocation for the full eighteen months.

Bevin had hardly presented his report to the Cabinet when the Services announced that they had underestimated the demands they would have to meet with the change from a defensive to an offensive strategy and required 314,000 more men and 59,000 more women than they had been allotted. Revised allocations for the remaining nine months to the end of 1943 would clearly have to be made and Anderson at once set to work with Bevin and Lyttelton on drafting proposals to be presented to the Prime Minister on his return from Washington.

After long discussions, Churchill gave a series of rulings which the War Cabinet accepted on 22 July 1943. The new demands of the Army and Air Force were heavily pruned: all three Services, however, received additional allocations above the figures fixed in December 1942. The most difficult problem was once again the aircraft industry. Cripps, now Minister of Aircraft Production, was prepared to reduce its demands for the remainder of the year from 359,000 to 212,000, but this he declared to be the limit in cuts, if the "realistic" programme of production he had introduced was to come anywhere near achievement. Faced with the unresolved disagreement between M.A.P. and the Ministry of Labour, Churchill tried to satisfy both by accepting Cripps's figure of 212,000 but extending it beyond the end of the period under discussion: by the end of the year Bevin was to do his utmost to find at least 115,000. To make this possible, the Ministry of Supply was instructed to run down its labour force by 185,000, Civil Defence was to give up an additional 15,000, and the proposed increase for other essential services besides munitions was cut back to the 1942 figure.[1] Even so, with demands almost halved, supply threatened to fall short of the numbers allocated by more than 50,000 men and women.

The crux of the calculations was clearly the figure for aircraft production. To help Bevin in finding the additional 115,000 he was called upon to provide, the Cabinet suspended the call-up of men from the aircraft industry, virtually ended recruitment for the Women's Services and extended registration for employment of women up to the age of fifty, hoping thereby to make easier the transfer of younger women to the aircraft factories. Every effort was to be made to plan the release of workers by the Ministry of Supply so that they could be absorbed by M.A.P. contractors; the Minister of Labour was to instruct his regional controllers to give priority to M.A.P. vacancies and to report monthly on the progress he had made in filling them. Finally, to leave no doubt of its determination to give the manufacture of aircraft precedence over every other claim, the War Cabinet laid it down that, if the supply of labour fell short of the re-allocations it had decided on, the deficit should not fall on M.A.P.

These decisions reflected Churchill's desire to increase the weight of the air attack on Germany at almost any cost. Bevin could not challenge a decision which lay outside his sphere of competence, but he was unshaken in his belief that aircraft production neither required nor could absorb so large a number of additional workers and he warned his colleagues in the Cabinet that so disproportionate an allocation of labour would have the same consequences on other branches of war production as the priority Beaverbrook had claimed in 1940.

Bevin found a simple way of making clear what he meant. He at once put the Cabinet's decision into effect—with a literal interpretation of his instructions which was well calculated to produce immediate reactions. He ordered the required priority to be given, but only to those items whose manufacture was directly controlled by M.A.P. Ball-bearings, tools and aircraft tyres, which came under the Ministry of Supply, were excluded. When this had been put right, the Ministry of Supply and the Admiralty continued to protest vigorously that they could not get labour for the most urgent items required by the Army and the Navy. The Minister of Labour was sympathetic, but regretted that he could not help: they had better complain to the

[1] See Tables 24 and 25 in Parker, pp. 206 and 209.

War Cabinet. It was left to Anderson to find a way out of the difficulty: this he did by a neat re-drafting of the original instruction[1] which satisfied honour all round. It was left to the Ministry of Labour and the Ministry of Production to agree which other products should be given equal priority with aircraft.

Having made his point, Bevin saw to it that his Ministry provided not just the minimum figure of 115,000 more workers for M.A.P. by the end of 1943 but another 50,000 on top: as a result the super-priority for aircraft manufacture could be dropped. The long controversy between the two Ministries thus ended with each side satisfied that it was vindicated. By absorbing 50,000 more in 1943 than had been allocated to it, the Ministry of Aircraft Production could claim to have proved that, without the special measures for which it asked, aircraft production would have been endangered for lack of manpower. The Minister of Labour, on the other hand, had the satisfaction of providing more labour than had appeared possible and at the same time showing that the figure of 212,000 which Cripps had stated as the irreducible minimum was well beyond the industry's real needs.

What is more striking than these inter-departmental disputes is that, with the exception of the group of essential industries and services other than munitions, not only M.A.P.'s but most of the other claims as well were substantially met by the end of 1943, each of the Services actually receiving more than the extra number of recruits allotted to them.

7

In the course of the summer of 1943 Sicily was occupied and the Italian mainland invaded, but pleasure at the overthrow of Mussolini was marred by the dispute which broke out over the Allies' deal with Badoglio. This marked another stage in the history of wartime politics. Hitherto political criticism of the coalition (as distinct from criticism of its conduct of the war) had been largely confined to

---

[1] Instead of directing the Minister of Labour to ensure that, if there was a short fall of labour, "no deficit should fall on the M.A.P.", Anderson altered the minutes to "the Minister should make every effort to avoid a situation" in which the deficit would fall on M.A.P.

domestic matters: it was now extended to foreign policy as well, to the character of the post-war settlement in Europe as well as at home. This was an issue in which Bevin was to become heavily involved before the war ended. At this time, however, he took no part in the discussion: the controversies of which he was the central figure in the second half of 1943 were still domestic, beginning with a brief but unexpected revival of opposition to his manpower policy.

At the end of July, Bevin announced in the Commons that, as part of its plans to release younger women for work in the aircraft factories, the Government intended to proceed with the registration of women up to the age of fifty-one. The direction of "grandmothers", as the popular press at once described it, roused strong opposition, particularly from the Conservative members—among them (*The Economist* noted) "many of those who welcome every opportunity of attacking the Minister of Labour".[1] The press was either hostile to the proposal or unconvinced of its necessity and in the course of August and September nearly 200 M.P.s signed a motion of protest.

No large number of women was likely to be made available for war work by the new order, but this was the first time there had been organised opposition to the registration of any age group. There were other signs of unwillingness to do more in a war that had now been going on for four years, and the Cabinet agreed with Bevin that the threatened parliamentary revolt must be met with firmness.

The debate came on towards the end of September 1943 (23rd and 24th), immediately after Churchill had defended the Government's handling of the Italian surrender against criticism from the Left. Bevin's critics were on the other side of the House but they were even less successful in persuading the Cabinet to change its policy. One reason for this was Bevin's own performance which the *Manchester Guardian* described as perhaps his best since coming to the House. "It was a clear exposition of the manpower problem with the sort of tidy marshalling of facts and arguments that one associates with highly disciplined minds like the Lord Chancellor's."[2] The opposition's appeal to emotion and vague exaggerations made a poor impression by contrast with Bevin's obvious mastery of his subject. His critics, *The Economist* remarked, "had apparently no real conception of what was going on and why"; while Bevin, by presenting the safeguards

[1] *The Economist*, 7 August 1943.
[2] *Manchester Guardian*, 25 September 1943.

that had been planned all along as a great concession, "rode off in triumph".[1]

Much was made of the dangers to health in directing women in their forties to take up employment but Bevin cut the ground from under the critics' feet by pointing out that they had made no protest when he registered nurses up to the age of sixty and cotton operatives up to fifty-five. More than a million and a half women over forty, he pointed out, were already in employment, over half a million between the ages of forty-six and fifty. "I am challenging the suggestion that there is a difference between the women who have always been to work and any other class in the country."[2]

When he made the closing speech on the second day, Bevin shifted his argument from the particular to the general. The nation was now on the eve of the decisive phase of the war, with the heaviest casualties still to come, and speaking with a sincerity of feeling which silenced criticism, Bevin reminded the House of what was at stake:

"Those who vote against complete mobilisation and full use of the manpower and the resources of this country at this moment should ask themselves what would be the result if others supported them. The result would be to carry on this war, and every minute it is being carried on, there is a greater sacrifice than calling women of 46–51 to serve in the factories, a far greater risk to the rising generation of this country, the most precious portion of the body politic. Everything you can do to turn out weapons of war helps to carry out the policy of the Government—that is, that metal is cheaper than men, and if we can use men and women in the factories to build up the most mighty equipment, the most powerful force, and with the sheer weight of that force shorten this war by a minute, or a day, or a month, then we shall be doing the most humane thing we can to end this holocaust."[3]

By the time the debate ended the House was so obviously on Bevin's side that the opposition did not press their motion to a division.

Bevin spoke with confidence because he was convinced that the opposition in the House did not represent opinion at large, certainly not among women. As a counter-blast, he organised a meeting of nearly 8,000 women (drawn from all over the country) in the Albert Hall the following week. He brought with him the most impressive representation of the Government to appear on a single platform

[1] *The Economist*, 2 October 1943.
[2] House of Commons, 23 September 1943, Hansard, Vol. 392, col. 467.
[3] House of Commons, 24 September 1943, Hansard, Vol. 392, cols. 674–5.

during the war: Churchill, the rest of the War Cabinet and fourteen other ministers. After the speeches, each minister in turn answered questions sent in beforehand, a further extension of the idea Bevin had developed in his three big industrial conferences of 1943, combining an opportunity for the Government to explain its policies with a chance for those most affected to air some of their grievances. So far as his own policy was concerned, Bevin was content to let the figures speak for themselves: out of a total population of thirty-three million between the ages of fourteen and sixty-four, twenty-two and three-quarter million were now engaged in one form or another of national service—a third of them women, without counting the million engaged in voluntary social services.

The conscription of women, and the lengths to which it was carried, remains one of the boldest acts of policy ever carried out by a demo-cratic Government—in Professor Postan's words, "a drastic act of total war, more drastic than anything done in the war of 1914–18 or anything that even Hitler could contemplate."[1] It was Bevin's greatest achievement as Minister of Labour, a policy which he not only persuaded the Cabinet to adopt in the face of opposition and much misgiving, but which he carried to success by the care he took to see that the millions of women who had to be registered and inter-viewed[2] were treated as human beings and that adequate welfare arrangements were made in time. What pleased him most was that, although his original appeal for volunteers failed, once compulsion was introduced it proved to be needed as no more than a framework, a guarantee that everyone would be treated alike.

"It is true," he told his audience of women in the Albert Hall, "there are sanctions in the background, but in the main you have responded because you felt the nation's need and have looked upon directions as determining where you should go rather than as a means of forcing you to go.... Behind it is the voluntary submission to discipline of a whole people."

[1] Postan, p. 148.
[2] More than eight and a half million were registered in 1941–2 and another million and three-quarters in 1943–5.

8

In June 1943, the Cabinet examined the results of the first year's working under the coal control. They were better than might have been expected the year before. The labour force had been maintained; output was five and a half million tons more than had been expected; cuts in consumption had saved 11 million more tons and stocks had risen by nearly four and a half million. Finally, the Greene Board, after giving a substantial rise in pay to the miners in June 1942, had gone on, in May 1943, to recommend the national settlement of wages which the miners had lost in 1926 and campaigned to get back ever since.

The prospects for 1943–4, however, and even more for 1944–5 were depressing. The gain in manpower in 1942–3 had already been lost through wastage and the labour force was beginning to fall. Fewer youngsters were going into the industry than ever and the proportion of older miners increasing. There had been little progress in the concentration of work on the more productive seams and pits, thanks to the stubborn opposition of both the miners and the colliery companies. Productivity per man was down and this, added to a falling labour force, meant a continuing drop in output: in 1943 nine million tons less were mined than in 1942, 30 million less than in 1940.

Higher wages and all the efforts to improve conditions had failed to remove the discontent from which the industry had suffered so long. Absenteeism was mounting. The percentage of shifts not worked because of absenteeism had risen from under 7 per cent in 1939 to 10.4 per cent in 1942 and over 12 per cent in 1943. Sickness and accidents accounted for part of these figures, but in two out of three cases the men had no good reason for staying away from work. Strikes were on the increase again and the tonnage lost through labour disputes, in defiance of the wartime ban on striking, had mounted from 340,000 tons in 1941 to 833,000 in 1942 and was to pass the million mark before the end of 1943.

Supplies for 1943–4 were probably safe enough, but there was a real danger, if the labour force and output continued to fall at the mid-1943 rate, of a coal crisis in 1944–5 when military operations would be at their height and when there would be a greatly increased demand for coal to supply the needs of the Services and to start up industry again in the occupied countries.

It was with all this in mind that Bevin went up to address the miners' annual conference at Blackpool on 20 July. He made a powerful speech, ranging from the progress of the war to the changes that were being planned in providing social services. Only towards the end did he bring in the mining industry, but what he said startled his audience and was received in complete silence. Shortage of coal, he told them, was the one thing that could prolong the war.

"At the end of this coal year there won't be enough men or boys in the in-dustry to carry it on. It is the one great difficulty in this war effort. . . . At the same time we are carrying out this invasion and every bit of territory we take from the enemy we have got to find coal for. . . . It is quite obvious that I will have to resort to some desperate remedies during the coming year. I shall have to direct young men to you."

And he went on to speak of seeing, if they had to send not only eighteen-year-old but sixteen-year-old boys into the pits, that they should have the same access to post-war educational schemes as if they had been through the Forces.

If Bevin's object was to shock the miners and the country into recognising how serious was the position in the coal industry, he succeeded admirably. The Miners' Federation called an emergency meeting of its Executive the same day and the press was full of Bevin's proposal.

The miners' suspicions were aroused on two points. Was the Government thinking only of directing boys from the mining villages into the pits? And was this intended as another device to evade the nationalisation of coalmining which was still, they insisted, the only way of putting the industry to rights?

On the first point at least Bevin was able to reassure them that he was thinking along more radical lines than they had given him credit for. In a statement to the Commons on 29 July he said that, if he failed to get enough volunteers to go to the pits and was driven to use com-pulsory powers, then (as had been the case with the direction of women) they would be applied equally to every class and area and would in fact relieve rather than increase the pressure on the boys in the mining districts.

On the second point, however, Bevin had nothing to say: nationalisation of the mines remained, as it had been for so long, the touchstone of British politics, an issue around which so much bitter feeling had accumulated that it had to be treated, not as an economic or technical, but as a political question.

On 29 August 1943 the *Sunday Pictorial* published an "Open Letter to Mr. Bevin" under the headline "Stop Trying to Fool Us". "Whether you like it or not," the *Pictorial* declared, "the nearer we get to peace, the more you are becoming the master architect of the future." The distinction between war and peace was unreal: any scheme that was going to raise production in the pits would have to provide guarantees for the future. Bevin should tell the Cabinet that before agreeing to use his powers to direct boys or men into the mines he must insist on nationalisation: "no conscription without nationalisation." If he did this, he would win 'hands down': if he failed to do it, he would be letting down not only the miners but the country. "The people can see no glimpse of the brand-new world in your handling of the coalmining industry."

The *Pictorial* article was reprinted approvingly by a number of Labour papers which urged Bevin to stand up to the mine-owners with the same firmness he had used towards the catering employers.

The Minister of Fuel and Power (Gwilym Lloyd George) had in fact come to much the same conclusion on different grounds. In the report which he presented to the Lord President's Committee in October 1943, he submitted, with the support of all his Regional Controllers, that the system of dual control set up in 1942 was an unsatisfactory compromise. The pit managers were responsible both to the colliery companies who employed them and to the Controllers appointed by the Government: they found it hard to serve two masters whose interests were frequently contradictory and this division of responsibility had undermined their authority in dealing with labour discipline as well as their efficiency. There was only one way, Lloyd George argued, to get effective day-to-day control of the 1,600 collieries nominally working under his control, to carry out the grouping of pits necessary to raise production and to establish new methods of labour management: the State must take over the ownership of the mines and become the employer of the managements for the duration of the war.

The Lord President's Committee passed the question to the War

Cabinet, which met to consider Lloyd George's and the miners' proposals on 8 October. What was said in the Cabinet is not known, but the result was very soon clear. If Bevin pressed the case for nationalisation, as seems probable, he was unsuccessful. All that the Minister of Fuel was authorised to say in the House of Commons was that a Cabinet Committee, including Bevin, had been set up to consider proposals for improving the control of the industry. Lloyd George's statement satisfied no one on the Labour benches: there was an admitted danger of coal running short, a danger which the Government took sufficiently seriously to propose calling up men of military age for service in the mines, yet all the Minister could offer was another committee "to consider improvements". An angry debate at once broke out between the two parties.

This time the Prime Minister did not leave it to anyone else to put the Government's case, nor did he allow time for a political crisis to develop. On the second day of the debate he intervened at once to state in the most disarming and yet, at the same time, authoritative manner the principle on which he believed it was alone possible to hold the coalition together. "Everything for the war, whether controversial or not, and nothing controversial that is not *bona fide* needed for the war." Applying this principle to the case of coal, he submitted that no case had been produced to show that nationalising the mines was a necessary step towards winning the war or would in fact raise output. "I certainly could not take the responsibility of making far-reaching controversial changes which I am not convinced are directly needed for the war effort without a Parliament refreshed by contact with the electorate." State control was to continue until Parliament decided what was to be done about the future of the industry, and he was willing to authorise discussions between the Minister of Fuel and the miners about the post-war period so that "the uncertainty and harassing fears" for the future "shall be as far as possible allayed".[1] But nationalisation, he made clear once and for all, was not a matter which he was prepared to discuss so long as he was Prime Minister of the coalition Government.

This unequivocal statement was surprisingly well received. Even more surprising, it settled the matter: for the rest of the war, nationalisation of the mines was not an issue.

[1] House of Commons, 13 October 1943, Hansard, Vol. 392, cols. 921, 924, 932.

9

Churchill's intervention, however decisive politically, still left Bevin and Lloyd George with the problem of how to get the coal and prevent production falling. A publicity campaign to secure volunteers for the mines proved a failure and in the debate on 12 October Lloyd George announced the decision to apply conscription. A month later Bevin had his scheme ready. It took the form of a lottery in which the selection of those required to go into the mines on call-up, instead of the Services, was determined by drawing numbers. Very few exceptions were allowed and Bevin made it clear that service in one of the training corps characteristic of the public schools would not entitle a boy to exemption. On the other hand care was taken to see that the "Bevin boys", as they became known, were given proper training before being sent into the pits.

Bevin's scheme, although fair enough in its procedure, was highly unpopular. Most of the young men drafted into the mines disliked the work and the conditions in which they had to live; there was more opposition than to any other form of conscription and a number preferred imprisonment to accepting direction.[1] In all, 21,800 were allocated to the pits by ballot before the end of the European war, not a large number, but enough to check the fall in the size of the labour force. In 1944, in fact, there were more men at work in the collieries than in any year since 1940, and if productivity and output continued to fall they would have fallen far more rapidly without the measures which Bevin took to provide additional manpower. It was reckoned in October 1944 that directions issued by the Minister of Labour had brought close on 100,000 men into the industry since 1941, nearly a seventh of the total number employed. The "Bevin boys" scheme was the last important step which the Government took —or needed to take—to bring men into the mines, and however unsatisfactory their record in attendance and discipline they helped to tip the balance in producing enough coal to ensure that the British war effort was never hamstrung for lack of it.

Bevin himself, although he did all he could to make a success of the

[1] Up to the end of October 1944 out of 16,000 youths picked by ballot, 500 had been prosecuted for refusal to obey the National Service Officer's order or for leaving their employment: of this total, 143 had been sentenced to imprisonment.

scheme, never had any illusions about it: he regarded it as an unsatisfactory expedient, a last resource to deal with a problem for which no satisfactory solution was possible, at least during the war.

In most other industries, greater willingness to accept direction of labour made it possible for him to use his compulsory powers effectively in the later years of the war, but in mining many of the men brought in remained unwilling recruits and the labour record of the industry provides strong evidence for Bevin's original belief that, where compulsion was unwillingly accepted, it produced a discontented labour force. It is ironical that his name should be associated with a scheme which necessity forced him to adopt against his own convictions.

When enough men had been found, by one means or another, to keep the numbers up, there remained the atmosphere of bitterness which hung over the industry like an acrid fog and seemed to condemn any attempt at improving the human relations on which in turn reorganisation and modernisation depended. A great deal was done to improve industrial relations in Britain during the war: coalmining was the outstanding failure. Not only was it the one major industry in which wartime production fell instead of rising but half the total time lost through strikes in 1943 (two-thirds in 1944) was lost in the pits. Bevin, however, was unwilling to accept the ingrained fatalism with which so many in the industry regarded its labour troubles and went on looking for a solution even if nationalisation had to be ruled out until after the war.

'I doubt very much," he wrote privately to Lloyd George, at the end of November 1943, "if the bulk of the miners are worrying about controls. It is more or less a peg on which to hang further demands. What the miner is concerned about, in my view, is the removal of uncertainty so far as his wages and conditions are concerned and this particularly applies to exporting districts."[1]

If this was so, then there was no need to wait for nationalisation before attempting to find a remedy and Bevin set out his ideas on what might be done in a letter of eight closely argued pages which Lloyd George described as "the most stimulating and comprehensive examination of the problems of the coal industry which I have read since I became Minister of Fuel".

[1] Bevin to Lloyd George, 29 November 1943.

Hitherto, Bevin reasoned, the coal industry had been able to avoid the necessity of making itself efficient because it had been able to supply the country with cheap coal at the expense of the miners' wages and conditions of work. To put these right and give the miner greater security would not only make it easier to carry out reorganisation (by reducing the opposition of the men) but would force the management —whoever they worked for, the State or private owners—to face up to modernising a notoriously out-of-date industry: for they would then have to win and keep their markets by technical efficiency, not as they had done in the past by cheap labour.

To begin with, Bevin urged breaking with the traditional system of fixing the miners' wages by dividing profits in a fixed proportion between the men and the owners. This was bad for the men because it left them uncertain what they would earn and led to inevitable disputes; it was also bad for the industry because it gave the owners an obvious incentive to set up subsidiary companies for developing the by-products of coal instead of "bringing in the newly created wealth to assist the cost of procuring the raw material in a similar manner to that done in other well organised industries".

First then, scrap the existing complicated system of wage calculation with its irritating deductions and substitute a guaranteed wage based on a five-day week, with any work above five shifts paid for extra. Second, scrap all bonuses, percentages and cost-of-living additions, substituting for them a new payments-by-results system "on a reasonable output yield of at least $33\frac{1}{3}$ per cent above the minimum".

Bevin's third proposal was to recognise that mining "is a dangerous, specialised, arduous and unpleasant occupation" which would have to be treated as a special case outside the framework of the normal social services. An example of this was workmen's compensation. The miner's risk of accident was so much greater that he could not be put on a footing with workers in other industries. Bevin suggested that a man should be paid a fixed percentage of his wage for all injuries which kept him away from work for more than seven days, leaving only compensation for minor accidents to be met from the ordinary social service fund. The same principle should be applied to pensions. Since the miner ran a greater risk of having to give up work at an early age, his pension ought to start earlier. Bevin suggested a graduated pension scheme based upon length of service and, as in the case of compensation, he added suggestions for dovetailing contributions and benefits with the general social service funds.

"The great gain to the industry arising out of these proposals would be that we should have a steady recruitment." It would remove the handicap of insecurity in one of the most dangerous of occupations. But the cost of the proposals which would have to be borne by the industry would add greatly to the price of coal and impose a heavy burden not only on the home consumer but on exports. How was this to be met? "I do not believe," Bevin wrote, "that we have really begun to tackle efficiently the production of coal or its distribution." And again he made a number of practical suggestions.

Taking distribution first, he proposed nationalisation of electricity and an enormous expansion in the consumption of electricity sold at a flat rate all over the country. This was at once a more efficient and a more economical method of supplying fuel than by the transport of the raw material.

Secondly, he called for intensive research, along the lines already begun during the war, into the more efficient and economical use of coal by industry. Cheap coal had hitherto discouraged fuel economy and there was much wastage which could be eliminated, as wartime experience proved.

His third proposal was to encourage the industry, once it had been reorganised, to run by-product plants of all kinds, entering the field of heavy chemicals, and so subsidising the cost of such raw coal as must still be burned.

To give time for the necessary reorganisation and make the men feel that their wages were no longer dependent on political pressure, Bevin suggested a guarantee of wages and conditions for a minimum of five years after the end of the war with Germany. This would provide the necessary foundation of confidence for a comprehensive settlement in place of a hand-to-mouth policy of expedients.

Had Bevin been allowed to put his ideas into practice, he might have been able to shorten the long-drawn-out crisis in the coal industry which was to have so grave an effect on the future of the Labour Government. Circumstances, however, denied him the chance; the successive expedients adopted by the Government, although influenced by Bevin's ideas, never added up to the comprehensive settlement at which he aimed. As for the miners' leaders, they had never liked Bevin's critical and independent attitude: later events, however, were to justify to the full his argument that their panacea of nationalisation, however necessary a change of ownership might be, would not by itself lift the industry out of its difficulties without drastic measures of reorganisation which the miners were as loth to face as the owners.

CHAPTER 9

Strikes, the Peak of Mobilisation
and Post-War Plans

I

THERE SEEMED a possibility in the summer of 1943 that the practice of unofficial strikes might be going to spread. Quite apart from the miners, there had been trouble in a number of other industries. In May 1943, 12,000 provincial busmen went on strike; in August 16,000 dockers at Liverpool and Birkenhead; in September 9,000 engineering workers in Vickers' works at Barrow; a little later, 16,000 in the Rolls-Royce aero-engine works at Hillington, near Glasgow.

At no time, of course, had the country been wholly free of strikes: it was inconceivable that it should be. The industries most affected were the two worst hit by the Depression, mining and shipbuilding, together with engineering (especially aircraft production), the industry most disturbed by wartime changes. The rise in the figures for 1943-4 is not difficult to explain.[1] The cumulative fatigue of four years of war put a strain on everyone's temper, especially in the months

---

[1] Working days lost through disputes:

Yearly Average								
Yearly Average	1911–14	:	17,707,000					
,, ,,	1915–18	:	4,230,000					
,, ,,	1919–20	:	39,769,000					
,, ,,	1936–39	:	1,983,000					
,, ,,	1940–July 1945		1,843,000	(A	508,000	B	163,000)
,, ,,	1941	:	1,077,000	(A	338,000	B	556,000)
,, ,,	1942	:	1,530,000	(A	862,000	B	526,000)
,, ,,	1943	:	1,832,000	(A	889,000	B	635,000)
,, ,,	1944	:	3,696,000	(A	249,500	B	104,800)
,, ,,	1945	:	2,847,000	(A	644,000	B	528,000)

A = Mining and Quarrying   B = Metals, engineering (including aircraft production), shipbuilding.

265

before D-Day. The rapid expansion in production and the many changes in production methods; the increase in piece-work and payment by results; dilution and the growing employment of women—these wartime developments all provided plenty of occasion for disputes over demarcation and wages, especially piece rates. The procedure for the settlement of such disputes, particularly in the engineering industry, was cumbersome and became slower in wartime when both management and trade-union officials were under heavy pressure of work: the issue which brought the Rolls-Royce factory at Hillington to a standstill for a month, the rating of women's work, had already been under negotiation for over a year without a settlement when the shop stewards decided to call a strike in order to get something done. Co-operation with the coalition Government placed trade-union officials in an awkward position, vulnerable to the accusation of "selling out to the bosses", while the recruitment of many of the most experienced for government work weakened communication between union headquarters and the rank-and-file membership. This gap in organisation was filled by the shop stewards, who were on the spot and knew what was happening, but were often inclined to show impatience with official policy and procedure and to take a more radical line of their own.

Looking back, it is easy to conclude that the danger from strikes was exaggerated and that the loss of production through disputes was negligible when compared with that due to other causes such as sickness, accidents and mechanical breakdowns.[1] But this was not a view which Bevin or anyone else could afford to take at the time. Morale in the factories was a constant preoccupation of the Government, and it was natural for them to feel anxiety lest industrial trouble might grow to such proportions as to hamper the build-up for the invasion of France in 1944.

The cornerstone of government policy was still Order 1305 of July 1940 which prohibited strikes and lockouts during the war and made arbitration compulsory. In making this Order Bevin had been as much concerned with creating a deterrent and giving a strong incentive to industry to settle its own disputes as with the formal machinery for arbitration which it established. The record justified

[1] K. G. J. C. Knowles reports that the loss of working time due to strikes in 1944 was estimated to be only one-fiftieth of the time lost from other causes, including sickness. (*Strikes*, pp. 270–1.)

this. In five years no more than 2,200 cases were reported to the Minister for arbitration under the Order, and of these half were withdrawn or settled by the parties themselves, sometimes after reference back by the Minister. A large number of other disputes were composed under the ordinary procedure of conciliation established by the Act of 1896 and by the efforts of the Ministry of Labour conciliation officers.

But strikes continued to take place and the question was naturally asked (not always without malice) whether the Minister of Labour was prepared to prosecute those who broke the law? Bevin's answer was—Yes, but only when he was satisfied that the purpose of the strike leaders was to make trouble and that he could count on the support of the responsible elements among the workpeople involved. The first prosecutions were instituted in April 1941, when six of the ringleaders of an engineering apprentices' strike in Lancashire were bound over for a year. In August sixty-two men on strike at a Dundee shipyard were served with summonses; when they thereupon went back to work without further trouble, they too were bound over. Similar prosecutions, sometimes involving fines and occasionally imprisonment, continued to be brought throughout the war, but Bevin was obviously reluctant to act except in flagrant cases and by January 1944, out of a total of one and a half million strikers up to that date (most of them out for no more than a few days), only 5,000 had been taken to court and less than 2,000 convicted.

To those who had little experience of industrial relations and reasoned from analogies with Army discipline, this looked like "being soft with the workers". A coal strike at Betteshanger, Kent, in January 1942, however, showed the limitations of legal procedure. This started with a dispute over the rates to be paid for working a difficult coal face. Only sixty men were directly involved but, with the backing of the local union branch, all the 1,600 workers in the pit came out on strike. After attempts to get them to go back had failed, summonses were taken out against the underground workers, more than a thousand in all. The Canterbury Bench sentenced the chairman of the local branch to two months' imprisonment, the secretary and a committee member to one month, and imposed fines on the rest of the men with a prison sentence if they failed to pay. After this, a compromise was found for restarting work and, early in February, the three officials were released from gaol on the intervention of the Home

Secretary who had received a good many protests at the sentences imposed. But it was decided not to remit the fines. The miners on the other hand had no intention of paying and, although warrants were taken out for their arrest, the justices were reluctant to put them into force. There was not enough prison accommodation to house a thousand men, and too few police to handle them, while to take the matter further would almost certainly lead to more strikes and a loss of production. Bevin, on being consulted, offered a judgment worthy of Solomon: the fines were not cancelled, but the warrants were held in abeyance, and with this the matter ended.

What the Betteshanger episode showed was that, if any large body of men was involved in a strike and stood firm, the sanctions provided by Order 1305 could not be enforced. This was the general conclusion, not only of a trade unionist like Bevin but of his opponent at the time of the Shaw Inquiry (1920), Sir Lyndon Macassey. Writing to *The Times* in January 1945, Sir Lyndon summed up the experience of the First as well as the Second World War in these words:

"Imprisonment for nonpayment of fines for illegal wartime striking, however juridically logical and theoretically justifiable, was under modern conditions (i.e. those of 1916) industrially ineffective and nationally undesirable—that in practice it operated to impair respect for the rule of law. You cannot imprison the whole of a large body of strikers. Alternatively, to pick out a selected few and make examples of them only elevates them into popular martyrs and brings in the always effective industrial rallying cry of 'victimization'."[1]

When he was confronted with the larger-scale strikes of the summer and autumn of 1943, Bevin's first concern was to insist on the constitutional procedures for the settlement of disputes and to give the unions time to get their members back to work. The one thing he would not do was to go over their heads. Thus, when 9,000 men in September came out on strike at Barrow in protest against an arbitration award, Bevin refused to intervene, leaving it to the union officials to persuade the men to return. When this failed in face of the men's conviction that they had been unfairly treated, he used his authority as Minister not to utter threats or bring the men into court but to back up the officials, telling the men that they were dishonouring their agreements and flouting the arbitration procedure to

[1] *The Times*, 27 January 1945.

which, through their unions, they were parties. At the same time, he brought the strongest possible pressure on the union leaders to end the strike and so avoid a humiliating confession of failure for the trade-union movement. This combination of tactics proved successful in getting the men to return, and most of the press, from *The Times* to the *Daily Worker*, supported Bevin, recognising that in the long run the loss in morale, and probably in production as well, was less than would have resulted from an attempt to browbeat men with a sense of grievance into surrender.

Where Bevin incurred criticism was less in his handling of the strikes than in his analysis of their causes. At the end of the Commons debate on manpower on 24 September 1943 he gave a warning on the danger of strikes, which he divided into three categories. One was the "last straw" type, where the remedy was investigation and con-ciliation rather than imposing penalties. A second was the strike "deliberately provoked for ulterior reasons by employers"; the third was the strike which was politically inspired by Trotskyites and other groups opposed to the war. It was on this last category that he laid most emphasis.

"While I will not be a party while I am in office," he declared, "to doing anything at all, under any circumstances, which will weaken the legitimate trade unions in any way—rather I want to strengthen them—I feel that steps must be taken to see that the war effort is not impeded by these activities."[1]

At Farnworth on 2 October he repeated his warning to the trade unions to be on their guard against "the anti-war people" who were trying to

"get hold of shop stewards and other positions in the movement, not for the purpose of prosecuting the war or of raising your standard of living but of trying to impede the war effort. . . . I call upon my friends in the movement not to leave their unions open to attack."

The press took Bevin to be putting the blame for wartime strikes on the insignificant Trotskyite group and, not surprisingly, refused to believe him. The most common explanation was that he had an obsession with communists and Trotskyites, seeing them everywhere, or found it less wounding to his pride to blame "subversive influences"

[1] House of Commons, 24 September 1943, Hansard, Vol. 392, col. 666.

than to admit that he could no longer command the loyalty of the workers.

A month later, however, Bevin offered an explanation of his own, which at least made his statement of the 24th sound more reasonable. Sir Archibald Southby, objecting to Bevin's charge that strikes were sometimes provoked by employers, raised the matter on the adjournment of the House of Commons and demanded names. These Bevin refused to give but he went on to tell the House that his remarks on the earlier occasion had been intended not so much as an analysis, as a warning in the light of the particular situation at that time.

"I made the statement, with very great deliberation, at a time when there was an earth tremor going on in the industrial world by which the war effort was likely to be seriously endangered. I proceeded to go through many of the inquiries held at the Ministry of Labour and to analyse what I regarded as the contributory causes of many of these disputes. . . . I think the proof of the pudding is in the eating. The statement I made that day . . . had a very salutary effect, as I think employers and trade unions in general will agree, and there died away what looked like a very difficult situation. . . . I am satisfied that the course I took had the effect of helping the executives on both sides to co-operate with me in a way which got over the most dangerous point we have come to in the four years of war. It was centred around one or two well-known disputes then in being."[1]

What all this was about, or what grounds Bevin had for his fears in September, remains a mystery. In the following year there was evidence of a Trotskyite group taking a hand in the Tyneside apprentices' strike, but it was hardly on a scale to justify Bevin's belief that political motives played a significant part in industrial discontent. Nor has any evidence to this effect come to light since. Perhaps the simplest explanation is that the wartime strain was beginning to tell on Bevin too—it would be surprising if it had not—and that this affected his judgment. It had more influence, however, on the explanations he gave than on his actions, and when the worst labour crisis of the war came, in the spring of 1944, although he took steps to strengthen the penalties against incitement, he showed clearly enough in his handling of the miners that he understood there were real grievances to be met, and acted promptly to remove them.

[1] House of Commons, 2 November 1943, Hansard, Vol. 393, cols. 626-7, 630.

2

In the summer and autumn of 1943, Bevin spoke at four of the biggest trade-union conferences, the T.U.C., the Scottish T.U.C., the Miners' and the T.G.W.U., and addressed a number of Labour rallies in Scotland and the North. Who had been right, he asked, the Government or the critics? There had been plenty of reverses, from Dunkirk to Tobruk, but these had been the result of "trying to do too great a task with insufficient force, an insufficient army and air force, insufficiently equipped".[1] While the critics railed, the Government had stuck to its plan, refusing to be diverted or provoked into attacking before they had built up forces which would be more than sufficient, which would bring to bear a superiority of power such as they had seen at Alamein. His own part in this had been to match the degree of mobilisation Germany had achieved by the methods of the police state.[2] "Well, we are a free people and as Minister of Labour I pinned my faith on the discipline of the voluntary organisations of this country."[3]

"In the early days of this mobilisation there were many critics—they are fairly quiet now—who pressed me to conscript everybody as though they were in the Services, to put a ceiling on wages, to set on one side for the duration of the war, the arbitration system and the joint negotiating machinery—forgetting one very important psychological factor, that a citizen . . . will accept one thing when in uniform, by tradition, but immediately he is out, he is an entirely different person."[4]

The power of the State had been used to strengthen, not to supersede, to harness not to suppress the free institutions of trade unionism, and no minister before had done so much to bring the trade unions into the centre of the picture, to give them an increased share of responsibility and to rely on their co-operation rather than giving

[1] Speech at the Biennial Delegates' Conference of the T.G.W.U., Edinburgh, 2 August 1943, reprinted as a pamphlet, *A Survey of the War Situation and Post-War Policies.*
[2] In fact, as we now know, in spite of the police state and their much vaunted gift for organisation, the Germans did not equal the degree of mobilisation achieved by the British.
[3] Speech at Wigan, 14 November 1943.
[4] Speech at Edinburgh, 2 August 1943.

orders. The implication was clear enough: now that the strain of the war was beginning to tell, he had a right to come to the unions and press them to use all their influence to see the nation and the Government through the last and most difficult period of all. When he turned to reconstruction after the war, Bevin went through the planning which had already been put in hand, for education, for giving effect to the Beveridge Report, for a national health service. But he was less concerned to defend the Government's record than to give a strong lead to the trade unions in the part they could play in the transition from war to peace.

He urged them to look on the partnership which had been created between management and unions not as a wartime improvisation but as a permanent pattern of self-government in industry.

"This does not mean," he argued to the T.U.C., "merely self-government in private industry; it means self-government in industry whether that industry is owned by the State or by private people. If there is to be a development of collective ownership, may I suggest, with the double practical experience I now have, you will need to maintain this principle of self-government in industry."[1]

The unions, he believed, should extend their responsibilities after the war, not fall back on the narrower conception of wages and hours as the limit of their interest or suppose that the formula of nationalisation would solve all problems.

One problem with which Bevin was already concerned—the resettlement of men demobilised from the Forces—illustrates the sort of responsibility in which he wanted the unions to take a share. The obligation to reinstate men in their employment, which had been written into the National Services Act, was of little practical value. "It is not worth the paper it is written on," he told the T.G.W.U. delegates at a private meeting before the T.U.C.

"I am making a new Bill which deals with the reinstatement business, one of the principles of which is that I bring you people into it. . . . The principle upon which I am working is this, that when the Government says go ahead, I shall charge the Joint Industrial Councils with reviewing the position of their own industries and putting up new proposals which I can work, very much on the lines of the Essential Work Order only in the reverse way. . . . Just as I

[1] Speech at T.U.C. Conference, Southport, 7 September 1943.

brought industries in to help me to mobilise, I want to bring industries in to help me to demobilise. It will, I think, give the unions a fairly good grip and hold on the resettlement of industry at the end of the war."[1]

There was another reason why Bevin wanted to get the unions to raise their sights and think in larger terms. He was convinced that the only way to avoid the experience of 1918–26 was to keep on controls for a period of three to four years after the war ended. Working-class folk would be as anxious as anyone else to throw off controls after the restrictions of wartime and might be taken in by the specious cry of "Freedom", unless the unions threw their weight on the side of the Government.

"The people who are screaming for taking off controls," Bevin told his audience at Wigan, "have no vision of the problems we have to tackle during the transition period. One of the most vital is the replenishing of our homes. There is hardly a home in this country that is not short of domestic utensils, sheets, pillow cases, blankets. . . . What is going to happen? I will tell you what I want to happen. I want to compel manufacturers to manufacture goods of proper quality, no rubbish, and replenish our homes. . . . Secondly, I would keep the controls on, but not down to the low standard they are now. If you don't, what will happen? Those with money, and what is more important, those with time, can go to the shops and will run prices up and make it more difficult for our working folk to replenish for a considerable time. . . . We must do it for the sake of the men who have married during the war, for two and a quarter million marriages have taken place since the war broke out and not more than 10 or 12 per cent have homes. They are living with their families or in furnished rooms. There must be control or these unfortunate people are going to be bled of their savings accumulated during the war. And the one essential thing if you are going to stop moral disaster after the war is to enable these young folk to start off under reasonable conditions of home life as quickly as ever you can."[2]

Without waiting for the end of the war, Bevin had urged the T.U.C., if they were eager to see improvements in social conditions, not to leave it to the Government to put schemes into operation but to take the initiative and practise self-help. He gave them two examples, the guaranteed week introduced by the Essential Work Order and education.

[1] Speech at the dinner of the T.G.W.U. Delegates to the T.U.C. Conference, Southport, 6 September 1943.
[2] Speech at Wigan, 14 November 1943.

The Essential Work Order had been subjected to much criticism during the T.U.C. debates. Bevin refused to apologise for it: it was necessary in wartime and it gave the men a guaranteed week.

"Does anybody," he asked the conference, "want to go back to the hourly payment? I cannot believe it. This standing on and off, this going to the factory door in the morning and 'Nothing doing, Tom, go home'—surely nobody ever wants to go back to that again. But could I make this suggestion, if you will allow me. Do not rely on the Government only to maintain it. Why not weave it into your collective agreements at the earliest opportunity? We are not anxious to have the duty of enforcing it by law. Do not turn the rising generation too much to the law and not enough to you."[1]

Education, his second example, had already been debated by the T.U.C. the day before Bevin spoke to the Congress. He picked out one particular provision in the new Bill of special relevance to the unions, continuation classes for the 14–18-year-olds who went straight into industry.

"Now the trade unions will say, the public will say, 'Will the Government keep its word or will Parliament carry it out when the war is over?' May I suggest to the trade unions that you begin to compel us now? You will be sitting down in joint conference week in, week out, fixing conditions for young people. Every new agreement you make, make it on the assumption that this is going to operate, and that industry has got to adjust itself to it. . . . See to it that it is woven into the hours of labour, time off and all the rest. And what better chance could you have than you have got now with this magnificent Joint Industrial Council meeting?"[2]

Some newspapers rebuked Bevin for talking on such occasions as a trade-union leader and forgetting that he was now Minister of Labour. Bevin himself denied that there was any conflict of loyalties. On the one hand, much of his authority as Minister of Labour derived from the support of the trade-union movement; on the other, all his advice was directed to persuading the unions—as well as the employers—to take a wider rather than a narrow sectional view of their responsibilities. Nobody thought it wrong for Woolton or Lyttelton to talk to businessmen and industrialists about the interests of business and industry in the post-war world; why was it wrong for him to talk to trade unionists about theirs?—except on the assumption that

[1] Speech at T.U.C. Conference, 7 September 1943.
[2] Ibid.

business interests were identical with those of the nation, while the trade unions represented only a class interest.

This was a view of the national interest that Bevin would never accept: after the war, the State could not return—any more than the trade unions—to the narrow conception of its obligations that had prevailed in the Depression years.

"It must be our determination," he told the Regional Council of the Labour Party in Manchester, "to compel the State to accept the principle that mass unemployment must be abolished. We have mobilised, organised and energised our people in the fight against Germany and in the same way in peace we must mobilise, organise and energise in this fight against mass unemployment. Private enterprise can no longer have the monopoly of employment. In the past the State has only been brought in when unemployment has risen to a dangerous level. I say that, if they turn to public enterprise when the State is in danger, then the State must assert its right to step in and prevent unemployment arising at all."[1]

The emphasis Bevin laid on the role to be played by the trade unions, and the fact that he said nothing about the political situation at the end of the war and the part to be played by the Labour Party, led some people to ask whether he was advising the trade unions to look to their own strength to secure the reforms they wanted, by collective bargaining rather than by political action. It was at this time that *The Economist*, as well as Nye Bevan, criticised him for a dangerous tendency to substitute the corporate for the Parliamentary State,[2] the former believing that "the biggest collective bargain of all" between the employers and the unions would be at the expense of the community and the consumer, the latter at the expense of socialism.

It has already been suggested that the right context in which to see Bevin's insistence on the right of the trade unions to be consulted on all important issues of economic and social policy is not the corporate state, but the post-war development, to be found in other industrialised countries besides Britain, of a managed economy in which Government is obliged, as a consequence of its intervention in the economy, to consult with and secure the co-operation of all the major interest groups affected by its policies, chief among them the

---

[1] 2 October 1943.

[2] *The Economist* expressed this view on several occasions, e.g. 31 October 1942; 16 January 1943; 20 March 1943; 23 October 1943; 13 May 1944. For Nye Bevan's attack on Bevin in the debate of 28 April 1944, see below, pp. 303–9.

trade unions and the employers' associations. Granted a managed economy, it can be argued, the traditional forms of parliamentary and party representation have to be supplemented (not replaced) by the sort of functional representation which Bevin was developing through his wartime practice of consultation. This of course was unwelcome to those who wanted to return as quickly as possible from the managed economy of wartime to a free economy with government intervention (and the power of the trade unions) reduced to a minimum. It was equally unwelcome to those like Nye Bevan who were not content with a managed economy but wanted to go further and establish a fully socialised system, the key to which they saw, not in the increased influence of the trade unions, but in the capture of a parliamentary majority by a party committed to carrying out a hundred per cent socialist programme.

Bevin did not share either of these views: he was implacably opposed to any return to a nineteenth-century free-enterprise capitalist society but had little faith in the immediate practicability of the socialist revolution which was the aim of the Left. What he was working towards was that combination of a Managed Economy and the Welfare State which the Labour Government established after 1945. His attitude, however, was easily misunderstood because of his unwillingness to talk about politics, the result of which was to leave the impression that he was interested only in the extension of collective bargaining and trade-union influence, not in political action.

The reason for this unwillingness was still the same: his determination to have nothing to do with party politics while he was a member of the coalition Government. In an open letter to his union members at the time of the 1945 election, when he was in the thick of the party fight, he still claimed with pride,

"During the five years I was in office, I did not make a political speech of any character, until I spoke at Leeds, and then it was necessary.[1] I was with the Government in every difficulty we had to face, including the terrible difficulties of Greece.[2] I took my corner without shirking a single issue".[3]

This claim was justified and by itself accounts for his refusal to raise political issues at a time when the strain of war was reaching its

[1] April 1945, see below, pp. 368–70.
[2] In December 1944, see below, pp. 340–47.
[3] *The Record* (Journal of the T.G.W.U.), Vol. XXV, No. 285, June 1945.

peak and any member of the Cabinet must have been anxious to avoid disruptive issues.

A second reason, however, which may well have played a part was uncertainty about his own political position and that of the coalition. He was now in his sixties and spoke at times as if he meant to retire when the war was over. The one thing which would keep him in politics, for a time at least, was his strong desire, which he had expressed to Arthur Deakin as early as May 1940, "to lay down the conditions on which we shall start again".[1] But that might well be accomplished more surely by a continuation of the existing all-party coalition for the period of transition from war to peace than by an immediate return to party politics which had little attraction for him and might in any case put the Tories in at the very moment when Bevin attached most importance to being in power himself. This was a question on which he had not made up his mind, nor did the Labour Party itself reach a decision until the autumn of 1944; it was the last question he wanted to raise in 1943 and this supplied an additional reason for saying as little as possible about politics until they were over the hump of the war. Once he felt free to speak out, however, as we shall see, no one was left in any doubt that he was as much convinced of the need for political action as for persuading the trade unions to use their bargaining power on wider issues than wages and hours.

3

In the meantime Bevin continued to show himself more active than any other member of the Cabinet in providing ideas and pressing forward with planning for the post-war period—ideas and plans which showed no disposition at all on his part to think of substituting reliance on trade-union strength for action by the State.

On 9 June he drafted a note to the other members of the War Cabinet urging the need for a firm decision on the financial commitments which the Government thought the country could afford to undertake after the war. If hostilities were to end suddenly, he argued, however much preliminary work had been done, none of it could be turned into legislation without a decision on finance: "and the Treasury and Government will be in no better position to forecast

[1] See Vol. I, p. 653.

the financial burden it is capable of undertaking then any more than now." Instead of waiting till the war was over and leaving the Government of the day to make up its mind at a time of confusion and conflicting demands, "the major decisions should be taken now as to the items we are prepared to see through."

The need to attack the poverty and low standard of living of the underdeveloped areas outside Europe and North America continued to occupy his mind as much as domestic reforms.

"It is not much good," he told the delegates at the T.G.W.U. Conference, "being wealthy and at the same time having to maintain armies and navies to protect yourselves against hungry folk. It is better, while having proper and adequate means of defence against aggressors, at the same time to work to remove the causes of discontent, most of which are economic, over the whole world."[1]

Thinking along these lines, he wrote another long letter to Eden in June 1943, urging him as Foreign Secretary to support the proposal for an I.L.O. Conference before the end of the year.[2] He was anxious, however, he told Leggett, not to see the I.L.O. saddled with the problems of relief after the war, a job better handled by U.N.R.R.A. The I.L.O. must not take a short-term view of what would be needed when the war ended, but look ahead and lay long-term plans for a permanent improvement in the standard of living of the poverty-ridden parts of the world.[3]

One obvious place in which to start was India, and Bevin continued to press the Cabinet not to remain content with the deadlock which had followed the failure of the Cripps Mission. Towards the end of June 1943 he dictated a note on India to serve as an aide-mémoire in arguing his case in the Cabinet.

"In my view," he wrote, "there should be an entirely new approach to the problem of the Indian continent and it should be visualised as playing an entirely different role in the future than it has hitherto done."[4]

He marshalled his arguments under four heads, defence, economic, political and constitutional.

[1] Speech at Edinburgh, 2 August 1943.
[2] This took place in London in December 1963, with Bevin welcoming the delegates on behalf of the British Government. See below, pp. 282–3.
[3] Note to Sir Frederick Leggett, 16 October 1943.
[4] Note dictated by Bevin, 21 June 1943.

"It has been customary in the past to look on the Balance of Power as being primarily situated and determined in Europe. My view is that the future balance of power will be largely determined in Asia where the Great Powers of China, Russia and ourselves will meet. . . .

"In my view we have accepted China [i.e. Chiang Kai-shek's China] as playing the prominent Asiatic rôle too readily. . . . China, I think, may well disintegrate again after this war and it will take more than the capital of the U.S.A. to hold it together, and there may be substituted Russia for Japan as a source of anxiety. India, therefore, from the point of view of a key position of defence of the East will play a very much bigger part than it has hitherto done."

Bevin's first proposal was to work out a new defence area to be held by the Commonwealth in partnership, stretching from Burma right round to the Persian Gulf. To supply the forces needed for its defence Indian industrialisation should be pushed ahead as rapidly as possible.

Bevin was in favour of greater industrialisation for economic as well as military reasons:

"I agree that 80 to 90 per cent of the Indians are not political but they are poverty-stricken. The population is increasing too fast, industrialisation is too slow and the standard of life far too low. Can India, therefore, be used with a drastic organisation drive behind it, to raise the whole standard of living in the East? In my view it can. For the first time she is a creditor nation. I would use the whole of these credit resources for capital goods to be used for the development of India both in her agriculture and her industry, all of which will give a higher standard of living and would be complementary to her defence."

In the third part of his argument Bevin turned to foreign policy, urging that India should at once be given the full status of a Dominion, and an Indian appointed as Minister of External Affairs.

"The India Office should be abolished; Indian affairs transferred to the Dominions Office and a special Indian Department set up in the Foreign Office. These steps would raise the status of India abroad and have a big psychological effect in India itself."

In this way, Bevin hoped to by-pass the obstacles to a constitutional settlement. "I do not believe," he wrote, "that India will attain satisfactorily a written constitution." He wanted, instead, to start fresh discussions on defence, economic development and India's place in the world, and so "begin to right our relationship now, without any further change in her political constitution, into a treaty as between a Dominion and ourselves in a full partnership". If this was done, it would then be much simpler later to introduce changes in the form of government as a result of experience.

In a coalition which had Churchill as Prime Minister and Amery at the India Office, there was little chance of Bevin's ideas being adopted; but the document quoted (together with a series of notes to Amery found among his papers) is proof that if the coalition provided no new approach to the Indian problem, it was not because Bevin or the other Labour members of the Government failed to recognise its importance or to press for something to be done.

Two other questions which he was turning over in his mind in the summer of 1943 were arbitration and public assistance under the Beveridge scheme. Bevin had always hoped to see the wartime procedure of arbitration continued after the war. In a memorandum to his department in June[1] he suggested strengthening it by appointing regional investigators who would start to make an inquiry into a dispute before it turned into a strike. Arbitrators, too, he believed, could with advantage be appointed regionally if only because local practice and customs varied so much.

In dealing with those who were unemployable or "work-shy", Bevin argued that there were people to whom the ordinary unemployment provisions of the Beveridge plan were not applicable. Before turning a man over to public assistance, however, they should try to discover how he had got into such a state:

"From my experience, it arises mainly from the exhaustion of a process in industry in which the craft has disappeared and a person is often too old to take up another and an adequate provision for training has never existed; or in the case of the professions when a person has found himself unemployed at an age at which it is difficult to reabsorb him."

Bevin believed that a good deal could be done to rehabilitate a man by training him for other employment and he wanted specific provision for this to be written into any Act. This would still leave a residue of men incapable either of retraining or of any form of employment. "I am strongly in favour," he wrote, "of expert study of how to

[1] Addressed to the Permanent Secretary and dated 7 June 1943.

deal with what is virtually a disease. Such a study may result in recommendations which would apply compulsion. That I should be prepared to face." But he wanted the inquiry first and held to the belief that, if remedial training were provided, the number who would refuse all work would be small.[1]

These were old interests of Bevin's; a more recent one was the needs of adolescents. In December 1941 the Cabinet had decided to register the sixteen- and seventeen-year-olds with a view to persuading them to join a youth club or organisation in which they could undertake some form of part-time national service. Nothing much came of the original purpose, but the registration brought to light for the first time some of the facts about the jobs which boys and girls of this age were doing, their pay, conditions and prospects.[2] In a great many cases they were pitchforked out of school on to the labour market, with little or no guidance about jobs, no training and no proper protection against being exploited. "The adolescents," as Bevin said, "have been neglected by all our institutions." It was his discovery of these facts which made him determined to get provision for the continuation of their education after leaving school written into the Education Bill. At the same time he took up with his Joint Consultative Committee the question of industrial training and technical education, persuading them to set on foot an inquiry into the provision made by different industries. As a first step he secured the appointment of the Forster Committee on juvenile recruitment for the mines and the scheme for building apprenticeships. Finally, in January 1945, he set up a committee under Godfrey Ince to draw up plans for an effective juvenile employment service after the war.

"The most important decision of a lifetime," he said in one of his last speeches as Minister of Labour, "is the choice of a career or occupation. A man's work is the background against which his inner life has got to grow. His conception of life cannot but be profoundly affected by the environment in which he has to earn his daily bread."[3]

The Ince Committee's report, submitted in December 1945, was put into effect by the Labour Government and led to the setting-up

[1] Note on Assistance dictated by the Minister, August 1943.

[2] Of the 835,000 boys who registered at the beginning of 1942, only 67,000 were still at school.

[3] Speech to the Standing Conference of National Voluntary Youth Organisations, London, 12 May 1945.

of the post-war Youth Employment Service along the lines which Bevin had suggested.

Apart from these practical steps, Bevin set to work to persuade employers, Local Authorities, education committees, trade unions and anyone he could get to listen, to pay more attention to the needs of these young people, including their health and facilities for the use of their leisure. In a speech at Farnworth in October 1943, which he devoted almost entirely to these needs, he declared:

"Industry must not regard the adolescent as a profit-making factor. It ought to recognise that the great producing age of the human being, upon which their economy must be based, must be the years from eighteen years onwards, that the period below that age must virtually be regarded as the school or training; and that the hours of labour for these juveniles and the combined educational facilities must be organised accordingly. Industry and the State will not be losers by taking that view and accepting the period of adolescence as one of investment. Its costs are maintenance costs—the maintenance of ability that can keep us in the forefront of industrial development . . . The idea that a youth can perform an industrial operation and do it cheaply must be discarded and the whole mind of the nation turned on a recognition that this is a preparatory age for industry, occupation and citizenship."[1]

When the Governing Body of the I.L.O. met in London in December, Bevin put care for the adolescent high among his priorities for social reform after the war. His speech[2] provides an interesting catalogue of the items he wanted to write into any programme of reconstruction:

Maintenance of controls to provide an orderly transition from war to peace and stop speculation.

Full employment—"Statesmen will not, on this occasion, as they did on the last, succeed in deluding the people with fine words which they fail to translate into the economic system afterwards."

A comprehensive system of social security.

Reconstruction of the educational system, with proper attention to technical education.

A comprehensive National Health Service—"Health is a great national asset that must not be allowed to deteriorate."

A prosperous agriculture with decent standards of living for the rural worker.

Rebuilding of the damaged cities and the abolition of slum housing.

Reorganisation—a euphemism for nationalisation—of the coal industry.

Rehabilitation of the disabled to make them fit for employment.

[1] Speech at Farnworth, Lancashire (George Tomlinson's constituency), 1 October 1943.

[2] Opening Address at the 91st Session of the Governing Body of the I.L.O., 16 December 1943, reprinted as a pamphlet.

All this, Bevin argued, could only be done on a planned basis. "Laisser faire will not do, nor must vested interests stand in the way.... The needs of the present age cannot be met with nineteenth-century economics." Drawing on his own experience in wartime, he urged the delegates to the Conference to think of a human as well as a financial budget:

"Year by year the Government should study prospective demand, taking into account failures of harvest, or anything that can be foreseen which would dislocate the world. With this an ordered economy could be planned so that if the trade of industrial countries contracted at home, it could expand abroad, or if it contracted abroad capital development could be turned on at home and so keep the measure of consumption stable. It is not impossible to deal with cycles of boom and depression if Governments have the facts before them in advance.... We must make our statistical forecasts in the form of the right use of manpower and not only of money."

To do this, Governments would need a constant flow of information about economic conditions from all over the world. This was the sort of job the I.L.O. could do, and in doing it they would be helping to give labour and economic questions their rightful place in Governments' outlook on international affairs, which had too long been limited to international law, defence and diplomacy.

4

By the time the I.L.O. Conference met, a new parliamentary session had begun. The political situation was all too familiar. Despite setbacks, especially in Italy, the Allies were now on top militarily: however hard they fought, the Germans and Japanese were being steadily pushed back and at some point in the next campaigning season the British and Americans would make their return to the European continent. After the Moscow Conference of Foreign Ministers in October, the first full-dress conference between the Allied leaders took place at Teheran (28 November–1 December 1943). The conduct of the war and foreign policy were debated in the House in mid-December, but the speeches were unremarkable and criticism muted.

Once again it was the Government's failure, or refusal, to give a lead on post-war domestic affairs that roused the critics. "Like the policemen in *The Pirates of Penzance*," *The Economist* wrote, "Ministers say 'We go, We go,' but like Gilbert's constables again, they do not go: no visible headway is made."[1]

Churchill was not to be moved from the line of policy which he had laid down earlier in the year, but he was anxious—so far as he could, without committing the Government to specific measures—to remove the impression that he was indifferent to the needs of reconstruction, and at the Mansion House, on 9 November, he repeated the promises of his four-year plan, pledging his Government to the provision of "Food, work and homes for all". Three days later he appointed Lord Woolton, who had won a big reputation as Minister of Food, to the newly created post of Minister of Reconstruction, with a seat in the War Cabinet.

Woolton's appointment was well received and once again the press celebrated the conversion of the Prime Minister to reconstruction and reform.

At the end of 1943 a new Cabinet committee on reconstruction was constituted under Woolton's reluctant chairmanship.[2] Among its members were Attlee, Bevin and Morrison, Anderson, Lyttelton and Butler. The committee took over the responsibility for government policy on the social insurance scheme and with it a host of other postwar questions in which Bevin was interested: employment, housing, the proposed National Health Service, physical planning and the location of industry. Bevin was not only interested; he supplied a great deal of the driving power and ideas for the committee, missing only six of its ninety-eight meetings before it came to an end in the spring of 1945.

The success of the Reconstruction Committee was due to the clear understanding by its members of the limits within which they had to work. To try to go beyond those limits would have meant either splitting on party lines and getting nowhere, or producing compromise solutions which would have satisfied neither side and made

---

[1] *The Economist*, 2 October 1943.

[2] Sir William Jowitt (first as Paymaster-General, then as Minister without Portfolio) had hitherto been responsible for post-war planning and had acted as chairman of the earlier Reconstruction Committee. For Woolton's reluctance, see his *Memoirs* (1959), c. 17. He was persuaded to accept the job of Minister of Reconstruction by Churchill on the grounds that, while still an Independent and not yet a member of the Conservative Party, he was known to be staunchly anti-socialist in his views.

the worst of both worlds. This meant, for example, stopping short of any attempt to agree on the post-war organisation of industry. But, having accepted this, the Committee wasted no time in recrimination but got down to the job of producing detailed plans on the subjects on which it could agree. What is remarkable is how far they succeeded in getting. It was the Reconstruction Committee which hammered out a common policy on employment and turned it into a White Paper which is rightly regarded as one of the milestones in modern British history. It was the Reconstruction Committee which took the Beveridge Report and converted it into a practical plan for a new system of social security. It was the same Committee which worked out the outlines of a National Health Service and even made some headway with plans for post-war housing and the intractable problems of compensation and betterment.

What they could not produce, and did not attempt, was more than plans, a common programme of action to be put into effect or at least started before the end of the war. Most people at the end of 1943 believed that in Europe at least the war would be over in 1944. The last chance for the coalition to produce such a programme was in the Parliamentary Session of 1943–4. But the possibility of doing this was less than ever. While Bevin was preaching the need for controls to continue after the fighting had ended, other ministers (Geoffrey Lloyd, Ernest Brown and Harold Balfour) were delivering speeches condemning planning and controls as incompatible with freedom. Lord Beaverbrook's newspapers had been conducting a vigorous campaign in the same vein all summer: an immediate end to all controls the moment the war was over. And Beaverbrook himself was back in the Government as Lord Privy Seal (28 September), brought back by Churchill with unspecified responsibilities, an appointment which mystified everyone. Any Cabinet which contained Bevin and Beaverbrook was going to find it hard to agree on a post-war programme.

For eighteen months the militants in the Labour Party had attacked the coalition for its failure to embark on major measures of reform. Now they were beginning to have second thoughts. They still wanted the reforms, but were by no means so sure that they wanted the credit for them to go to Churchill and the coalition, when a coalition led by Churchill might be the post-war alternative to a straight fight in a general election and the chance for Labour to form its own Government. The opposition too, therefore, was as unsure of its objectives as the Government.

A large section of opinion, however, still expected the coalition to produce an agreed list of measures for the transition from war to peace and was critical of the Government for failing to do so. The King's Speech (24 November 1943) setting out the Government's programme for the session did nothing to satisfy this section of opinion and received a poor press. Lord Hinchingbrooke, a leader of the Tory Reform Committee, described it as "cleverly drafted to allay some fears but calculated to rouse few hopes and no enthusiasm". The Government was reproached—and not only by members of the Labour Party—with the long delay in making decisions on nearly every one of the issues, from social insurance to town planning, which would dominate the period of reconstruction. For the Labour members of the coalition, in particular, it was evidently going to be a difficult session.

The first controversy in which Bevin became involved had nothing to do with the King's Speech or post-war policy. On 17 November, Herbert Morrison, as Home Secretary, announced that he had decided to release Sir Oswald Mosley from further detention on the grounds of ill-health. Looked at from this distance Morrison's action appears sensible and courageous, but at the time it made many people angry. The Labour Party Executive passed a resolution expressing regret, the National Council of Labour and the General Council of the T.U.C. disassociated themselves from Morrison's action. Bevin was furious. This showed, he declared, that he had been right about Morrison, and that he was not to be trusted. If the House of Commons let him get away with it, they were fools. He demanded that the precedents should be looked up for a member of the Cabinet abstaining from voting and spoke openly of resigning. It was left to Brendan Bracken to talk him round, a task which he performed with great skill, producing only a single case of abstention, from MacDonald's National Government (a precedent which he knew Bevin would dislike), and appealing strongly to his personal loyalty to the Prime Minister[1] for whom his departure from the Government would be a grievous blow.

Whether it was Bracken's intervention or his own better judgment which calmed him down, Bevin said no more about resigning and, on

[1] Brendan Bracken to Bevin, 30 November 1943.

1 December, voted with other ministers in support of Mosley's release. The open quarrel, however, between two of the three Labour members of the War Cabinet did not help to improve matters in a divided Party. Morrison had been making a series of speeches which attracted a lot of attention and were an impressive attempt to work out a Labour policy for the post-war period. It was all the more unfortunate that at just this moment Bevin's suspicions of Morrison should have been fortified, leading him to see in Morrison's independent initiative only a bid to replace Attlee as the post-war leader of the Labour Party and removing any prospect of reaching agreement between the three Labour ministers on a common programme.

Still ruffled and angry from the Mosley affair, Bevin succeeded in making a good many other members of the House angry the following day, when he announced the arrangements for young men to be chosen by ballot for service in the mines. When members began to ask questions and some to call for a debate, he retorted with unnecessary heat that the responsibility for the measures proposed was legally his, that the House could certainly debate the matter if it wanted to, but that this had nothing to do with him and that he meant to put the scheme into operation at once, a way of handling the House which stirred up immediate opposition.

When the debate took place a fortnight later, Bevin had recovered his temper, but now puzzled his audience by his inability to express himself clearly. The occasion was remarkable for the unanimity of mine-owners and miners' M.P.s in welcoming a government measure. Most speakers recognised the fairness of Bevin's scheme and there was little disposition to criticise its details. When he rose to reply Bevin said bluntly, "It is no secret to say that the Government cannot leave the coal industry where it is after the war . . . and expect it to survive."[1] But the hopes which he expressed—based in fact on the proposals he had made to Lloyd George[2]—were wrapped in such obscure and cryptic language as to leave the House quite unable to penetrate his meaning. It was on this occasion that the *Manchester Guardian* made the comment: "Like all oracles he speaks in riddles."[3]

[1] House of Commons, 17 December 1943, Hansard, Vol. 394, col. 1856.
[2] See above, pp. 261–4.
[3] *Manchester Guardian*, 18 December 1943.

5

The two Bills for which the Ministry of Labour was responsible in the 1943–4 session show Bevin in yet another light. One was the Disabled Persons (Employment) Bill, embodying the recommendations of the Tomlinson Committee; the other was the Reinstatement in Civilian Employment Bill. Neither was of major importance, but the first was very close to Bevin's heart, and the second provided an unexpected test of his skill as a parliamentarian.

Bevin entrusted both Bills to his parliamentary secretaries, Tomlinson in the first case, McCorquodale in the other, allowing them to make the opening speech in the Second Reading, to take the lead in committee and sum up for the Government on the Third Reading. He stayed in the background, holding himself in reserve for any difficulties but allowing his junior ministers to have most of the limelight. After the quarrels he had had with his Party he might well have decided to play the leading part in promoting two Bills which were certain to be popular in the Labour Party, and the fact that he chose not to do this attracted favourable comment from a number of speakers in the debate.

The first of the two Bills was a good example of Bevin's skill in using the needs of wartime to carry through a social reform which he had urged in vain as a trade-union leader. Bevin was perfectly frank about this, refusing to claim the war as the justification of his scheme for re-habilitating the injured or to limit its provisions to ex-servicemen. Its real justification, he told the House, was the cost to the nation, and to the individual, of neglecting men injured in war or industry:

"The terrible loss of skill which was involved was impressed upon me when I took office and found that nearly 200,000 people had been written off in cold blood by local committees as no good to society and a permanent charge on the Assistance Board. These men were marching up twice a week to parade their misery and their suffering and yet by measures of this kind in this war that number has now been brought down to 18,000."[1]

Arguing that, whatever the cost of rehabilitation to the State, it was an investment in restoring human skill, Bevin added: "It is not enough merely to say that this country is going to be poor. . . . This

[1] House of Commons, 10 December 1943, Hansard, Vol. 394, col. 1346.

country will not be able for the next fifty years to afford an unemployed man or to allow a man to be kept away from industry
because he is unfit or injured."[1]

The Bill was given a second reading with the warm support of all
parties and of the press. When the committee stage was reached in
January 1944, however, opposition was raised on two counts. A
group of Conservative M.P.s tried to amend the Bill to give preference for training and employment to "war-disabled" as defined in
the Pensions Appeal Tribunals Act of 1943, and to limit the Minister
of Labour's power to make administrative regulations. Bevin stepped
in to handle both amendments and did so adroitly.

He was opposed in principle to making any distinction between
those injured while serving in the Forces and those injured while
serving the country in civilian employment, and he was not the only
one to think that the movers of the amendment were trying to make
political capital out of championing the cause of the ex-serviceman.
But he also recognised the strength of popular feeling and proposed to
create a preference by administrative action. This did not satisfy the
Tories who pressed for the distinction to be written into the Act.
Bevin therefore produced, off the cuff, an amendment of his own
which met their objection without sacrificing the flexibility of
administration which he was anxious to keep.[2]

The second objection, to the Minister's power to issue regulations
under the Act, was part of a general campaign conducted by Conservative members to establish more effective control by the House of
Commons over delegated legislation. Bevin and other Labour
members were annoyed at the attempt to make a test case out of this
particular Bill. "I think it is most unfortunate," he said, "that in the
case of the halt, the lame and the blind, a political fight should have
been started."[3] The amendment he had himself suggested, however,
defining but not curtailing his powers under the Act, divided the
opposition and although the issue was pressed to a division Bevin

---

[1] Ibid., col. 1349.

[2] Bevin's amendment promised that in selecting names for training or employment, "the Minister should so exercise his discretion" that, "so far as is consistent
with the efficient exercise of his powers", preference would be given to men and
women who at some time had served in a branch of the Armed Forces or in the
merchant marine. The original amendment had involved a cumbersome reference
to the Pensions Act which Bevin avoided.

[3] House of Commons, 27 January 1944, Hansard, Vol. 396, col. 1008.

successfully defended his position by a majority of seventy-two to eighteen. The Act was finally passed on 1 March 1944.

Between July 1941 and September 1945, when the permanent scheme for rehabilitation came into effect, 426,000 disabled men and women had been interviewed by Ministry of Labour officers. Of these not less than 310,806 had either been placed in employment or given training; another 28,500 were awaiting training or placing; another 67,000 had been found to require no assistance. They are surprising figures. No one would claim that the 1944 Act was a major reform; it aroused little controversy, stirred no enthusiasm amongst those who would not be content with less than a radical transformation of society, and, once passed, was taken for granted. It was one of the measures, however, which gave Bevin the greatest satisfaction to have had a hand in passing into law, and added one more item to the long list of practical humane measures which give him his place in the history of British social reform.

The second Bill with which he was concerned in the opening months of 1944 was of a different character. A clause had been written into the National Service Act of 1939 guaranteeing those called up under the Act reinstatement in their previous employment. As it stood, this clause was unworkable for a number of reasons. No machinery for enforcement had been provided; the employer's legal obligations were not defined and new provisions had to be made to take account of the fact that as many as five or six men then in the Services might have held the same job at some time during the war. The simplest plan would have been to scrap the original clause but this would have been represented as a dishonouring of the Government's pledge. On the other hand, it was difficult to postpone action until the general plans for demobilisation and resettlement were ready, for cases of men discharged from the Services were cropping up daily and the Ministry of Labour needed workable legislation to deal with them. The Bill, in fact, had a very limited purpose; it was not intended as a substitute for, or even as the first instalment of, the Government's major scheme for resettlement at the end of the war.

All this was said clearly and repeatedly by McCorquodale and Bevin in the debate which accompanied the second reading of the Bill (3 February 1944). None the less a number of members, on both sides of the House, persisted in treating the Bill as if it were the Government's only proposal for dealing with the very much bigger problems of demobilisation and post-war employment. Peter Thorneycroft, in particular, startled the House and stirred Bevin to anger by a bitter speech denouncing the Bill as misleading, a near-swindle, and a political manœuvre. Bevin did not disagree with Thorneycroft's view that it was impossible to fit the problems of 1944 into the framework of 1939, and attached little importance to the Bill, except as the redemption of an out-of-date pledge. But he resented the implication of dishonesty and let his anger appear only too plainly in a reply which brought down upon him the old reproof that he was far too inclined to take criticism personally.

Once his anger had blown itself away, however, Bevin showed the same patience he had displayed during the debates on the Catering Wages Bill in meeting objections and carrying opinion with him. This was particularly true of the committee stage (17–18 February) in which his interventions showed his mastery and flexibility as a negotiator. When he had to deliver a set speech at the opening of a debate Bevin could, on occasion, score a real success.[1] In replying to a debate he was frequently at his worst: he expressed himself poorly, his ideas were disorganised and his temper liable to carry him into ill-timed outbursts. It was in the more informal atmosphere of a parliamentary committee that his true gifts shone out: his rich and varied experience, his grasp of the issues under discussion, his resourcefulness in finding alternative means of achieving his object and his gift for cutting through legal terminology to the real problems of real people. The Reinstatement Bill is a good example of this: by the time he had reached the end of the committee stage, Bevin had so far won over the opposition that the Reinstatement Bill, in the end, passed without a single division and with virtually universal goodwill.

6

In September 1943 the British reached the peak of their wartime mobilisation. At that point over twenty-two million men and women out of a population of thirty-three million between the ages of fourteen and sixty-four were serving in the Armed Forces and Civil Defence or were employed in industry, an expansion of three and three-quarter

[1] For example, his speech on manpower in September 1943 and his speech on employment policy in June 1944.

million in four years. As impressive as the scale of the expansion was the size of the transfers which this had involved. The Armed Forces, less than half a million strong four years before, now numbered over four and a quarter millions. The munitions industries had grown by close on two million workers, the less essential industries had been reduced by more than three and a quarter million.

This was the limit of the country's resources in manpower. What that meant was seen as soon as Bevin presented the autumn 1943 manpower survey to the Cabinet. A gap between manpower demand and supply was familiar enough, but the situation this time was different. Demand was as high as ever, a total of 1,190,000 men and women required by the Services and industry in 1944, but this time there was no supply at all: in 1944 the total intake from all sources would not suffice to make good the ordinary wastage from the labour force and, even if no men or women were called up for service in the Forces, there would still be a deficit (or, to put it another way, a fall in the occupied population of the Kingdom) of 150,000. Looking at the numbers asked for, Bevin wrote in his covering memorandum:

"These demands cannot be met. The standards and amenities of the civil population cannot be further reduced. A fresh review must be made of the uses to which the available manpower is to be put."[1]

This was a disquieting report on the eve of the great battles which could be expected to mark the invasion of France, with greater losses of men and material than ever and no additional resources of manpower from which to make them good. Everything turned upon the duration of the effort which was now required to defeat Germany and end the war in Europe. If this could be done by the end of 1944, then it would be a reasonable risk to take men for the Forces from the munitions industries, relying on the stocks already built up and American aid to see that the Army did not run short of equipment and supplies. If the war lasted into 1945, then not only American equipment but additional U.S. forces would be needed to make up for the declining strength of the British Army. At a meeting of ministers on 5 November 1943, presided over by the Prime Minister, it was decided to assume that the European war would be over by 31 December 1944 and plan accordingly.

[1] Quoted, Parker, p. 211.

With the date for which they were to plan settled, the allocation of cuts was handed to a two-tier committee in which a team of officials worked out proposals and put them up to a group of four ministers, Bevin, Lyttelton and Cherwell (Paymaster General) meeting under the chairmanship of Sir John Anderson.[1] Their plan, accepted by the War Cabinet on 1 December, provided more men for the merchant navy, transport and mining, but not much more than a third of the numbers originally asked for by the Services; and that could only be done by taking more than 300,000 men from the munitions industries. Even the aircraft industry had to accept a reduction of 69,000 in its labour force, and the Ministry of Supply more than 200,000.

While the Ministry of Labour set to work to find the numbers called for in these allocations, the Minister was already deep in plans for the reverse operation, demobilisation. In the course of the autumn session, Bevin had repeatedly stated the Government's intention of avoiding the mistakes of 1918 by publishing such plans well in advance. In November 1942 the War Cabinet had agreed that demobilisation should be based upon a combination of age and length of service in determining the date of a man's release, with exceptions limited to those urgently needed for reconstruction work, returned prisoners of war, married women and cases of great personal hardship. There the matter rested until the summer of 1943 when a new complication was introduced by the likelihood that the war against Japan might continue after the end of the war in Europe. The possibility of a two-stage ending to the war threw the whole plan back into the melting pot and the Prime Minister proposed the appointment of a new committee, composed of a number of parliamentary secretaries under Sir William Jowitt's chairmanship, to produce a fresh scheme.

The decision to set up a new committee (in which the Ministry of Labour was not represented) had been taken while Bevin was away in Scotland on the first holiday which he had been persuaded to take during the war.[2] When he got back and heard what had happened,

[1] Anderson succeeded Kingsley Wood as Chancellor of the Exchequer in September 1943, but retained his personal responsibility for manpower questions.

[2] It seems to have been on the occasion of this trip to Scotland that Bevin was persuaded to pay a visit to the Fleet. He was received in great style by the First Lord of the Admiralty, A. V. Alexander, whom Bevin had known from his Bristol days but who now appeared to Bevin's scornful amusement in the uniform of an Elder Brother of Trinity House. Not content with this, the First Lord paced the

he was furious. He had no intention of letting anyone else handle demobilisation and this may well be the occasion on which for the only time during the war he is reported to have lost his temper with Churchill. It hardly needs saying that when the unfortunate demobilisation committee produced its report on 22 November, it came under heavy fire from the Minister of Labour. Bevin took exception to any idea of talking about demobilisation until the end of the war with Japan as well as with Germany; he criticised the exceptions to the principle of release on age plus length of service as too complicated and too numerous, and he persuaded the Cabinet to allow him to make alternative proposals. When he got back to his Ministry after the Cabinet meeting he sent for Godfrey Ince and told him to put the proposals they had discussed into a paper. When Ince asked how soon the Minister wanted it, he got the reply, "Yesterday." Without waiting for a typist, Ince wrote out Bevin's and his own alternative scheme by hand, calling the two categories of release Class A and Class B, labels which immediately stuck. Having taken the precaution of securing the agreement in advance of the Secretary of State for War (Grigg), Bevin brought it to the Cabinet in December and got it accepted in place of the Committee's report. After that there was no further question of who was to have the responsibility for carrying it out.

Instead of a plan for demobilisation (a word all too likely to be misunderstood both in the U.S.A. and at home), Bevin proposed that the needs of the Services for the Far Eastern war should continue to be met by the call-up of eighteen-year-olds and those whose military service had been deferred, while a "re-allocation of manpower" took place between the Services and industry, releasing men for return to civilian employment on the basis of age and length of military service. A second and much smaller group, limited to men with

---

bridge of the destroyer on which they were being taken out to the Fleet and summoned the Yeoman of Signals to send several rather pompous and unnecessary messages to "their Lords of the Admiralty", the Commander-in-Chief of the Fleet, etc. When these had been dispatched, Alexander suddenly saw a battered and dirty collier passing close by. He again called for the Yeoman of Signals and instructed him to signal her: "A message from the First Lord of the Admiralty: God Speed and God Bless You." This was too much for Bevin. "Here," he called to the Yeoman, "send them another message: From the Minister of Labour, would you like a bob a day more?"

special qualifications (Ince's Class B), would be released for re-construction work, but they were to have a shorter period of paid leave (three weeks instead of eight), and would be subject not only to direction by the Ministry of Labour but to recall by the Forces if they failed to comply with the directions they received.

The scheme was finally approved by the War Cabinet on 2 February 1944. To have told people at the time, however, that Britain had passed the peak of her mobilisation and the Ministry of Labour was now working on the plans for demobilisation would only have provoked a flood of angry comment. January 1944 was the fifty-third month of the war: the First World War had lasted fifty-two. Alamein was more than a year ago; Hitler had survived the over-throw of Mussolini and blocked the advance in Italy; the Allies had still not landed in France, had still not launched their main forces against the enemy. Everyone was certain that the Allies would win the war: why didn't they get on with it? In the early months of 1944 a large part of the British nation was in one of its periodical moods of disillusionment and irritation with its Government.

One sign of this was the by-election results. In January 1944 a Commonwealth Party candidate at Skipton defeated the Conser-vative despite Government support and a Conservative majority at the last election of 5,000. In February, at Brighton, one of the safest Tory seats in the country, the former Conservative majority of 40,000 was reduced to under 2,000 by an Independent challenge, and a fortnight later Lord Hartington, standing for the Cavendish family seat of West Derbyshire, was beaten by over 4,000 votes. The official Labour candidate won at Kirkcaldy on 18 February but with a majority considerably reduced by the anti-Government vote. These results improved nobody's temper. In a debate on the war situation (22 February) Churchill attacked those who felt "that the way to win the war is to knock the Government about, keep them up to the collar and harry them from every side." In another reference to the In-dependent candidates at by-elections, he spoke scathingly of "the little folk who . . . frolic alongside the Juggernaut car of war, to see what fun or notoriety they can extract from the proceedings."[1]

During March 1944 the House of Commons was the scene of a series of angry debates over reconstruction plans. Housing, the future of civil aviation, the White Paper on the National Health Service, the

[1] House of Commons, 22 February 1944, Hansard, Vol. 397, col. 700.

Education Bill—each in turn provided the occasion for open clashes between the parties and for much criticism of the Government, from the Right (over the Health Service, for instance) as much as from the Left. The Education Bill was the one major piece of legislation for the post-war period which the Government had so far presented to the House: its slow passage through committee and the divisions to which it gave rise provide some justification for Churchill's view that a controversial programme of post-war reforms—even if they could have been agreed on by the coalition Cabinet—would have dangerously divided and distracted Parliament and nation. An attempt to fix the date for raising the school-leaving age to sixteen produced a vote as high as 137 against the Government. A second division, on equal pay, a week later (28 March) led to the Government's first defeat. The vote, 117 to 116, was unrepresentative of the House and most of those who voted against were taken aback at the result. Greenwood rose at once to plead with the Government not to take this defeat too seriously, arguing that the vote had not been intended as censure of the minister responsible (Butler) or of the Education Bill as a whole, still less of the wartime coalition. Churchill, however, was in no mood for appeasement. He had already expressed his scorn for those who treated the Government as "a set of dawdlers and muddlers unable to frame a policy, or take a decision or make a plan and act upon it".[1] Determined to assert his mastery of the House, he insisted on a vote of confidence on the specific issue on which the Government had been defeated: the amended clause was deleted from the Bill and the original reinstated by a vote of 425 to 23.

This episode aroused fresh criticism of the Prime Minister. Those who were impatient for reforms and frustrated by Churchill's refusal to embark on them in wartime argued that the coalition had a right to demand down-the-line settlement only on issues directly connected with the war and that Parliament should be given much greater freedom to discuss domestic matters.

"The leadership of the war is not in question," *The Economist* wrote; "but for every one elector who, two months ago, suspected that the Government was needlessly obstructing reform or who doubted whether Mr. Churchill is the man to lead the country in peace as well as in war, there must now be three or four."[2]

[1] Broadcast of 26 March 1944, in *The War Speeches of Winston Churchill*, Vol. III, p. 111.

[2] *The Economist*, 8 April 1944.

There was renewed talk of Churchill's "dictatorship", and of the rights of Parliament in face of the Executive's encroachments.

Everyone was aware that the nation was on the eve of great events. Nobody living in the southern half of England in those months could fail to notice the signs that the preparations for the assault on the mainland of Europe were at last coming to a head. Instead of the Dunkirk spirit of 1940, however, impatience at the delay combined with anxiety for the result to produce a state of irritable tension. The House of Commons was particularly sensitive to this atmosphere as the frayed tempers of members and the exaggerated language of charges and counter-charges show.

During March other ministers had borne the brunt of criticism in the House. Bevin had taken part in the divisions, and had not concealed his anger with the members of the Labour Party who consistently voted against the Government. Apart from carrying the rehabilitation and resettlement Bills through their final stages, however, he had only spoken in reply to questions.

Two matters, in particular, brought a continual stream of questions directed to the Minister of Labour: "Bevin boys" in the mines and domestic service, both issues which were bound up with the explosive subject of class distinction. There was no doubt that service in the mines was very unpopular and resented not only by the boys but by their parents, especially if they came from middle- or upper-class homes. Their grievances, some of which, like rates of pay, were well founded, were given undue publicity and Bevin had to stand up to a good deal of criticism especially from the Tory benches. Domestic service was bound to suffer from the increased demand for women in industry, and again the Minister of Labour came in for harsh words. A flood of letters to the press and questions in Parliament recounted the difficulties to which his policy was subjecting overworked housewives and mothers,[1] and arousing middle-class resentment.

---

[1] Bevin was in fact very much interested in the problem and had already asked Miss Violet Markham and Miss Florence Hancock to take his ideas and make them the basis of a report on the organisation of domestic service on a new, professional footing. This could only apply to the post-war period, however, and their report was not published until the summer of 1945, when it led to the setting up of the National Institute for Housecraft with the object of securing proper training, pay and status for domestic work. Miss Markham made a spirited defence of the Minister in a letter to *The Times*, 6 April 1944.

7

Far more serious than these grievances over conscription for coal-mining and lack of servants was an increase in the number of strikes which not only taxed Bevin's skill in handling industrial disputes but plunged him into the thick of political controversy.

The centre of the trouble, as it had been so often before, was in the coalfields. In the winter of 1943–4 the Miners' Federation put in a claim for a new minimum wage. The award by the National Reference Tribunal (known from its chairman as the Porter Award) granted a rise in the minimum, but fixed it at £1 a week less than the miners had asked for. This the Federation accepted. Much more likely to create trouble, as the miners' leaders warned the Government, was the tribunal's refusal of the claim for an increase in piece rates. For the raising of the minimum without a corresponding increase in piece rates threatened to wipe out the accepted differentials between one man and another, especially in the more poorly paid coalfields like South Wales; in particular it affected the earnings of the men who worked at the coal face and the skilled craftsmen on whom output depended.

In an effort to remove the anomalies created by the Award, miners and owners hurriedly negotiated new piece-work rates in one district after another, expecting the Government to meet the additional cost from the Coal Charges Account. Alarmed at the prospect of a steep rise in the price of coal, the Government first hesitated and then, on 11 February, announced that it would do nothing of the sort.

This announcement was not followed by any sharp increase in the number of strikes: fewer miners, in fact, were on strike in February than in January after the original Award. But there was every prospect of real trouble if the Government did nothing more. The National Reference Tribunal in making its award had spoken of it as a temporary expedient pending the thorough review of the wages structure of the industry "which is long overdue". No one believed this more strongly than Bevin and he now persuaded the Cabinet that, however much they might have hoped in 1942 to keep the Coal Control's operation of the mines separate from questions of pay, they could no longer avoid taking the initiative in pressing the industry to reform its complicated and highly unsatisfactory wage system.

Nothing less, he argued, would give the mining industry any hope of stability and nothing but a strong lead from the Government would bring either the owners or the miners to make the fresh start which was needed.

Bevin not only carried the Cabinet with him but, working closely with Lloyd George, the Minister of Fuel,[1] played the leading role in the negotiations which began, on the Government's initiative, in the second week of March 1944. The owners, true to form, adopted a sceptical attitude from start to finish; they agreed to sign the agreement when it was completed but said openly that they did not expect it to bring peace, stability or any other benefit to the industry. Bevin ignored them, concentrating his attention on the miners' representatives. He secured the Cabinet's consent to offering them a national agreement to last for four years, thereby committing the Government to maintain control of the coal industry after the end of the war. The "district ascertainment" dividing the proceeds of the sale of coal between wages and profits, a practice which had been the cause of much bad feeling and continual disputes since its introduction in 1921,[2] was to be suspended; the district output bonus was to be abolished. Instead a national minimum wage, consolidating the wartime advances, was to be paid, fixed at a higher rate than in any other industry, while piece workers were to be given proportionate increases. Finally, no further general change in the rates, once they had been settled, was to be sought by either side for the four years during which the agreement ran.

As the miners' leaders recognised, when the negotiations were completed, this was the biggest step ever taken to meet their claims for fair treatment. But agreement was not easily reached and the miners' negotiating committee manœuvred to secure further concessions. Bevin was too experienced a negotiator to be put off by such tactics, but discussion of the different points took time and, since the miners were organised in a Federation, every one had to be referred back to

[1] On 6 April Lloyd George wrote a personal note to Bevin: "I feel that before the day is out I must thank you again most warmly for all your help and support during the last few days."

[2] A major objection to the ascertainment system was the exclusion from the district proceeds of the profits of coke and other by-products, a practice to which Bevin took strong exception. (See p. 262 above.) For the details of the 1944 negotiations and settlement, cf. Court, c. xiv, and R. Page Arnot: *The Miners in Crisis and War*, pp. 392–403.

their Districts which were far from unanimous in their views. By the end of March, two months after the Porter Award, nothing had been settled and an angry mood was growing among the miners who were as critical of their own leaders as of the Government. In the first half of the month, 15,000 men had been on strike in the Scottish pits, 100,000 in South Wales. Before they could be got to go back, 120,000 came out in Yorkshire and stayed out from 16 March to 11 April—in all a total loss in output of 1,850,000 working days since 24 January.

Nor were strikes limited to the miners. An unofficial Tyneside Apprentices' Guild which had been set up in the shipbuilding yards started a national strike of apprentices against their inclusion in the ballot for "Bevin boys". 6,000 came out on Tyneside; 5,000 on Clydeside and another 1,000 in engineering works at Huddersfield. A few days later, nearly 30,000 workers were reported out in the Belfast shipyards and engineering factories, a protest against the imprisonment by the Northern Ireland Government of five shop stewards.

The Prime Minister and the other members of the Cabinet looked to Bevin to get the situation under control quickly. If he failed, Churchill was not the man to hesitate in ordering military measures. If, on the other hand, he lent himself to too drastic steps against the strikers, he would lay himself open to the charge of strike-breaking and attacking the workers. There was no sharper test for a trade-union Minister of Labour during the whole war.

Bevin met the challenge boldly. Without waiting for the press or Parliament to demand that the Minister of Labour should do something, he determined to act first and so keep the game in his own hands.

On 4 April he was due to speak at a lunch given by the Civil Engineering Construction Conciliation Board. He startled his audience by saying that the policy of relying on joint industrial co-operation which he had persuaded the Government to adopt in place of military control was in danger of being wrecked at the eleventh hour by a single industry.

"I would say to the miners: You do not live alone. If you pursue this policy of wrecking industrial agreements properly arrived at and destroy the National Arbitration Court for which you have worked for thirty years and have now got, and continue to throw over your leaders, then you will reduce the position to anarchy. And it is not the miners alone who will suffer; it is the great mass of the working class.

"What has happened this week in Yorkshire," Bevin added, "is worse than if

Hitler had bombed Sheffield and cut our communications. It is the most tragic thing that in Britain you can do more harm by thoughtless action and lack of discipline than your enemy can do to you."

Instead of trying to conceal, he deliberately underlined the implications of the situation—"whether my policy survives or whether other steps have to be taken". The liberties of the people depended upon the smooth working of joint industrial relations. "I do not want everything from the top. I believe in the struggle, but I believe in an ordered, disciplined, methodical, regulated struggle."

The miners had suffered more than most industries between the wars, but since 1940 more had been done to remedy their wrongs than for any other group of workers.

"Apart from the political issue of nationalisation I can stand with my hand on my heart and say the miners have achieved everything in the last four years that they have fought for since 1912. I think therefore we are entitled to claim from them good output and loyalty and at least to have our lines of communication open and our soldiers supported.

"Enough of that. I am bound, however, to take this opportunity of saying it because in the next few days great decisions on this question of industrial relations have got to be taken. We are not going to lose this war, whether it is apprentices or miners or anybody else. We are not going to have this country let down. We cannot afford it. Too much is at stake."

The next day Bevin's speech not only captured the headlines but won strong support in the press.

"The Minister of Labour," wrote the *Daily Mirror*," is facing a difficult problem with courage and firmness. . . . There is of course a great deal to be said on both sides, but there is no point in arguing about these matters at the present time. The eve of the great attack upon Europe is emphatically the wrong time for strikes, unofficial or otherwise."[1]

Even the *Daily Worker* urged the Yorkshire miners to go back to work.

The same morning, 5 April, Bevin went to Transport House and took the unusual step of attending a meeting of the General Council of the T.U.C. With a South Wales miner, Ebby Edwards, in the chair, he pressed the T.U.C. to issue as strong a statement as possible condemning the strikers. If the trade-union movement did not master the crisis, the Government would be forced to take emergency action, and

[1] *Daily Mirror*, 5 April 1944.

all that the T.U.C. and he had done to make the unions equal partners with employers and the State would be thrown away. There was little disposition to disagree with the view he put and the meeting ended by endorsing a statement signed by Citrine and Ebby Edwards which gave Bevin the full backing of the trade-union movement.[1]

During the next week Bevin held on to the initiative he had gained by his speech of the 4th. He met the Miners' Executive on the 6th and pressed them too in the strongest possible terms to get the Yorkshire strikers back to work and to sign the new agreement on wages. He got both. The Yorkshire colliers were working again by the 12th and on the same day a miners' delegate conference meeting in London gave the Federation Executive power to accept Bevin's wage proposals. After further negotiations the agreement was signed on the 20th.

At the same time Bevin drafted an amendment to the Defence Regulations which, without resorting to emergency measures, strengthened the Government's hands in dealing with anyone who attempted to foment or exploit a strike in an essential service. The penalty for incitement was raised to a maximum of five years' penal servitude or a fine of £500 or both—a sufficiently strong threat, Bevin hoped, to make anybody stop and think before starting trouble. He argued the case for it at a meeting of his Joint Consultative Committee on 11 April and secured the support of both the employers' and the unions' representatives. On the 17th an Order in Council was published adding a new section to this effect—1AA—to the Defence Regulations.[2]

---

[1] Text in the *Daily Herald*, 6 April 1944. The I.L.O. was due to meet in Philadelphia in April and Roosevelt had written to Churchill urging him to send Bevin to lead the British delegation and spend at least a day with him at the White House. It was an invitation which Bevin would dearly have liked to accept—he had never met Roosevelt—but he had no doubt that it would be a mistake for him to leave the country at such a time and he told Churchill that he must refuse (6 April).

[2] Prosecutions had already begun of three men and one woman, members of the insignificant British Trotskyite party, who were implicated in the apprentices' strike. This action was taken under the 1927 Trade Disputes Act, not under the Defence Regulations. The chief figure in the Trotskyite group was Jock Halston who described himself as the national organiser of the British section of the 4th International. The group published a monthly, *Socialist Appeal*, from an address in Harrow Road which was raided by the police. On 19 June the four accused were found guilty of acts in furtherance of a strike but not of inciting the apprentices to strike. The men were given sentences ranging from six to twelve months, the woman thirteen days, which meant her immediate release. The sentences were subsequently quashed on appeal.

By the 20th Bevin could report to the Cabinet that the strikers were back at work, the law suitably strengthened and the new national agreement on wages signed by the two sides of the mining industry. He had mastered the industrial crisis and done so without impairing his policy of co-operation between employers, unions and the State. But he had now to face a storm of parliamentary criticism over the methods he had used.

8

Parliament had risen for the Easter recess on 6 April and did not reassemble until the 18th. The Order in Council publishing the new regulations was laid on the table of the House in the normal way but this did not satisfy those who maintained that the Minister of Labour had deliberately slighted the House of Commons by consulting the T.U.C. and the Employers' Confederation without coming to the House or giving it the opportunity for a debate. Aneurin Bevan took the lead in drafting a "prayer" to annul the Order and in pressing the Government to provide time for a debate, a demand which Churchill finally agreed to accept.

In the ensuing week pressure was brought to bear by Labour ministers and by union leaders to avoid another split in the Party's ranks. A recommendation to support the Minister of Labour was accepted by the Parliamentary Labour Party at a meeting on the 27th which Bevin attended for the first time since the Beveridge row. None the less, when the debate took place the following day the House heard a bitter indictment of the Minister of Labour and the trade-union leadership, delivered not by a Tory but by the Labour member for Ebbw Vale, Aneurin Bevan, and supported by half a dozen other members of the Labour Party.[1]

Two immediate issues were involved: the character of the strikes in the early months of the year and the procedure by which Bevin had introduced Regulation 1AA. Bevan accused the Minister of Labour of working up a campaign of calumny against the miners through the press and of trying to cover up the failure of the Government's industrial policy, especially in the mines, by starting a witch hunt for

[1] These were Pritt, Silverman, S. O. Davies and the three Clydesiders, Kirkwood, Buchanan and McGovern.

agitators. In his reply Bevin dismissed this charge of an "inspired" press campaign as a lie, pointing out that he had waited three weeks to see if the unrest would die down before saying anything at all and arguing that, if he had remained silent in the face of the strikes spreading, he would have been held to account by the House. The most surprising omission in his defence was any mention of the strenuous efforts he had made to cut through the confusion created by the Porter Award and to secure for the miners the national wages settlement which went a long way to meet their grievances. Regulation 1AA was in fact only one of the steps he had taken to deal with the strikes, and by allowing the opposition to fasten on it in isolation from the rest he failed to make the best of his own case.

In one of the more effective passages of his speech Bevin pointed out that much of the criticism was directed, not against Regulation 1AA, but against Order No. 1305, which had established compulsory arbitration in place of strikes and had been accepted by the unions as the basis of wartime industrial relations. To demand an unfettered right to strike, as the Opposition now appeared to be doing, war or no war, was to invite chaos.

To some extent, Bevin's defence suffered from the timing of the debate. By 28 April the strikes were over and the crisis past. The fears of the first weekend of April appeared exaggerated. Nor did his earlier references to Trotskyites help him: none of his critics was prepared to take their part in the strikes seriously. But there was evidence (to which Greenwood referred in the debate) of strikes being started by irresponsible minorities who either cared nothing about the consequences of their action or were out to make trouble. However distasteful it might be to Labour's traditions, Greenwood argued, no Government could tolerate such activities at the stage which the war had now reached and it was on these grounds that he accepted the case for granting additional powers.

Bevin himself said in the course of the debate that he hoped the new regulations would never have to be used, and this proved to be the case: there were in fact no prosecutions under clause 1AA of the Defence Regulations. It is indeed unlikely that Bevin ever meant there to be: what he wanted was a deterrent, something with which to impress anyone minded to stir up trouble that the Government was not to be trifled with. But this could obviously not be said in public if the threat was to prove effective.

The sharpest criticism of Bevin was over the procedure he had followed in drafting the new regulation in consultation with the T.U.C. and the Employers' Confederation and then publishing it without giving the House an opportunity to debate it first. It was for Parliament, and Parliament alone, Bevan thundered, to decide whether a case had been made out for further restrictions on the liberty of the individual, not for the T.U.C. or the Employers' Confederation, bodies unknown to the constitution and responsible to no one but themselves. "All these are matters for the House of Commons, not for outside bodies. There was therefore no reason why the right hon. gentleman should conspire behind our backs and make laws in our absence."[1] This was not merely to strike a blow at parliamentary institutions: "I say that it is the enfranchisement of the corporate society and the disfranchisement of the individual."[2]

This part of Bevan's indictment clearly took Bevin by surprise. As Greenwood reminded the House, if the T.U.C. had not been consulted, Nye Bevan would have been the first to protest and Bevin, never very sure of his ground when it came to parliamentary procedure, argued that consultation with interested parties was a normal practice which had been followed with the Roman Catholics and the Church of England in the case of the Education Bill without anyone suggesting that Parliament was being slighted. When Bevan interrupted to ask why the Minister had not proceeded by a Bill instead of by an Order in Council, Bevin replied, convincingly enough, that he had not wanted to introduce legislation which, once it was on the Statute Book, might be difficult to remove. The Defence Regulations were limited to the wartime emergency and the powers in dispute would lapse once the war was over.

Mixed up with the criticism of the way in which Bevin had treated the House of Commons, and giving it much of its emotional force, was indignation at the way in which he had treated the Parliamentary Labour Party by carefully consulting the T.U.C., and even the employers, but neglecting to say anything to the Labour members of Parliament. It was the Labour Party, Bevin was told, which represented the working classes (including the millions of non-unionists), not the T.U.C., which, Bevan added, was out of touch with

---

[1] House of Commons, 28 April 1944, Hansard, Vol. 399, cols. 1063–4.
[1] Ibid., col. 1072.

rank-and-file opinion in the unions and represented only the trade-union officials.

"Do not let anybody think that he is defending the trade unions; he is defending the trade-union official . . . who has become so unpopular among his own membership that the only way he can keep them in order is to threaten them with five years in gaol."[1]

Not content with asserting the political party's autonomy, the opposition carried the attack into the trade-union camp. Bevin had provided safeguards against anything said at a trade-union meeting being used for the purposes of a prosecution under Regulation 1AA and thereby laid himself open to the charge that he was creating a privileged position for the unions from the protection of which the thirteen million workers who were not trade unionists were excluded. What had the Law Officers to say, Silverman demanded, about such discrimination against the ordinary citizen? Bevin's reply that he intended nothing of the sort but was simply safeguarding rights which the law had accorded to the trade unions since the Act of 1875 did not satisfy the opposition.

Aneurin Bevan went further still. The Minister of Labour's policy throughout the war had been to channel all the workers' grievances through the machinery set up by the trade unions. If they tried to redress their grievances in any other way—by striking, for instance—they were threatened with imprisonment. This claim to a monopoly by the trade unions deserved the attention of Parliament: were the trade unions willing to have all their rules submitted to scrutiny by Parliament? "Of course not: they dare not." But Parliament ought not to be content with anything less before handing over the citizens of the country, including the millions of workers who were outside the unions, to the protection of a self-appointed group of trade-union officials responsible to nobody but themselves.

Whatever the faults of the trade unions—and nobody in the House of Commons was ignorant of these—Bevan had overreached himself in attacking not simply their shortcomings but the principle of trade-union organisation.

From the national point of view, without the much abused union officials and the T.U.C., leave alone Bevin at the Ministry of Labour, the organisation of war production and the mobilisation of the

[1] Ibid., cols. 1071-2.

country's manpower to anything like the point achieved in 1943–4 would have been impossible. The position of the trade unions might be open to abuse but Bevin was on strong ground in arguing that the only alternative to the system of industrial relations he had built up with the support of the trade unions was the military control of labour.

From the working-class point of view the one way in which the workers had ever been able to improve their position had been by organisation. This was the basis of the trade-union movement. "But I know," Bevin remarked, "that some hon. Members would rather the working classes went to Hell through chaos, than that they won a victory by organisation."[1]

The decision to give up the strike weapon during the war in return for arbitration had not been taken by the General Council or by union officials: it had been made by a conference of nearly two thousand representatives from the union Executives, the great majority of them lay members.

"I do not believe," Bevin added, "that the working classes of this country have lost by that decision. The evidence . . . of the wage increases and the changes and improvements is the greatest evidence of the wisdom of that decision. The miners themselves have not lost by the new agreement which has recently been arrived at. . . . I believe that if that great conference were called tomorrow and that they were asked whether they wanted to give it up for the rest of the war, there is not an executive who would do so."[2]

Even from the party point of view, while no political party which aimed at winning a majority and forming a Government could accept the view of the Labour Party as the political instrument of the trade-union movement, to carry repudiation of this to the point of openly attacking the trade unions on their own legitimate ground of industrial policy was to break up the Labour Movement. The Parliamentary Labour Party might be "sick to death" (Buchanan's phrase) of being told by trade-union members like Glanville (the Durham miner who vigorously defended Bevin in the debate) that "the trade-union movement is not the infant, but the parent of the Parliamentary political Labour Party,"[3] but Glanville had some grounds for his retort that neither the member for Ebbw Vale nor

[1] Ibid., col. 1131.
[2] Ibid., col. 1124.
[3] Ibid., col. 1113.

most of the other Labour members would be in the House at all if it had not been for the trade-union movement. This was not only true historically: without trade-union contributions (89 per cent of the Party's central income in 1944[1]) and without trade-union organisation to back it, the Labour Party would have been little more effective than the I.L.P. The Parliamentary Labour Party had a legitimate grievance against the way Bevin had treated it on this and on other occasions, but Bevin had also grounds for complaint in the persistent attacks to which he and other Labour ministers had been subject from a section of the Party throughout the war, in flat contradiction, to Bevin's way of thinking, of party solidarity and often of the majority decisions of the Parliamentary Group. The movement would only survive if the trade-union and political wings learned to live together on terms of mutual toleration. This was not going to be made easier by violent attacks such as Bevan's on the personal integrity of the trade-union leaders, "a speech," Arthur Greenwood told him, "of an anti-trade-union character the like of which I have never heard from the most diehard Tory in this House or outside."[2]

Bevan's attack went too far for most members but the strength of the feeling which existed in the Party is shown by the voting. The Government's majority in repudiating Bevan's "prayer" was convincing enough, 314 to 23 (16 of them Labour votes), but only 56 out of 165 Labour members voted in support of Bevan. And the division within the Party was underlined by the steps which both sides took to carry the quarrel outside the House. The day after the debate Bevin spoke at the annual Bristol Festival of the T.G.W.U. in the Colston Hall. He described the half hour he had spent listening to Bevan's onslaught as one of the saddest experiences of his life.

"I could understand it from a Conservative or a reactionary, but as I sat on the Front Bench I said to myself, more in pity than in anger, 'Why do men curse the ladder over which they have climbed to position? Why do they? Why do they want to kick it away?' ...

"Friends, what has the Labour Party been, what has it done? It is the fashion now for men like Harold Laski, Aneurin Bevan, and Silverman, the intelligentsia, of those people who claim to be members of our Party, to ridicule us, to denounce us, to say we are slow, to say we are conservative, to say we are reactionary.

[1] See Table 13 in Martin Harrison: *Trade Unions and the Labour Party since 1945* (1960) and the discussion in c. 2.

[2] *Hansard*, Vol. 399, col. 1118.

"After all, we are the Labour Party. In this very city I have stood, for week-end on week-end, outside St. George's Park gates on Sunday nights, preaching socialism to our people, without any hope of being anything or anybody. . . . Why, when we have done it, do they then turn upon us in a way which can throw disrepute and discredit upon us?"

Tribune retorted by declaring that

"a free vote throughout the Trade Union movement would kill the Regulation and dispose utterly of the claim of Mr. Ernest Bevin, which he has palmed off on the credulous Churchill, to speak as the Labour Emperor whose lightest whims are honoured law among the working people of this land."[1]

The issue had now become much wider than Regulation 1AA. "Recent events in which I have been involved," Bevan told his constituents, "were the result of attempting to force Tory policies down the throats of Socialists." And he went on to speak of the need to be vigilant unless they were going "to submit to the corporate rule of Big Business and collaborationist Labour leaders."[2]

The parliamentary leadership of the Party demanded Nye Bevan's expulsion, but despite the backing of the Administrative Committee, and the clear recommendation of both Attlee and Greenwood, a majority voted for Shinwell's amendment to refer the matter to a joint meeting of the Administrative Committee and the National Executive. The result of this was a statement deploring Bevan's action in ignoring decisions of the Parliamentary Party and causing disunity within the ranks of the Party: he was not expelled from the Party but required to give an assurance that he would in future abide by the Standing Orders of the Party. Bevan gave the assurance but the gloss which he put on it hardly promised a peaceful future for the Party. If he gave it, he declared, it was

"because I believe that there are elements in the Party which wish to continue association with the Tories when the war is over [and] I refuse to allow myself to be manœuvred out of the Party and thus leave them with a clear field in which to accomplish the ruin of the Labour Movement."[3]

---

1 *Tribune*, 5 May 1944. The same number contains a reply by Laski to Bevin's attack on him.
2 Quoted by Michael Foot: *Aneurin Bevan*, Vol. I, 1897–1945, p. 461.
3 Ibid.

CHAPTER 10

More Post-War Plans, Demobilisation— and Greece

I

THE ROW over Defence Regulation 1AA[1] was followed, rather surprisingly, by an improvement in the relations between Bevin and the Labour Party. One reason for this was a general feeling that Aneurin Bevan had gone too far and a reaction in favour of the man he had singled out for attack. Another was a greater willingness in the Party to recognise the size of the job Bevin had done at the Ministry of Labour and even to take pride in the fact that it had been done by a Labour minister.

But the most important reason was a change in the role Bevin himself began to play. Although he had been active for a long time behind the scenes in planning for the period of reconstruction, in public he had held back from saying much in order to avoid stirring up controversy. A good many of Bevin's disagreements with the Labour Party could have been removed if he had felt free to say what he thought and not felt obliged to defend coalition policy even when he disagreed with it.

In 1944, the situation changed. The task of mobilisation was over; in June the invasion of Europe was successfully launched and the war had clearly entered its final phase. The discussion of post-war plans could no longer be avoided, and there was no longer the same reason for avoiding it. In the final year of the war the Prime Minister and the other members of the coalition adjusted themselves as best they could to the task of preserving sufficient agreement to see the war through as a united Government while leaving themselves sufficient freedom to express their differences over post-war policy.

The result was to show a different Bevin to the country—and to the

[1] S.R. & O. 1944, No. 461.

Labour Party—a man as interested as ever in the improvement of industrial relations, but one who could no longer be represented, in the way Aneurin Bevan had tried to represent him, as a man who could only look at post-war problems from a trade-union point of view, a man with an interest, vigorously expressed, over the whole field of government policy.

He made no more attempt now than he had before to win support in the Labour Party or trim his sails to catch popular favour. He said what he thought, whether it found favour or not, and in December, for instance, went out of his way to support Churchill's policy in Greece against the tide of feeling at the Labour Party conference. But much of what he said, and the force with which he said it, had a strong appeal to many members of the Party who had been irritated by his stubborn defence of coalition policy in earlier years. This is not the sort of change which can be dated at all precisely, but it is a change clear enough from a comparison of his 1944 speeches with those he made in 1943. By the election of 1945 it had at last established Bevin as one of the leaders of the Labour Party and, in the eyes of many, as a national leader too.

The question which brought out the change most clearly was the reconstruction of industry. Plenty of other people were interested in plans for starting up British industry and trade again after the war: Bevin's contribution was his insistence on the social purpose of industry, particularly the provision of employment. He interpreted his responsibility for demobilisation to mean not simply releasing men from the Forces or war industry but finding them jobs. And that in turn meant seeing that there were jobs for them to find.

He was convinced that, if the older industries like cotton and coal were to compete in post-war markets, they would need to modernise not only their equipment but their ideas. He said something of this when he went to Manchester in January 1944 to open an exhibition at the Cotton Board's Design Centre.[1] He made his meaning plainer still in a private talk with Dalton, the President of the Board of Trade, early in March. His list of the reforms which he believed necessary to make cotton an efficient industry ranged from bulk purchase of the raw material, fixed transport charges and standard prices from spinner to manufacturer to the need for compulsory amalgamation, double shift

[1] His speech to a conference of Cotton Industry Representatives in the Houldsworth Hall (7 January 1944) was reprinted as a pamphlet by the Cotton Board.

working, the extension of automation and improvements in sales organisation based on the British Overseas Cotton Corporation.

One point on which Bevin laid emphasis both in his talk with Dalton and in his speech in Manchester was the change that would have to be made in the industry's labour force. "Lancashire has hitherto depended on children and married women. I doubt very much whether married women will go into the cotton mill as they did in the old days and it is obvious that with the development of the education scheme there will not be the availability of children."[1] The cotton industry, he told its leaders in Manchester, would not get back the 175,000 operatives it had lost during the war unless employers changed their ideas drastically about amenities and the conditions of work. They would also have to introduce a simplified wages system and sweep away anomalies carried over from the past. Cotton, like coal, was an old industry, much burdened with out-of-date traditions and habits of mind, both on the unions' and the employers' side. It needed to re-discover the radical, innovating temper which had made it a leader of the earlier industrial revolution.

He said the same thing to the Institution of Production Engineers at Cardiff:

"A hundred years ago you were very enterprising people in tinplate, steel, copper and coal; you were in the vanguard. . . . Can anyone tell me how much research is going on in Wales now? . . . I have a feeling that there is not enough localised research and imaginative action going on down here . . . I have been wondering what the great enterprises in Wales have been doing . . . whether they are looking to the Government to do it all or to the big combines outside Wales. I tried to look up the list of inventions in the last ten or twenty years to see what the outstanding Welsh inventions are. . . . If there have been any they have either been sold to London or the inventors have hidden their light under a bushel."[2]

Bevin continued to urge the need for more research and scientific development in the coal industry in particular, writing to Attlee in March 1944 that the rise in the cost of coal as a result of the new wage agreement provided an incentive for technological progress. This was an argument he had already used in Manchester:

---

[1] Note of a discussion with the President of the Board of Trade on Cotton, 2 March 1944.
[2] Speech at the Engineers' Institute, Cardiff, 24 March 1944.

"I am convinced," he said then, "that if coal had not been so cheap, our power situation in this country would have so scientifically developed that the men could have been paid properly and power could have been much cheaper long ago."[1] .

He told Attlee that because coal had cost so little he believed not much more than 25 per cent of the potential value was being extracted or turned to full account. It took a pound of coal to produce a unit of electricity: suppose this could be cut by half, it would have an immediate effect not only on the electricity supply industry, but on the manufacture of generating plant and other electrical equipment. Greater efficiency in the use of fuel could be applied to steam no less than to electrical plant. "If we can get an immediate and vigorous study of these problems we should not only improve the efficiency of industry here but we should develop a great market for our products overseas."[2]

But Bevin was interested in something more than plans for the revival of particular industries; he wanted a clear government commitment to maintain full employment as an overriding object of state policy. For months a committee of senior officials had been discussing how far the Government could go in this direction.[3] In 1944 the question came before the Reconstruction Committee and Bevin pressed strongly for a government statement in unequivocal language. Besides the Keynesian proposals to check a slump by maintaining total expenditure on goods and services—a programme for keeping up purchasing power of which he had long been in favour—he wanted to see the Government given powers to secure the location and development of industry in accordance with the social needs of the nation, above all the need to provide employment.

With the depressed areas of the 1930s very much in mind, he asked if the State could afford to leave it to private enterprise to decide where new factories were to be started. Would this produce a balanced distribution of industry in those parts most vulnerable to unemployment?

---

[1] Speech at the Houldsworth Hall, Manchester, 7 January 1944.
[2] Bevin to Attlee, 17 March 1944.
[3] The Economic Section of the War Cabinet Office had produced a paper on full employment after the war as early as 1941.

"My view is that if public and private industry as we have understood it hitherto, that is public trade services together with private manufacture, can provide a balanced industry, then irrespective of other considerations which may arise, I should have no objection to giving them a fair chance to do it. . . . But where private industry either does not, will not or cannot establish a balanced industry, then steps must be taken to do it. I think private industry is placing its claims far too high when it takes the line that even at the expense of social wastage and human deterioration they must have a monopoly of manufacture. As long as they serve the social purpose in accordance with the State requirements, there can be no quarrel, but when they cease to do that, surely the right thing to do is create organisms which will do it."

This quotation is taken from a lengthy note which Bevin dictated for his own use at the beginning of February 1944. He went on to marshal a number of cases in support of his argument.

One was the Lanarkshire coal pits. These were nearing exhaustion and the men employed in them ought to be transferred, before Lanarkshire became a distressed area, to the neighbouring Fife coalfield which was capable of development. This could not be done because the two coalfields were owned by different companies.

"Or take transport. I hold that the essential thing for the East coast is to develop a great dock undertaking in the Tees . . . if that area is to survive. Yet is it likely that with the Railway Company, with its miserable Middlesbrough Dock and the multiplicity of wharves all up the river and the lying outside of big ships for days on end, [this] will be remedied by waiting for private enterprise to adjust its differences or by any appeals that may be made by the Board of Trade or anyone else?"

South Wales provided another example. The greatest success of pre-war policy in the depressed areas had been to get Richard Thomas's to build their new steel strip mill at Ebbw Vale. But the consequence of this was to ruin the tin-plate industry of West Wales and create unemployment there. To whom was West Wales to appeal for help to prevent itself becoming another depressed area?

The answer, Bevin concluded, was self-help: to set up publicly owned corporations, backed by state capital and associated with the Local Authorities, which would run industries in districts threatened with unemployment. If private firms were willing to do the job, they should be given every opportunity, but the community ought not to be left without some form of defence to preserve its social capital, the industrial skill of its people.

"I am certain," Bevin added, "when the shipbuilding industries closed down at Jarrow, had the Corporation of Jarrow been allowed to build ships it could not only have built them efficiently but could have sold them and probably assisted the shipbuilding industry, instead of merely looking for a method, as the capitalists were at that time, of keeping prices up."

The more he heard of talk in business and Conservative circles about their plans for starting up the post-war economy, the more certain he became that the Labour ministers had got to press for the public interest not to be subordinated to the profit of private interests. In a letter to Attlee (18 May 1944), he criticised the Party for failing to put a distinctive socialist point of view on a whole range of questions from land development to the nationalisation of transport.

"I am very much concerned," he wrote, "about this question of compensation and betterment. . . .
"Here is our Party with some of the greatest Local Authorities of the country under Labour control, with tremendous issues at stake affecting [property] development and the Party doing nothing at all in the matter . . . I think that as Leader of the Party you really ought to take some vigorous steps to get some opinion on the matter.
"Another matter that is worrying me a good deal is civil aviation.[1] All the vested interests are on the rampage.
"The Prime Minister has taken the line that he will not agree to nationalise anything during the war. We must await a general election. Yet it looks as if Max Beaverbrook and all the forces associated with him are attempting to de-nationalise what we have got, and I think that there has got to be a pretty strong Labour view put to the Cabinet on the question of civil aviation.
"Leathers is very carefully avoiding putting up any proposals on transport at all . . . I had lunch the other day with the General Managers of the railways and they left me in no doubt of their assumption that the railways are just going to be handed back and they talked with such a degree of confidence that it left me with a feeling that there must be some understanding that we know nothing about.
"There is also the question of road transport—they seem quite comfortable as to the post-war position . . .
"I plead for a clear lead," Bevin ended, "and that if we have to stand up to it, we must. But if we cannot, as a coalition, carry any nationalisation of mines, railways or electricity, surely the Party must make its position clear and keep its hands clean for the Election; and in Uthwatt, public ownership and transport, denationalisation of the air service there is a clear-cut division and I really think we ought to face it."[2]

1 This was the subject of debates in the House of Commons on 29 February and 14 March 1944.
2 Amongst Bevin's papers are three other letters to be taken with that of 18 May. The first, to Dalton, dated 15 May, strongly urged him to trim down the tin-plate

Besides putting up a fight to prevent control of the economy being handed back to private interests, at least before a General Election, Bevin attached great importance to getting a statement which would bind both parties—whichever won the election—to a policy of full employment. The final draft of the White Paper on Employment Policy[1] corresponded with what he wanted and Bevin was the chief spokesman for the Government in introducing it to the House. Chapter II of the White Paper made out an unanswerable case for maintaining controls and rationing in the transition from war to peace. Chapter III was an elaboration of a sentence (para. 26) which might well have come from Bevin's notes: "It will be an object of government policy to secure a balanced industrial development in areas which have in the past been unduly dependent on industries specially vulnerable to unemployment." Chapter IV contained the key promise: "The Government are prepared to accept in future the responsibility for taking action at the earliest possible stage to arrest a threatened slump. This involves a new approach and a new responsibility for the State." Chapters V and VI set out the proposals for carrying this promise into effect, including varying the rate of social insurance contributions and even taxation in order to keep up purchasing power, and Bevin's favourite plan for continuing the manpower budgets into peacetime so that the Government could foresee what measures would be necessary to maintain employment.

Woolton and Bevin held a joint press conference when the White Paper was published and Bevin recalled the Bank Act of 1844, exactly a hundred years before. Under the free trade system which had been operating since that time, "whenever the exchanges went wrong, you rectified them by restricting credit and producing unemployment."

"The first check that made that almost unworkable was when Governments began to introduce social service, because you could not starve the people quickly enough to rectify the exchanges, and that was a problem that was

---

industry's proposal for a redundancy levy on the grounds that it was designed to protect firms with obsolescent and inefficient plant. The second, to Anderson, now Chancellor of the Exchequer (16 May), protested against speculation on the Stock Exchange and asked if he could not tax the gains which were made in this way. (Anderson's answer was "No".) The third, to Attlee, dated 25 May, again urged him to make an issue in the Cabinet of the proposal from Beaverbrook to hand over civil aviation to private enterprise.

[1] May 1944 (Cmd. 6527).

facing us very much . . . in 1929 on the Macmillan Committee. To-day this plan just leaves the nineteenth century behind and it says in effect that, instead of the human being having to fit himself into an exchange system, the exchange system has to fit into human requirements. It reverses the policy we have been following ever since the industrial revolution."[1]

2

Ten days later, the invasion forces made their assault on the Normandy coast. It had taken four years exactly from the Dunkirk evacuation to mount an invasion in sufficient strength to break through the German defences and drive the enemy out of Western Europe. They had been four interminably long years in which the group of men at the centre—the War Cabinet, the Chiefs of Staff, the senior civil servants—had stuck to their plans despite setbacks, criticism and a strain on resources and morale which few Governments have had to face. Many had contributed to the planning and organisation which finally launched the armada on its journey across the Channel: with the exception of Churchill, few had contributed as much as the Minister of Labour on whose judgment and steadiness of nerve had depended so much of the industrial effort and the co-operation of labour needed to equip the forces which now set sail.

As a gesture of recognition for the part he had played, Churchill invited Bevin to join Eden, Smuts and himself in his special train at Droxford, near Portsmouth, from where they could go to see the men embarking. It was not the most convenient accommodation: Eden complained that there was only one bath and one telephone and "Mr. Churchill seemed to be always in the bath and General Ismay always on the telephone". The Prime Minister was outraged to learn that Bevin cleaned his own shoes and made a great fuss about the man who was responsible for mobilising the manpower of the entire country being left to look after himself. He instructed one of his aides to see at once that a Royal Marine was sent to act as the Minister of Labour's batman. Bevin was embarrassed: "I wouldn't like you to do that, Prime Minister," he said, "I get such splendid ideas when I'm cleaning my boots."

More serious was the Prime Minister's quarrel with de Gaulle. Much moved by his sense of history, Churchill had advanced to greet

[1] Verbatim notes of the M.O.I. press conference, 26 May 1944.

the Frenchman with arms outstretched, only to discover that other great man in one of his most unyielding moods and bitterly offended at the way he had been treated, particularly by the Americans. The result was to exasperate Churchill, who finally declared that if it came to the point he would always side with the United States against France. Eden did not like this pronouncement nor did Bevin, who said so in a loud aside. His sympathy with the French and his understanding of the difficult state of mind created by the humiliation of defeat and occupation were already evident. His opposition was not lost on Churchill who later took Eden furiously to task for stirring up Bevin and Attlee and trying to break up the coalition.[1]

As they stood watching the men file past and crowd into the landing craft, some of the members of his own union recognised him and called out: "Look after the missus and kids, Ernie." Bevin's sense of humanity was as strong as Churchill's sense of history, and all too conscious of what awaited the men he had just spoken to when they reached the other side of the water, he walked back to his car in tears.

The preparations for invasion had produced unexpected demands on manpower which had been hard to meet. "Mulberry", for instance, the prefabricated floating harbour, called for several different types of labour all in short supply. To build the concrete caissons (Phoenix) required a force of 20,000 men drawn from the reduced building and civil engineering industries. To assemble the prefabricated parts of pier heads and pontoons (Whale), 1,600 men were needed, 600 of them electric welders of the highest proficiency who could only be found by loans or withdrawals from war production. A third part of the project—Bombardon, the erection of steel structures to form a shelter for ships in the deep water anchorage outside the Phoenix breakwater—required 6,000 men engaged on prefabrication of the parts and another 2,000 (including 500 skilled men) on the site at Southampton chosen for the assembly.

"Our manpower had already been allocated," Bevin later told his Ministry staff, "and then they came to us for an additional 20,000 men. Every man was supplied. I had a wonderful interview in that little conference room at St. James's Square. I decided that day that I would argue with nobody. So they made up the list of the people they wanted and I said to the Army—you must give me so many carpenters by next Wednesday, to the Navy so many steel benders, and to everybody else the rest, and the meeting ended. One

[1] *The Eden Memoirs: The Reckoning*, pp. 452–3.

representative went back to his department and his Minister asked him what had been arranged. He said, 'I have to deliver 400 carpenters by next Wednesday.' 'And what did you say?' he was asked. 'I said nothing—I was told,' he answered."[1]

In each case the men were found and each part of the Mulberry programme finished on time.

Quite apart from emergencies, the manpower planning involved called for careful organisation several months in advance without giving away what was happening. The operating staff on the railways, for instance, was raised by over 9,000 men in the first six months of 1944; 13,500 drivers were found for road transport and the number of registered dockers brought up from 70,000 to 77,000, many of them being transferred from London to the main invasion port, Southampton. At the critical moment the unions and their members had not failed the country: little direction had been needed, most men only asked to be told where they were wanted. Bevin was justifiably proud, both of the job of organisation carried out by his Ministry and of the voluntary response from the working class on which he had pinned his faith. He called a press conference to tell the country something of what had been done by the Ministry of Labour during the war but stepped aside himself and insisted that it should be taken by Godfrey Ince, the Director-General of Manpower.

Two days later, on 21 June, Bevin moved the adoption of the White Paper on Employment Policy by the House, opening for the Government and making one of his best speeches since he had entered Parliament. He spoke with unusual lucidity, turning the Treasury language of the White Paper into terms everyone could understand, yet showing a sure grasp of the changes in economic policy which it proposed. He also brought a sense of historical occasion to the debate, no mean feat when everyone's attention was drawn towards the battles being fought in Normandy.

Bevin linked the two in the most natural way in the world by telling the House what the men of the 50th Division had said to him on the eve of embarkation: "Ernie, when we have done this job for you, are we going back to the dole?" Conservative members, shocked at the familiarity, interrupted: "Ernie?" they queried. "Yes," Bevin answered, "it was put to me in that way because they knew me

[1] Speech to Headquarters staff at the Ministry of Labour, 6 March 1945.

personally. They were members of my own union and I think the sense in which the word 'Ernie' was used can be understood."[1]

Without any appeal to the emotions or any recrimination, he succeeded in making the whole House understand how deeply the experience of being unemployed had burnt itself into working-class feeling and how much it meant to him and to the men and women he represented that, after more than fifty years of protest, the State should accept the obligation no longer to alleviate but to prevent unemployment. From the time of the industrial revolution down to the present war, unemployment and deflation had been regarded as the automatic correctives for any lack of equilibrium in the nation's economy. It had been not only a callous but a remarkably wasteful system.

"From 1922 to 1939 we lost 250 million days of production through strikes and lock-outs alone. Over 60 per cent of these disputes arose from the need for adjustments due either to deflation or Gold Standard adjustment, and were outside the control of industry. . . . With all that loss of 250 million days, wages went down, wages went up, went down again and went up again. What was the nett result at the end? The change in money wages over the whole field of sheltered and unsheltered industries . . . was only five points. We had all these fights and struggles going on throughout the country, with all the consequent difficulties, and the adjustment was five points. . . .
"In that same period of 17 years, we had an average of 1,700,000 unemployed and we paid out a total of £1,260,000,000 in benefits and assistance. That payment helped to keep the consuming market going and to that extent probably prevented unemployment from being worse, but we had not a single pennyworth of production for all that expenditure. I do not think that that was good for the country."[2]

Now, instead of the automatic corrections of deflation and unemployment, it was proposed to introduce conscious direction of the economy for the first time. And this direction was to follow the pattern expounded by Keynes and now accepted by the Treasury: "to meet the onset of any depression at an early stage by expanding and not contracting capital expenditure and by raising consumption expenditure and not reducing it."

As *The Economist*[3] remarked, the Beveridge Report had not been

[1] House of Commons, 21 June 1944, Hansard, Vol. 401, cols. 212–13.
[2] Ibid., cols. 215–16.
[3] *The Economist*, 10 June 1944.

much more than a projection of familiar principles; the White Paper on Employment Policy represented a much more revolutionary change in state policy. Yet the first was greeted with "a chorus of hosannas"; the latter "now causes hardly a ripple". Why? Part of the answer lies in the dates of the two documents. In 1942–3, at the time of the Beveridge Report, the nation was still eager for a united lead from the coalition on post-war policy. Now, when the Government produced an agreed statement representing a radical break with past policies, and when Bevin in particular pleaded for a national approach to a common objective, the mood had changed.

Many who had led the demand for the coalition to take action on the Beveridge Report in 1943 had now altered their minds, pinned their faith on the return of a Labour Government at the end of the war and were wary of anything which might be used to keep the coalition in existence once the war was over. Every speaker on the Labour side, Greenwood and Shinwell as well as Nye Bevan, made the same point: how far could two parties, one of which believed in private enterprise, the other in socialism, travel together in carrying out a policy of full employment?

The issue, Bevan argued in one of his best speeches, was fundamental. Bevin's pragmatic approach was incompatible with socialism. If you were a socialist, how could you believe that the problem of unemployment could be solved without the "transfer of economic power" by the nationalisation of finance and the key industries.

"The subjects dealt with by the White Paper," Bevan declared, "represent all the matters which distinguish that side of the House from this. The question of how the work of society is to be organised, how the income of society is to be distributed, to what extent the State is to intervene in the direction of economic affairs—all these are questions which first called this party into existence. . . . Indeed, I will go so far as to say that, if the implications of the White Paper are sound, there is no longer any justification for this party existing at all."[1]

Bevan's conclusion was that the coalition Government had gone outside its terms of reference and that the issue, instead of being covered up, should be left to be decided by a general election.

Bevin had tried to meet this argument in advance. Rejecting the

[1] House of Commons, 23 June 1944, Hansard, Vol. 401, col. 526.

claim of the Tory press that the Labour ministers had abandoned their belief in the public ownership of industry, he argued that "the question of how you can give effect to the decision as to who will own industry is not prejudiced by this White Paper. . . . What we have tried to do is to devise a plan which, however you may decide the ownership of industry, . . . seeks to attain the objective [of full employment]."[1]

When a member interrupted to ask if he was in favour of "a continuous coalition", he replied "No" without hesitation. What he wanted was a new code in the conduct of industry, whether privately or publicly owned, which would accept full employment as "a common objective nationally" as it had accepted winning the war. He added, with some insight into the workings of Government, "It is in the attitude of mind, the direction of Government policy in the whole of the Civil Service, as well as ministerial support, that this problem must be faced."[2]

As the previous section has shown, Bevin was not blind to the conflict between socialist and Tory views on the future of industry, but he believed that the reforms which had "stuck" in British history were those which had been accepted by all parties and still thought it worth the effort to get the principle of full employment adopted by both sides. In the long run he was to be vindicated: whatever the differences in emphasis and conviction, the policy of full employment has been accepted, has had to be accepted, as common doctrine by all parties and Governments since 1944. But in the short run he was swimming against the tide of reviving party feeling: in the summer of 1944 no one on the Labour side of the House was prepared to believe in the conversion of the Conservative Party to the principles of the White Paper, and without the massive Labour victory in the 1945 election to drive home the argument it is arguable that their scepticism would have been justified.

3

Although there was a good deal of controversy in July over town and country planning—a White Paper and Government Bill—as well as

[1] Ibid., 21 June 1944, col. 214.
[2] Ibid., col. 213.

over housing, Bevin was not directly involved. The one debate in which he took part was arranged to allow George Tomlinson to report on the I.L.O. conference in Philadelphia. The attendance was small, most of the speakers friendly and the occasion would be hardly worth recalling, had Bevin not been drawn into expressing his views on the international settlement after the war.

A Conservative member suggested that one disadvantage of the I.L.O. was the fact that, while Britain carried out the conventions she ratified, other nations did not. Bevin's reply that they were not going to get very far by approaching other nations in "a spirit of unctuous self-righteousness" started a dispute about the possibility of ever trusting the Germans, the Italians and the Japanese. Was Bevin prepared to do this?

"Certainly," he answered. "When this war is over and they are back in the comity of nations . . . you will have to deal with these nations somehow or other. . . . We cannot ignore the existence of 80 million people anywhere in the world."[1]

This had already happened in the case of Italy: they had accepted Italian signatures on the armistice and recognised the new Italian Government. If the Foreign Secretary negotiated a treaty and then was told by the House of Commons, "We know that we are going to keep that treaty, but will the other people?," the conduct of foreign and commercial policy would become impossible. Labour conventions would have to be treated on the same footing. He would not accept the argument that, because the League of Nations had failed, it was an illusion to suppose that any international organisation could ever succeed. The League represented a first attempt at carrying out "the biggest thing in the world". Its failure and the fact that this had led to a second war had created greater support for any future attempt. It was necessity, not sentiment, which would force the nations to renew the attempt and go on until international organisation was made to work.

A great advantage of the I.L.O. in Bevin's eyes was its method of organisation. "If you can bring together people working in similar occupations, they soon forget their race and are talking their trade, and you produce a friendship which it is very difficult for war or anything else to break."[2] He was particularly anxious to get

[1] House of Commons, 26 July 1944, Hansard, Vol. 402, col. 853.
[2] Ibid., col. 582.

the Russians back into the I.L.O. It had been a mistake to force them out and it would be an even worse mistake to let the I.L.O. become involved in international politics and diplomacy.

"I do not want to paint the lily," Bevin concluded, "and say that the I.L.O. has done wonderful things. What I do say is that it is the duty of this generation to hold on to every international organism which has survived this war. We shall need them."[1]

Everything else that summer, however, was overshadowed by the tremendous news from the war. Immediately before the House rose for the summer recess, on 3 August 1944, even Churchill relaxed his usual caution so far as to admit that "victory may come perhaps soon". The next six weeks were among the most dramatic of the war: the break-out from the Normandy bridgehead and the landing in the South of France followed by the liberation of Paris and the greater part of France and Belgium; the sweeping victories in the East clearing Russian soil of the invader and carrying the Red Army into Poland and Rumania.

The progress of the war and the belief that the end was in sight were at once reflected in politics. As early as 11 September *The Times* reported that the Labour Party was virtually agreed on ending the coalition and fighting the post-war election as an independent party. *The Economist* welcomed this as the right decision for the future of British politics, but one more than likely to condemn the Labour Party to several years in opposition. Apart from Churchill's prestige, *The Economist* pointed to "the striking failure of the Labour Party to produce any figures of real eminence save for the special case of Mr. Bevin".[2] The *Manchester Guardian* made the same point, adding that

"personal jealousies even inhibit the party from capitalising the achievements of its two outstanding ministers, Mr. Bevin and Mr. Morrison. No Labour man goes about claiming for the credit of his party that Mr. Bevin has performed the greatest administrative feat of the war in mobilising the nation's manpower. What would the Tories not have made of Mr. Bevin's achievement, had he been their man? Mr. Bevin is left to thank God on those days when someone in his party is not actually attacking him."[3]

[1] Ibid., col. 589.
[2] *The Economist*, 23 September 1944.
[3] *Manchester Guardian*, 19 September 1944.

In these circumstances the publication of two White Papers[1] setting out the Government's plans for implementing the Beveridge Report took most people by surprise and left the supporters of the Report astonished. After all the anger created by the Government's original mishandling of the Report and the bitter reproaches heaped on its Labour members, Beveridge's proposals emerged unscathed and in certain respects strengthened. This was particularly true of the second White Paper which went further than Beveridge in sweeping away the old workmen's compensation and proposing a national scheme for insurance against industrial injuries. Bevin had from the first preferred such a scheme to Beveridge's, had borne the leading part in getting it accepted and regarded it with satisfaction as complementary to the measures he had already introduced for the rehabilitation of the disabled. The *Manchester Guardian* called it a great reform and rightly gave the credit for it to the Labour members of the coalition. Both reports were approved by the House of Commons in November and although the coalition ended without any attempt to translate them into legislation—the condition on which Churchill had always insisted—it is none the less true that preparations for turning the Report into legislation had been completed ready for any Government returned at the election to put into effect.

A fortnight after the White Paper on Social Insurance was published, on 11 October 1944, Bevin introduced the second reading of a Bill providing for a 20 per cent increase in unemployment insurance benefits. This was an interim Bill to meet the difficulties of the transition period between war and peace. As he pointed out in the debate, you could not mobilise nearly twenty-five out of thirty-six million people and then swing them back to peace-time occupations without disturbance and some temporary unemployment. The Bill brought the level of benefits up to the scale proposed in the White Paper and was to operate only until the new Social Insurance plan came into effect.

Bevin, however, never brought a Bill dealing with the welfare services to the Commons without being subjected to violent criticism by a section of the Parliamentary Labour Party—and this Bill was no exception. The prospect of any unemployment at all produced an intense emotional reaction and he found himself under attack

1 The first, on Social Insurance, was published on 26 September 1944; the second, on Industrial Injury Insurance, on the 27th.

because the increased benefits, without any increase in contributions, were still too low to live on.

This was a real issue between Bevin and other members of the Labour Party. They believed that a subsistence standard of living should be provided out of insurance funds: Bevin did not. Insurance, he held, ought not to provide subsistence, but a contribution towards it: the rest should come out of general taxation through the Assistance Board.

"I want to state a principle," he told the House, "I happen to be a Socialist and I am still a Socialist in spite of the chains that join me to the Coalition. . . . As a Socialist, I am never going to admit the principle that insurance is the right way in total to deal with unemployment."

After drawing a distinction between the views of the I.L.P. ("the Little Bethel of Socialism") and those of the Social Democratic Federation on which he had been raised, he added:

"I take the view, having gone into this insurance, that it was after all not a Socialist measure. It was a Liberal measure, a Liberal device . . . to avoid the actual steps that ought to have been taken to deal with unemployment. I have never departed from that principle . . . I am being asked to forsake all my Socialist principles and come down to the I.L.P. philosophy that the dole is the solution for unemployment and I am not going to do it."[1]

In the end, only six of those who had attacked the Bill as "reactionary" and "contemptible" voted against it; but the Labour Party showed no signs of conversion to Bevin's S.D.F. principles, the clash of views being repeated in the debate on the Social Insurance White Paper a few weeks later.

4

There was one question which could no longer be left without an answer: whether the coalition was to continue after the end of the war with Germany. There were many in the country who would regret the break-up of the wartime administration; and the making of the post-war settlement might well seem to require nothing less than a national all-party Government. In practice, however, for the

[1] House of Commons, 13 October 1944, Hansard, Vol. 403, col. 2139.

coalition to continue meant one of two things, either prolonging the life of a Parliament which had already sat for nine years without a general election, or holding an election in which the coalition would campaign as a united front. Neither was really a practicable course. When the possibility of a coalition appeal on the lines of 1918 was considered by the Cabinet, Bevin said later, everyone present declared himself opposed.

Without waiting for a Government pronouncement, the Liberal and Labour Parties settled the matter for themselves. Following a Liberal statement on 5 October 1944, the Labour Party Executive announced on the 6th that it would fight the first post-war election as an independent party. The statement was drawn up by Attlee and, after approval by the Executive, had been agreed with the Labour ministers in the Government. It vindicated the decision to join the coalition in 1940 and justified the loyalty of Labour ministers to the common cause not only in waging war but in preparing for the post-war problems any Government would have to face. Nothing must be allowed to conflict with the paramount necessity of bringing the war against Japan as well as Germany to a successful conclusion. But once that was said there was no equivocation about the future. "Despite malicious whisperings to the contrary, no responsible leader of Labour has ever toyed with the idea of a coupon election," a denial which Bevin repeated with emphasis in April 1945. The Labour statement added the hope that the coalition might be brought to an end "with dignity and good feeling", and the wish that "so great an adventure" should not end in "squalid bickerings".[1]

When Churchill came to move the prolongation of Parliament on 31 October he accepted, regretfully but without attempting to dispute them, the arguments which had moved the Labour Party to reach its decision. "Indeed I have myself a clear view that it would be wrong to continue this Parliament beyond the period of the German war."[2] This was accepted by everyone, without debate, as definitive. There would have to be an interval after the end of the war in Europe before the general election could be held, but no one proposed extending the coalition's term of office as long as the end of the Japanese war, until Churchill himself made this suggestion in May 1945.

With this settled, both parties began to make their preparations.

[1] Text in *Manchester Guardian*, 7 October 1944.
[2] House of Commons, 31 October 1944, Hansard, Vol. 404, col. 667.

Ralph Assheton, Bevin's former parliamentary secretary, was appointed chairman of the Conservative Party Organisation, and on 29 October 1944 the Labour Party Executive held a special meeting with the three Labour members of the War Cabinet, Attlee, Bevin and Morrison. A brief statement reported that the matters discussed included the Party Conference in December and "questions which the party will have to deal with when the progress of the war makes a general election possible".

Bevin's attendance at the meeting with the Executive was another important step in his rapprochement with the Party; so was his concurrence in the decision to end the coalition. He was not by temperament much of a party man; his loyalty was much more to the Labour Movement than to the political party as such and his only experience of politics had been as a minister in the coalition. What mattered to Bevin was to have a hand in framing the post-war settlement, in "laying down the conditions on which we start again", and he saw a strong case in favour of this being done by the same sort of National Government which had conducted the war and which, it could be argued, was just as necessary to cope with the difficulties of the transition from war to peace.

"National Government" and "coalition", however, were ambiguous terms and, in view of the Labour Party's earlier history, highly charged with political emotion. The Tories later put about the story that Bevin was prepared to join a National Government *even if Labour decided against it*, and this was eagerly accepted by those in the Labour Party who found fault with his loyalty to the wartime coalition. If an approach of this sort was made, there is no evidence that he ever considered it and an emphatic denial by Bevin himself that he would even have looked at such a suggestion.[1]

But a continuation of the existing coalition between the parties for a limited period after the war ended was a quite different proposition. In his memoirs, Eden recalls a conversation with Bevin on their visit to the embarkation ports in June 1944:

"Bevin appeared to think that it might be necessary to continue the National Government into the immediate post-war period and asked me if I knew what Churchill's intentions were. . . . He said that if the old man

[1] At Leeds, 7 April 1945. See below, pp. 368–70.

were to retire then he and I, he was sure, could work together, if that were the right thing to do. . . . He would not care which office either of us held. I replied, 'Neither would I.' There was however one thing that he must have. 'What?' I asked. 'The nationalisation of the coalmines.' The trade unions would have to have that."[1]

By the autumn, however, Bevin had come round to the majority view in the Labour Party and accepted it without demur. This still left open, as 1945 was to show, the question of *when* the General Election was to be held, and this was to cause further disagreement in May 1945. But the important decision, that Labour should leave the coalition before an election and fight it as an independent party, was not re-opened.[2]

In announcing its decision the Labour Party was at pains to stress its loyalty to the National Government as long as the war with Germany lasted. Even Nye Bevan, speaking in the debate on the prolongation of Parliament, had been prepared to concede that "it would be foolish to break up the National Government now. I think our representatives are entitled to say that, having gone so far, they must complete the journey and remain in the Government until Germany is defeated."[3] But as Churchill remarked, the odour of dissolution was in the air and the clash of interest and opinion was to be openly expressed from now to the end of the war.

The first big row of the autumn was over the compensation clauses in the Government's Town and Country Planning Bill. Bevin had spent much effort on finding a compromise which both parties could accept, however reluctantly, and Churchill went out of his way to praise his Labour colleagues for the concessions they had made and Bevin for his efforts "in the cause of unity".[4] To Churchill's undisguised irritation, however, Conservative ministers at the beginning of October decided that they could no longer accept the compromise clauses and reopened the whole question. This plunged the House into an angry dispute which lasted for several weeks. Some sort of Planning Bill had to be passed if the housing programme was not to

---

[1] *The Eden Memoirs, The Reckoning*, pp. 453–4. When Eden told Smuts that Bevin wanted nationalisation of the mines if the coalition was to continue, Smuts commented: "Cheap at the price."

[2] See below, pp. 374–77.

[3] House of Commons, 29 October 1944, Hansard, Vol. 404, col. 682.

[4] Churchill's speech in the House, 6 October 1944, Hansard, Vol. 403, col. 1372.

be held up indefinitely, and Bevin set himself to work out, and then persuade Labour to accept, a further compromise which finally allowed the Bill to pass.

The V1 and V2 attacks left no doubt that housing was the most urgent of post-war priorities: since 14 June something like a million houses had been destroyed or damaged, most of them in London. After the disputes of the summer, the Government had no difficulty in passing both its housing Bills[1] through their remaining stages at the end of September, despite strong feeling in the House that they were inadequate. At the same time Bevin met the last of the war's emergency demands on manpower by drafting more building labour into London to carry out immediate repairs. A force of 28,000 men already engaged on repairing earlier air-raid damage was raised to 60,000 by August, and to 130,000 by the end of 1944. The Ministry of Labour not only had to find the men and get them to London, but to take over the responsibility for housing them, no small task in a city where accommodation had been drastically reduced by the damage they were called in to repair. No one was satisfied with the speed of the repairs, yet by early 1945 the Ministry had brought no less than 40 per cent of all the building labour in the country into the London area, an indication of the extent to which even an industry as important as building had been stripped of its labour.

It was not only the building industry which was short of men. Even after the cuts made in December 1943, Bevin's Ministry found it difficult to meet the manpower allocations for 1944. A report to the Cabinet in June 1944 put the gap likely to separate supply from demand as high as 225,000 men and women. The ministerial committee on manpower was re-assembled and in two reports to the War Cabinet (6 July and 1 September 1944) revised the figures yet again, cutting everybody else's allocation but raising the Army's in order to meet the high casualty rate. The Forces actually ended the year 1944 with 80,000 more than they had been allotted in December 1943, an

---

[1] The first dealt with long-term, the second with emergency plans. Bevin told Edgar Harris, of the T.U.C., that Churchill had urged him to become Minister of Health and take over responsibility for the post-war housing programme. Bevin, so he told Harris, was willing, provided it did not mean surrendering his seat in the War Cabinet: he had no intention of giving up his wider interest in reconstruction. When Churchill told him that, if he went to the Ministry of Health, he could no longer be a member of the War Cabinet, Bevin turned the suggestion down and Churchill did not press him.

extraordinary last squeeze on the hard-pressed resources of man-power.

But the hardest job of all was to see what would be needed in 1945. Manpower calculations in the last year of the war proved, in fact, to be more difficult to make and more unreliable, when made, than at any time since manpower budgeting had begun. This was hardly surprising since no one could say when the European war would be over, how many men (and munitions) would then be needed for the Far Eastern war and how long that, in its turn, would last.

The first attempt at a guess was made by the Joint War Production Staff of the Ministry of Production working on figures produced by the Chiefs of Staff. The result was not impressive. When the Minister of Production presented them to the War Cabinet he suggested that the estimate should be cut by a million men. Bevin at once moved to repel this invasion of his territory by the Ministry of Production. He wrote a strong letter to the Cabinet in which he had little difficulty in questioning the competence of the Joint War Production Staff to carry out such an investigation or the arbitrary assumptions made by the Minister of Production in presenting them to the Cabinet. A compromise settlement by the Cabinet in April left it to the Ministry of Labour to work with the J.W.P.S. and the Chiefs of Staff in pro-ducing a better estimate of manpower needs after victory in Europe.

The result was scarcely more impressive. The estimated length of the Japanese war was cut from three to two years, but even that left a gap between supply and demand of a million and three-quarter men.[1] In fact, at no time before the Far Eastern war abruptly ended did the planners succeed in reconciling the proposals of the Chiefs of Staff with the manpower which industry would need to start on a partial reconstruction programme.

Even without the complications of the Japanese war, Bevin's officials were finding enough difficulties in devising a budget for the first six months of 1945 when, it was assumed, fighting in Europe would still be going on. The first draft of the autumn manpower survey, finished in October 1944, had to be revised before it was ready to present to the Cabinet in December. It proposed reinforcements of 140,000 young men for the Forces, but this was well below the estimated casualty rate while, even with the munitions industries

---

[1] See Parker, c. 14, for the details.

giving up more than 600,000 workers, the net reinforcement of civilian industry was not more than 400,000.

The ministerial committee made some improvements. They found more men for the Army, but their proposals for cutting down the munitions industries in order to hasten reconstruction ran into strong resistance from the supply ministries, particularly M.A.P. Two months after the committee first presented its proposals to the War Cabinet, it had to come back again, on 12 February 1945, and ask for a ruling between the irreconcilable interests. This they were given, but by 12 April 1945, when the figures were settled finally, the period under review was more than half over. The last manpower budget of the war (for the first half of 1945) had little more than an historical interest.

The same day that it presented its final report, the ministerial committee was asked to address itself to a manpower budget for the Japanese war on the assumption that victory in Europe would be won by 31 May. They were still arguing with the Chiefs of Staff about the size of Britain's military commitment when the coalition broke up and a general election supervened. Before the new Government had time to grapple with the problem, the A-bomb had been dropped and the war in the Far East was over.

5

Despite the untidiness of its final stages—few of its stages had been anything else—the wartime mobilisation was already accepted as a remarkable achievement. But Bevin had still to face the most severe test, at least politically, of his reputation as Minister of National Service. For demobilisation, if clumsily handled, could have produced the biggest political storm of all: no other issue so closely affected every man and woman in the Forces, and every family in the country. And the need to carry it out in two stages, starting with the end of the European war while the war in the Far East was still being fought, created problems which had not been present in 1918.

Bevin's proposals had been ready, and the necessary White Paper in draft and approved since the spring of 1944. When to publish it was a question to make any Government hesitate, especially with heavy fighting still in progress and the war, even in Europe, far from over.

By September 1944, however, Bevin was certain the time had come, and on the 12th of that month wrote a powerful memorandum for his colleagues in the Cabinet arguing for immediate publication of the full plan. The Cabinet accepted his advice and he presented the White Paper to a crowded press conference on the 21st.

Few state papers have had a more long-winded title: "The re-allocation of manpower between the Armed Forces and Civilian Employment in the interim period between the defeat of Germany and the defeat of Japan."[1] This was deliberate. Bevin wanted at all costs to avoid using the word "demobilisation" until the war in the Far East as well as in Europe was over, and he made it clear that the only basis on which the release of men from the Forces could begin was the continued call-up of new classes to replace those who had already served for long periods. If the title was clumsy, however, the proposals themselves were admirably simple. Up to the very last moment Bevin had had to resist suggestions to complicate them by bringing marriage or overseas service into the calculation. But he refused to depart from his original formula, arguing that this alone would produce what was needed, a scheme which could be grasped at once and recognised by everybody as fair. Two things only were to be taken into account, age and length of service, with two months' service the equivalent of a year in age. Those with the highest score on the two counts combined were to be released first. The only exception was a number of men (Class B) urgently required for reconstruction work (notably, house building) who were to be released out of turn. These were to be retained on the active reserve, subject to recall if they left the jobs to which they were directed, and were to receive three instead of eight weeks' paid leave as a gratuity. Nor would the numbers placed in Class B delay anyone's chances of release under the main scheme (Class A) since their places would be filled by calling up additional men who had hitherto been held back in reserved occupations.

The first reaction was favourable but guarded. Everything turned on the way in which the proposals were regarded by the men in the Forces, and the debate in the House was delayed until 15 November to allow time for this to become known. When the debate finally took place, it was abundantly clear, as Bevin had always been confident it would be, that they were widely and generally approved. This was

[1] Cmd. (1944) 6548.

the verdict of two serving Members (Lt.-Col. Profumo and Major Nield) who had been flown back from the front for the occasion, and it was borne out by all the other evidence. Attempts were made during the debate to persuade Bevin to recognise the claims of special groups but his answer in every case was the same: his overriding aim was to keep the plan simple and to defend the claims of the ordinary soldier in Class A against all comers.

Bevin's opening speech lasted over an hour and a half. His mastery of his subject was obvious and by the end of the debate it was clear that he had largely convinced the House that to add to the original scheme would weaken it. What might have been a matter of bitter controversy was thus removed from the agenda of politics, and when the time came to put it into practice the scheme worked so well that it was continued without modification for the full-scale demobilisation at the end of the Japanese war.

In the course of the debate Bevin remarked that most of the trouble after 1918 had come not from those who had stayed at home but from the ex-soldiers disillusioned on their return to industry. With that in mind he laid as much stress upon the measures proposed for re-settlement as on the arrangements for demobilisation.

Some of these had already been given legislative form, in the Disabled Persons (Employment) Act and the Reinstatement in Civil Employment Act (both dated March 1944). The White Paper on Employment Policy might also be regarded as part of the same policy, together with the Bill which Bevin had introduced to provide increased unemployment benefit during the transition from war to peace. To these were added during 1944 three (ultimately four) schemes to help the large number of young men and women whose training had been interrupted by the war. They were now to be given the chance to complete it with grants and training facilities provided by the State. For those wishing to go on to a university or prepare for a professional career there was the Further Education and Training Scheme; for those wanting to learn a skilled trade, the Vocational Training Scheme; for those who had begun but not completed apprenticeships the Interrupted Apprenticeship Scheme. The fourth, instituted in 1945, was a scheme for three months' full-time training in business methods, for which Bevin secured Sir Frederick Hooper as director.

To make sure that the existence of these facilities was widely known the Ministry of Labour not only distributed millions of leaflets and made teams of lecturers and films available throughout the Services, but set up Resettlement Advice offices in all the principal towns. Through these, men and women looking for jobs were put in touch with the local employment exchanges of the Ministry and, where necessary, with a new service for higher appointments which Bevin started to provide advice on careers and bring together employers and candidates for technical or administrative posts.

All this was a world removed from the pre-war conception of the Ministry of Labour's offices as principally concerned with the payment of the "dole". The inspiration and driving force throughout was Bevin's: it was he who saw the problem of demobilisation and resettlement as a whole, secured the support of employers as well as trade unions, linked it with the provisions of the new Education Act and gave the nation for the first time a comprehensive service in which the State helped its citizens to train for and find employment at every level from the industrial apprentice to the university graduate. This was an administrative reform which proved its value ten times over in easing the adjustment from war to peace and helped to establish a new attitude in post-war Britain towards the training and employment of skilled manpower. And Bevin made it very clear that even from the Opposition benches he would see to it that his plans for reallocation and resettlement were faithfully carried out.[1]

The day after the Commons debate on the scheme for releases from the Forces, the Government published a second White Paper[2] dealing with the parallel problem of civilian manpower between the end of the European and the end of the Far Eastern war. While part of industry would already be switching to production for reconstruction, it was necessary to keep sufficient industrial and manpower resources in hand to supply the Far Eastern fronts. This posed difficult problems of allocation and the question how far the Government was to retain its wartime controls.

The White Paper was only published after lengthy talks with both sides of industry. It established an agreed order of priority in the release of workers, beginning with Class K. This consisted of women with household responsibilities; women wishing to rejoin their

[1] See Bevin's speech in House of Commons, 16 May 1945, Hansard, vol. 410, col. 2531.
[2] Cmd. 6568. For a summary, see Parker, pp. 261-3.

husbands on release from the Forces; older women over sixty and men over sixty-five. At the same time, the scheme made provision for the release of men of military age to the Services as soon as they could be spared from reserved occupations and set out the controls—including the Essential Work Order—which the Government proposed to retain in the interim period. Registration of young women was to continue in order to replace older women in civilian work, and a comprehensive Control of Engagement Order required all men between eighteen and fifty and all women between eighteen and forty to continue to obtain employment through a labour exchange.

These proposals attracted much less attention than those for men serving in the Forces and were not debated by the House of Commons until May 1945. In practice they worked equally well and were retained, without serious modification, for the second stage which followed the abrupt end of the war with Japan.

In the course of the November debate, Bevin said:

"It was quite obvious that we should have to wind this country up to a point at which it had never been wound up before in terms of manpower. The great anxiety which we have had all the time is, having wound it up to such a point, can we unwind it in an orderly way, or will it snap? If it snaps there will be chaos."[1]

It did not snap and there was no chaos. Several million men and women moved from the Forces and war industry to peace-time occupations without any repetition of the breakdown which followed the 1918 war. Bevin's double achievement, the mobilisation and demobilisation of an entire nation, was complete.

6

The government programme for what proved to be the last wartime session of Parliment promised only two Bills of any importance: Bevin's Bill to set up wages councils and Dalton's to plan the distribution of industry. For the rest the King's Speech (29 November 1944) merely said that progress would be made "as opportunity serves". "Opportunity will not serve," the *Manchester Guardian* com-

[1] House of Commons, 15 November 1944, Hansard, Vol. 404, col. 2026.

mented, "and the Government knows it."[1] After the impressive output of White Papers and Reports this was a meagre result: the only part, for example, of the revised Beveridge scheme to be translated into law before the coalition broke up was the Family Allowances Act.

The reason for this is not hard to find. The coalition was still held together by the common purpose of winning the war, but on most domestic questions was sharply divided, each side watching the other suspiciously. Bevin thought that they must get through to victory as best they could and saw no point in trying to force a trial of strength on domestic issues before the election. This view was not shared by Morrison, who argued that they could not leave the reorganisation of the principal industries until the election and wanted the three Labour members of the War Cabinet (Attlee, Bevin and himself) to put forward far-reaching joint proposals.

Morrison had first made this suggestion in October when he drafted a memorandum outlining "a great national plan of capital re-equipment and technical reorganisation."[2]

Bevin was non-committal: no plan that came from Morrison was likely to rouse his enthusiasm and he suspected him of playing for the party leadership. He made no comment on the detailed proposals but ended his letter of acknowledgment with the pointed comment: "When the Leader puts a policy before all of us we shall have to give it our serious consideration."[3]

Morrison was not put off by this snub. He wrote back the next day:

"This episode reminds me once again of something I so often think about—we miss many chances by failure to discuss a line of action beforehand. When the two of us (or three of us) happen to find ourselves of one mind, as in the case of the iron and steel proposals, we really get somewhere. . . . Surely it isn't beyond our wit or will to make the necessary arrangements to keep together for the things that really matter?"[4]

Bevin did not reply, but when the matter came up again in November it became clear that he had other objections to such proposals besides his inveterate suspicion of Morrison. This time Attlee's own office had produced a draft paper—The Immediate Future of

---

[1] *Manchester Guardian,* 30 November 1944.
[2] Morrison to Bevin, 17 October 1944.
[3] Bevin to Morrison, 24 October 1944.
[4] Morrison to Bevin, 25 October 1944.

Financial and Industrial Planning[1], but Bevin was no more impressed by this than he had been by Morrison's. He gave his reasons in a long letter to Attlee on the 22nd:

"In my view," he wrote, "the presentation of these papers to the Cabinet now, with an election pending, cannot be any more than a demonstration. If, as a matter of tactics, it is desired to put forward these views in order to have them aired in a debate of the type we usually have in the Cabinet on these matters, with the ultimate view of showing what disagreement there is, I can understand that more, but I do not recommend it . . .
"If we attempt to set up the forms of organisation suggested in your paper, are we not thereby prejudicing the whole socialisation of industry? I myself am being forced to the conclusion that a country run by a series of London Transport Boards would be almost intolerable."

Bevin had never forgiven Morrison for his part in setting up the London Transport Board,[2] and to leave no doubt how strongly he still felt he added: "Boards like this would be unrepresentative, unresponsive and unlikely to pay much attention to the public interest."

But the old quarrel with Morrison was secondary. Bevin gave the real grounds of his objection in the next paragraph, and in doing so showed the extent to which he accepted the Labour Party's political objectives.

"In my view," he wrote, "the only real way to bring these big basic industries to serve the public is not to apologise for the State but to come right out for state ownership, but this is not the time and this is not the Cabinet to take that course. We must have the general public behind us and a clear idea of what we are going to do if we are to face the most formidable opposition we shall undoubtedly meet."

He used the example of coal to reinforce his point. Nothing effective could be done to put the mining industry to rights under its existing organisation, and hundreds of millions would have to be spent on its technical re-equipment. Labour argued—rightly in Bevin's view—that the industry would have to be nationalised: "but to attempt to do anything in this Cabinet and to enter into further compromises pending an election will, it seems to me, be a serious handicap to our people in putting their point of view when the time comes."

[1] Dated 11 November 1944.
[2] See Vol. I, p. 459.

Victory in Europe. Churchill, flanked by Bevin and Anderson, gives the Victory Sign to the crowd in Whitehall.

Britain's wartime leaders at Buckingham Palace on VE Day. *In front*: Churchill, H.M. King George VI, Bevin, Anderson. *Behind (left to right)*: Sinclair, Woolton, Bridges (Secretary of the Cabinet), Ismay, Lyttelton, Portal, Morrison, Cunningham, Alan Brooke.

"For these reasons," Bevin concluded, "I have been reluctant to give my support to these papers or to put forward anything further in this Government. The time at our disposal is now so limited that it is very unlikely we shall get even the social legislation referred to in the King's Speech. To have a show-down now on these other matters seems to me to be the wrong tactics, which will be seen through quite easily."[1]

Bevin's opposition settled the matter. Further talks took place between the Labour ministers in December 1944, but it was clear that Morrison, if he meant to take the initiative, would have to do so on his own, and on the 20th he wrote to Attlee to say that he would not press his proposals further.

Bevin's refusal, however, to engage prematurely in a political fight over the whole front of industrial policy did not mean that he had given up efforts to push through specific proposals. Nor were these efforts limited to legislation in which he had a personal interest (such as the Wages Councils Bill): he joined forces with Dalton to get the latter's Distribution of Industry Bill through Parliament before the coalition broke up.

Bevin regarded such a Bill as an essential part of the commitment to full employment which the Government had accepted. He discussed with Dalton the proposals to be put forward, defended them stubbornly in the Distribution of Industry Committee set up by the Cabinet (of which he was chairman), and when it was decided to give the responsibility under the Bill to the Board of Trade (rather than the Ministry of Labour) helped Dalton to strengthen his departmental organisation.[2]

Once the outline of a Bill was ready, it had still to be approved by the War Cabinet. There was Tory opposition to a measure interfering with the right of a firm to put its works where it liked, and in his letter to Attlee of 22 November Bevin spoke of "a great struggle on Distribution of Industry". No mention of it had been made in the first draft of the King's Speech and only a strong *démarche* by Bevin and Dalton secured its inclusion.[3] On 3 December Churchill wrote to Bevin to propose that, as the discussion of the matter in the Cabinet had shown political issues to be involved, the Distribution of Industry

[1] Bevin to Attlee, 22 November 1944.

[2] For example, Bevin used his influence to help secure the transfer from the M.A.P. of Sir Charles Bruce Gardner as Chief Executive of Industrial Reconversion.

[3] Dalton, *The Fateful Years*, p. 448.

Committee (three Labour and two Conservative ministers) should be strengthened by the addition of Beaverbrook, the Lord Privy Seal.

Bevin's reply was diplomatic but firm. He pointed out that the question of policy had already been settled by the Cabinet's adoption of the White Paper on Employment Policy and that the only question under discussion was the legislation needed to carry this out. The five ministers named as members of the Committee had been joined at each of their four meetings by other departmental ministers without regard to party, and if the nucleus was felt to need strengthening it would be best to add one of these with a special interest in the matters under discussion.

Bevin did not, however, evade the personal issue:

"Your specific proposal in regard to the Lord Privy Seal," he wrote, "is a matter of some delicacy so far as I am concerned. I have tried to work with the Lord Privy Seal in Cabinet, but I do not wish to preside over a committee of which he is a member. You will appreciate that I was in this position for nearly fifteen months[1]—an experience which has left me with very strong feelings, which in the interests of unity within the Government I have always endeavoured to suppress. But I would have great difficulty in accepting responsibility for the arrangement you suggest and I sincerely hope that you will not press it."[2]

This was the end of the proposal that Beaverbrook should be brought on to the Committee, but not of Tory opposition behind the scenes. It was not until 7 February 1945 that the Bill was approved by the War Cabinet and it actually received the Royal Assent on 15 June 1945, the day Parliament was dissolved. While much of the credit for the Act rightly went to Dalton, he would never have succeeded in placing it on the Statute Book without Bevin's determination to get it through in face of strong Tory pressure on the Prime Minister to drop it.

7

December 1944 brought another of those sharp passages of disagreement between the public and the Government which punctuated the history of the war. The disagreement this time was over Greece, the first row over foreign affairs for more than a year.

[1] As chairman of the Production Executive.
[2] Bevin to Churchill, 6 December 1944.

As the Germans were driven out of one country after another, the Allies had to face the same question that had embarrassed them in Italy, which of the rival groups they were going to recognise or allow to take over power. During 1944 this was raised not only in the case of Italy but in Yugoslavia and the countries liberated by the Red Army, Poland, Rumania, Bulgaria and Hungary. The answer to the question was complicated, particularly in the case of Poland, by the differences between the Allies themselves. The contest for power in post-war Europe, both locally and internationally, had already begun. This was not, however, recognised, still less accepted, outside the handful of people concerned with the negotiations, a fact of great importance in understanding the public reaction to events in Greece.

Bevin was not, of course, directly concerned with these questions which were handled by Churchill and Eden. But he took more interest in them than most other members of the Cabinet, read all the papers and was kept in touch with what was happening by Eden who continued to consult him frequently on foreign affairs. An illustration of his interest is provided by a letter which he wrote to Eden on 22 September 1944 about the situation in the Middle East where another storm was blowing up, this time between Britain and France, over the independence of Syria and Lebanon.

"I have been following with very close attention," he wrote, "the telegrams that are passing about the Levant.
"I am aware that, while we have agreed to independence, it was always recognised that the French had a special interest. But it does seem to me now that we are in danger of creating a situation very similar to that at the end of the last war. Having made use of these people during the war we are now deserting them and forcing them to accept the French."

Bevin went on to ask if the whole situation in the Levant could not be reviewed and a treaty negotiated between Britain and France to regulate the matters in dispute. Eden replied on 2 October explaining at some length the problems which had to be dealt with and the steps he was taking.

In October 1944 Churchill visited Moscow and reached agreement with Stalin on a rough division of interest in the Balkans. In return for recognition of Russia's major interest in Rumania and Bulgaria, Stalin agreed to leave Britain and America a free hand in Greece and to recognise an equal interest with them in Hungary and Yugoslavia.

341

The Greek Communists, however, were not a party to the Moscow agreement and, within a short time of landing in Greece, British troops found themselves involved in a civil war. The charge was at once made—and, after Italy, readily believed by many in Britain as well as the United States—that the Prime Minister's preference for monarchical regimes was leading to the use of British forces to impose an unpopular right-wing Government on an unwilling nation and to suppress any challenge from the republican Left. Churchill, on the other hand, believed that he had to act immediately and, if necessary, ruthlessly to prevent the seizure of power in Greece by armed Communist bands already fighting their way into the centre of Athens.

This was the first time that public opinion in Britain or the U.S.A. had been confronted with the possibility of a Communist *coup d'état* and the first reaction was one of angry incredulity. Without reading again the speeches and press comments of the time (American no less than British), it is hard to realise how difficult people found it not merely to believe that Churchill was right but to admit that he could be right. The idea that Communists might use resistance movements and the opportunity of liberation to instal themselves in power was too novel to be accepted by most people even as a possibility. The threat to freedom in the public mind was still (as it had been for a decade and more) from the Right, not from the Left, and Churchill was furiously denounced as a reactionary.[1]

The trouble came to a head on 3 December when fighting broke out in Athens between the Greek police and former Resistance bands (E.A.M.—E.L.A.S.) under Communist control. On the night of the 4th–5th Churchill, believing the situation to be desperate and acting on his own authority, ordered the British commander, General

[1] This charge was not only made by the Left. Accusing Churchill of reviving "the Bolshevik bogy", *The Economist* wrote (9 December 1944): "Never once, since its first appearance, has the Red scare led to a revolution of the Left. For twenty years the revolutions have come from the Right. How can British foreign policy be based on the Anglo-Soviet alliance and on the Red scare at one and the same time? . . . A British policy that was openly and blatantly hostile to the forces of the Left would put this country in the position of Metternich rather than of Palmerston—and without any Holy Alliance to support us."

A fortnight later *The Economist* was still unconvinced by Churchill's defence of his policy: "The real issue is not Communism at all. It is the monarchy." *The Times* and the *Manchester Guardian* were no less critical. For an historical account, which gives a very different view from that of the contemporary press, see D. George Kousoulas, *Revolution and Defeat, The Story of the Greek Communist Party* (1965).

Scobie, to hold the capital at all costs. The news that British troops were in action against former Resistance fighters brought a plain statement of disapproval from the American State Department; the American as well as the British press thundered against Churchill's intervention, and in the House of Commons a group led by Sir Richard Acland, Seymour Cocks and Nye Bevan, tabled an amendment challenging the Government's action.

This was a dangerous moment for Churchill. There was widespread anger in Britain and a threat of American disassociation from his policy. His speech in the Commons on the 8th, the uncompromising attitude he adopted and the vehemence with which he spoke provoked his critics and was ill-received by the press. With the Labour Party conference meeting on the 13th, the unity of the coalition was under heavy strain. The Left were up in arms, calling on the Labour Movement to express its sympathy with the democratic forces of the Greek resistance which British troops were being used to suppress. They were helped by the fact that an American journalist had secured and published the text of Churchill's drastic telegram of the 5th to General Scobie and by the continued street fighting between British soldiers and Greek irregulars in Athens.

To avoid the danger of the Party Conference passing a direct vote of censure on the Government and its Labour members, the National Executive put forward a resolution calling for an armistice without delay and the resumption of talks to establish a provisional National Government in Greece. This was moved without enthusiasm by Arthur Greenwood and attacked by Jack Benstead, the railwaymen's general secretary, as milk-and-water compared with the resolutions the N.U.R. and the miners had submitted. Bevin had used his influence to secure the block vote of the major unions for the Executive's resolution, but Benstead as well as the Miners' spokesman made it clear that the vote would be given out of loyalty rather than conviction, and one speaker after another got up to repudiate Churchill and his policy amid loud applause. It was obvious that some more effective defence would have to be made of the Government's policy, but it was not Attlee, the Deputy Prime Minister and Leader of the Party, who undertook the task. Attlee was too shrewd a politician to involve himself in a head-on collision with the Party Conference: he left the job to Bevin. Hunching his shoulders and sticking his hands in the pockets of his jacket, Bevin walked out on to the platform to face the concentrated hostility of the audience crowding Central Hall.

Unmoved by his reception, Bevin took care to avoid the flamboyant phrases with which Churchill had goaded the opposition in the House of Commons. He kept his temper in the face of interruptions and argued his case with skill, but he was no less resolute, no more inclined to appease the critics than Churchill had been.

"No one," he began, "can complain of any criticism that may be offered to the Government—whatever Government it may be—in those great and vexed problems which will arise in the resettlement of Europe. All that I would say is that, if we win at the next election, as I hope we will, we shall find that we cannot govern the world by emotionalism; hard thinking, great decision, tremendous will-power will have to be applied, and the Labour Movement will have to learn to ride the storm as these great issues arise from time to time."[1]

Benstead had asked why the Government had not got agreement with Russia and the United States. The answer was that they had:

"Russia undertook the main problem of Rumania, we undertook the main problem of Greece, in agreement between the two Governments [i.e. Britain and the U.S.A.], and when the plan for dealing with Greece was worked out here by us, it was taken to Quebec, submitted to the President of the United States, agreed and initialled."

In a voice that could still be heard over the interruptions, Bevin declared:

"These steps which have been taken in Greece are not the decision of Winston Churchill, they are the decision of the Cabinet. . . . I took part with my Labour colleagues in the whole of these discussions, going on for nearly four years, trying to work out the best way of handling those terrific problems that would arise at the end of the war, and I say to the public—boldly, I hope, because I am not going to hide behind anybody—that I am a party to the decisions that have been taken, and looking back over all the efforts that have been made, I cannot bring it to my conscience that any one of the decisions was wrong."

Bevin told the conference that he would have preferred to see a similar organisation to A.M.G.O.T.[2] in Italy take over control in Greece, feeding the people and providing a breathing space before anybody began playing politics.

[1] Speech at the Labour Party Conference, 13 December 1944.
[2] Allied Military Government of Occupied Territory, first set up in Italy.

"But what happened? Our own party in the House of Commons denounced A.M.G.O.T. for all they were worth, and in the end we gave way and A.M.G.O.T. went by the board. Yet we were condemned last week because we did not have an A.M.G.O.T. to put into Greece instead of the Papandreou Government."

The only alternative to an Allied organisation was a provisional Greek Government formed in anticipation of the liberation of the country. The British stipulated that this must be representative of all the six main parties, including E.A.M.—E.L.A.S., and this condition was secured when the Papandreou Government was first set up. To help in the difficult period until a stable regime could be established, the British offered to provide food and armed forces; but they were only willing to do this if every party in the provisional Government gave written approval. "Every party signed, and it was on that agreement and that signature that we went into Greece. I do not think Labour members in any government could take greater precautions than that."

The programme which the Cabinet had envisaged was, first, to start the distribution of food and break the black market; then, as soon as order was established, to hold a general election with precautions to see that it was not rigged; and finally to organise a plebiscite on the question of a monarchy or republic.

"Now, I ask this Conference, could anybody on a democratic basis, have a better layout than that? That is what we agreed to. . . . We never expected E.L.A.S. or any other of the armed bands to go back on this agreement. We believed in the Cabinet that, once that signature was given by all parties, the signature would be honoured."
A delegate jumped up to shout: "What did Churchill think?" "I do not care what Churchill thought," Bevin roared back, "or what anybody thought; that is what I thought when the signature was put on and I believe in honouring signed agreements."

Of course, Bevin admitted, the British Government could have washed its hands and left the Greeks to fight it out. If they had followed this course, the conference would be discussing a different sort of resolution. "You would have been on this platform denouncing us for cowardice and for refusing to face our responsibility." In fact, now that British forces had been attacked, the Cabinet had decided that it could not give way. That was too dangerous, the sort of backing down which allowed Hitler to get where he was. But nobody wanted the fighting to go on a minute longer than was necessary. The Cabinet had not abandoned its plan for the feeding of the people and an election followed by a plebiscite. That still stood and they were ready, once the fighting could be stopped, to sit down round a conference table and discuss how to get the broad-based Greek Government for which the Executive resolution called—not necessarily under the present Prime Minister, Papandreou.

Bevin did not ask the conference to vote either for or against the resolution—he had already taken steps with the trade unions to see it safely carried—but he ended his speech with a vindication of his own position.

"I am here as a member of the War Cabinet. I am here as one of your nominees in that Cabinet. I am here, I hope, loyal to this Labour Movement, and I am here to justify the action I have taken as a representative of that movement in the Cabinet, believing that . . . whether the judgment was right or wrong, at least it was arrived at on the basis of facts, and with the sole objective of trying to get Greece and other countries back on their feet, with . . . their eyes turned to the future instead of upon the old quarrels that have defeated them hitherto."

Bevin did not convince the conference. Nye Bevan got the loudest applause of all when he climbed on the platform and promptly attacked the speech they had just listened to as "garbled and inadequate when it was not unveracious." The only people in the world to go on record in support of the account they had heard were Fascist Spain, Fascist Portugal and the Tory majority in the House of Commons. Disclaiming any desire to break up the coalition on this issue, Bevan added: "But, remember, we cannot be carried very much farther along this road." If the Labour ministers in the Government could not exercise a more decisive influence on the conduct of affairs, they should leave the Tories to do their own work.

Bevan got the applause, but, thanks to the trade-union block vote, Bevin got the decision (2,455,000 to 137,000)—to the relief of Churchill who never forgot the debt he owed to Bevin for his intervention at a moment when an adverse vote by the Labour conference could have seriously shaken his authority. The fighting in Greece lasted until mid-January 1945 and a settlement was only reached in February. But, once past the Labour Party Conference in December, the Government had weathered the worst of the political crisis at home.

It was the first time that Bevin had taken the lead on an issue of foreign affairs and when the dust settled in the first of many bitter controversies over the Communists and post-war Europe it was evident that, far from losing, he had added considerably to his reputation as a national leader by his readiness to assume responsibility at the cost of popularity. In the next six months it was to occur to many, both inside and outside the Labour Party, that if Labour was looking for a leader capable of standing comparison with Churchill, it would come nearest to finding him in Ernie Bevin.

CHAPTER II

Resettlement

I

THE FIRST half of 1945 at last brought victory over Germany, but at the same time growing anxiety about what was going to happen in Europe when the war was over. Churchill was depressed not only by the difficulties with Russia, but by the differences between his own and the American reaction to Russia's attitude. Roosevelt's death on 12 April and his replacement by an unknown and untried President added to the uncertainty about American as well as Russian policy.

The Prime Minister did his best to play down these difficulties in public and, with the exception of a debate on Yalta which showed the House much troubled over Poland, successfully avoided the discussion of foreign affairs in Parliament. But he did not hide his anxieties from the Cabinet, and Bevin in particular is reported by Eden to have been as impressed as Churchill by the evidence that Russia meant to establish control over as much as possible of Eastern and Central Europe.

After his intervention at the Labour Party Conference in December, Bevin made few references to foreign policy until the next party conference at Blackpool in May. But there are two at least which are worth noting.

The first was made in the course of a speech to delegates to the World Trade Union Conference in February 1945. In the course of the discussion, the Russian delegation had been understandably severe in its demand that the German people should make full reparation for the suffering and destruction they had caused. Bevin was in no mood to be "soft" with the Germans: they would have to go a long way in making reparations. "But," he added,

"the Labour Movement will have to be very careful in working out the methods of its approach to the problem. It would be only too easy to make

60 million people in the centre of Europe a submerged labour force which, if not handled correctly, can bring down the standards of all other countries."[1]

Particularly interesting are some remarks Bevin made in a speech at Leeds in April. The greater part of the speech, delivered at the Yorkshire Regional Conference of the Labour Party, was taken up with an outspoken attack on the Tories which raised a storm of political controversy.[2] The reference to foreign policy came at the end and was overlooked by the press until a letter to *The Times* from Ivor Thomas drew attention to it.

Bevin warned his audience that it was going to be "a difficult job" to hold the alliance of the three Great Powers together.

"There are suspicions. There is bias. We have to be patient and tolerant and press on with our task, however many attempts we have to make. . . . War has changed. The coming of the V1 and the V2, and the knowledge which all States have of their possible development makes the prospect not very reassuring, but very disturbing."

In the passage that followed, he made it clear that he spoke only for himself, not for the Party:

"I cannot help feeling that on the question of our defence, our foreign policy and our relations with other countries, there is an imperative necessity for the will of the nation as a whole to be expressed, and for a combined effort to be made. . . .

"I would not object to all the relative documents being made available to the Leader of the Opposition in order that all Parties may be informed of the facts and come to right decisions. If that happens the Government of the day would take their own decision. But . . . I feel that a complete knowledge of the facts is essential to both the party in office and the party in opposition."[3]

When Bevin's remarks had been hunted out and published, they attracted a good deal of comment. The fact that they came at the end of a speech in which he had gone out of his way to underline the differences between the parties on domestic affairs made all the more

[1] Speech at a lunch to the delegates of the World Trade Union Conference, 9 February 1945.

[2] See below, pp. 368–70.

[3] Speech to the Yorkshire Regional Council of the Labour Party, Leeds, 7 April 1945.

striking his plea for foreign and defence policy to be put on a different footing outside the party conflict. To the Left this was one more example of Bevin's naïveté or worse, a pointer to the battles ahead over the demand for a distinctive socialist policy in foreign affairs. At the time it was made, however, his Leeds speech reflected the impression that the difficulties with Russia and the United States had made on Bevin, an impression which, as we shall see, strongly influenced his attitude towards Churchill's proposal to continue the coalition until the end of the Japanese war and was the background to his assumption of the responsibility for foreign affairs, still unforeseen, less than four months later.

2

While the political difficulties after the war were now beginning to overshadow the other aspects of international relations, Bevin had not given up his efforts to bring social and economic questions more fully into the picture. When the I.L.O. Maritime Commission met in London in January 1945 he put all his influence behind the attempt to draw up a Seamen's Charter which would regulate the wages and conditions of merchant seamen throughout the world. When the Governing Body of the I.L.O. arrived for its 94th session later in the same month, the British Government—"which," the *Manchester Guardian* remarked, "in this case means Mr. Bevin"—proposed that the same principle of international joint industrial councils should be applied to all the other major industries as well as shipping.[1]

The Cabinet was at this time considering the Bretton Woods proposals for an International Bank and Monetary Fund. One of the more tantalising items amongst Bevin's papers is a letter to him from Amery, dated 26 January, which begins:

"My dear Ernest,

I am most grateful to you for the way in which you spoke up about Bretton Woods yesterday. The scheme may not be one for a rigid return of the gold standard but all the same it is a scheme whose avowed purpose is to bring about exchange stability for the sake of promoting international trade and investment on nineteenth-century lines. That is bound to cut

[1] Bevin delivered the opening speech, welcoming the I.L.O., on 25 January 1945.

across the right of nations to maintain the stability of their internal price level with all that it means for stability of wages and stability of employment. We have no guarantee whatever that these two interests will coincide in the post-war period which, as you pointed out, may well be much longer than five years before things settle down."

There is enough in this letter and other scraps of evidence[1] to indicate that Bevin took a keen interest in the future framework of international trade and finance and its effect on domestic policy, although without access to Cabinet records it is impossible to document this fully.

Parliament, in the meantime, with a general election in prospect and the parties turning their attention to the constituencies, had lost much of its interest, even for the politicians. Bevin attended dutifully, answered questions and voted when necessary, but only once held the centre of the parliamentary stage.

The occasion, the introduction of the Wages Councils Bill in January 1945, rounded off Bevin's efforts, as a trade-union leader as well as Minister of Labour, to create a comprehensive system of industrial relations. The Bill was a successor to the original Trade Boards Act introduced by Churchill at the Board of Trade in 1909, and this fact no doubt helped to account for Bevin's success in getting his proposals accepted by the Cabinet. Its other predecessors were the Road Haulage Wages Act and the Holidays with Pay Act of 1938—in both of which the General Secretary of the T.G.W.U. had played a leading part[2]—and the Catering Wages Act of 1943 which Bevin had carried single-handed through Parliament. The principle common to all this legislation was the statutory regulation of wages and conditions in trades and industries where the organisation of one side or the other was not adequate to settle these satisfactorily by collective agreements negotiated between trade unions and employers.[3] The method adopted avoided state or parliamentary regulation of wages by setting up independent boards representative of both sides of the industry and leaving these to fix the minimum rate of wages which were then to be enforced by statutory authority.

[1] E.g. his letter to Eden of 24 April 1942 (quoted above, p. 204) and a note on the proposal for a U.N. Reconstruction Bank written in June 1944.

[2] Cf. Vol. I, pp. 601, 618–19.

[3] The Holidays with Pay Act of 1938 empowered statutory wage-regulating authorities to make provision for holidays with pay.

Parts I and II of the 1945 Bill, while retaining the same procedure, changed the name of the old Trade Boards to Wages Councils and extended their powers. In future the Councils were to have power to fix not only maximum hourly rates but other forms of remuneration including a guaranteed weekly wage. Besides questions of pay, they could also advise the Minister on such matters as training, recruitment and conditions of work.

Under existing legislation the Minister of Labour had the power to establish a statutory procedure for fixing wages where no adequate voluntary negotiating machinery existed. This power he retained but the new Bill also allowed him to act ,in cases where satisfactory machinery had been established during the war under pressure from the Ministry (e.g. joint industrial courts) but where there was reason to fear that, with the return to peace and the sudden expansion of such restricted trades as retail distribution, these arrangements might collapse. In the words of the Bill, he could act where "voluntary machinery is not and cannot be made adequate or does not exist or is likely to cease to exist or be adequate". Where a breakdown in collective bargaining appeared possible—and the best of the employers were as alarmed at the prospect as the trade unions—Bevin argued that it was unreasonable for the State to stand aloof and wait until the standard of wages and conditions had fallen to a point where intervention would be justified. He proposed therefore that joint industrial councils which feared that voluntary organisation might not be adequate should be given the right to apply to the Minister for statutory machinery under the Wages Councils Act to enforce minimum standards on all employers in the trade. And if no one else would take the initiative, the Minister himself was given the power to hold an inquiry to decide whether there was a case for a Wages Council.

Bevin still believed that the best way to settle wages and conditions was by voluntary negotiation. But even in cases where both sides were organised and negotiation a well-established practice he thought it right to give the State's backing to the enforcement of collective agreements at least for a period. This had been done during the war by Part III of Order No. 1305, which required any employer in a trade where wages and conditions had been settled by negotiation, to observe the rates and terms agreed upon—whether he was a party to the agreement or not. Bevin wanted this preserved for five years after the war in order to make sure that unscrupulous employers should not secure a competitive advantage by undercutting the wages paid by those who honoured an agreement.

Why five years, Bevin was asked. Because, he replied, it would take five years to get over the greatest upheaval in the history of British industry, five years in which it would be essential for any Government, whoever was elected, to count on stability. He was opposed to any attempt to peg wages. "What we are really doing is allowing the voluntary machinery to operate unhindered either up or down during that five years."[1] There would be no state inspection or legal penalties (as there would in the case of rates fixed by a Wages Council): it would be the job of the unions and the employers' federations to see that the agreements were kept. But under the new Bill they could go to the Industrial Court to establish that the agreement was one that ought to be honoured and, once that was done, it became a part of the contract of service and an employer who disregarded its terms could be sued in the ordinary courts.

In Bevin's mind the three parts of the Bill fitted together into a single scheme to preserve the advance in industrial relations made during the war. He had put his proposals to the Joint Consultative Committee a year before (January 1944) and had encountered considerable misgivings, not least from the T.U.C. representatives who feared that to extend the statutory enforcement of wages might jeopardise voluntary negotiations and weaken the unions. Bevin, however, thought they were taking too narrow a view, and, although he modified his proposals in detail to overcome some of their objections, he persisted with his Bill. When he introduced the second reading in January 1945, he encountered none of the organised opposition which had threatened to wreck the Catering Wages Bill two years before, and the Wages Council Bill was given a warm welcome both by the House of Commons and the press. Bevin had attained a position of authority in the field of labour relations which no one was prepared to challenge openly. *The Economist* described it as "one of the most important pieces of labour legislation ever laid before Parliament", adding,

---

[1] Bevin's speech introducing the second reading of the Bill, 16 January 1945, Hansard, Vol. 407, cols. 69–84.

"If it were to be the last measure to be enacted by the present administration, the Bill would be a fitting swan-song, for in spite of stresses and strains, the wartime coalition has achieved real progress in the standards of labour."[1]

The credit for this was rightly given to Bevin, but what really satisfied him was to see passed, with the minimum of debate,[2] a measure which guaranteed that progress against the process of reversal which had followed 1918. A quarter of a century later he was in a position to finish the wartime job of putting into effect the recommendations of the Whitley Report which had made so deep an impression on him at the time. If the Wages Councils Bill was carried, he told the House, it would mean that the wages and conditions of work of fifteen million men and women, the overwhelming majority of the working population, would come under the protection of negotiated agreements or statutory regulations.

The hopes Bevin placed in his Wages Council Act were not illusory. By the time he died (April 1951), the number of wages councils had risen to sixty. In retail distribution alone, a group of trades notoriously difficult to organise, one million and a quarter workers had been brought under the protection of the Act, as many as those covered previously by all the trade boards. If agriculture and catering were added, the total number of workers whose wages and conditions were subject to statutory regulation had reached four and a half million[3]—and this without any weakening of the principle of collective bargaining or of the trade-union movement.

3

In February 1945 Bevin announced the last of the Government's re-settlement schemes, providing grants for men and women leaving the services who wanted to start up small businesses of their own. The best of plans, however, was only of value, he insisted, if those affected knew of the opportunities open to them. The Ministry of Labour therefore took special pains to give its release and resettlement schemes the

[1] *The Economist*, 13 January 1945.

[2] After a brief committee stage, the Bill received the Royal Assent on 28 March 1945.

[3] *The British Economy 1945–50*, edited by G. D. N. Worswick and P. H. Ady (1952), p. 115.

widest possible coverage: this included distributing five million copies of a specially written booklet which was sent to every member of the Forces. In a broadcast explaining the programme, Bevin repeated his promise, "There will be no wangling"; and at the end of April, Godfrey Ince, his Director-General of Manpower, announced: "If victory in Europe comes tomorrow, we shall be prepared to meet the strain."

Not content with these official steps, Bevin devoted his main energies in his final months of office to a personal campaign up and down the country in order to explain and win support for his resettlement plans, particularly from employers. Thus, to take his programme for only one of these months, January 1945, he visited Manchester for a conference with the Recruitment and Training Committee of the Cotton Board; spoke to the newly formed Joint Industrial Council for the motor retail and repair trade: addressed the National Federation of Building Trade Employers, and went down to Birmingham to talk to the Rotary Club, and afterwards to a special meeting of the Chamber of Commerce, attended by more than two thousand members, in Birmingham Town Hall.

One of his themes was the difficulty of settling back into civilian jobs several million men and women fresh from the very different experience of service in the Forces. He felt a strong sense of personal obligation to those he had directed into the Services and he was determined to make others recognise the same debt.

"It will take weeks, possibly months, before they will have acclimatised themselves to citizenship again. This is where every firm in the country, from the charge hand to the Director, must be prepared to exercise tolerance and patience. . . . Do not expect the men will come back to the bench or the counter as if nothing had happened. If a man is sacked because he is awkward we shall be creating for ourselves a very difficult situation indeed. I would plead therefore for everybody to make a very close study of the psychology of the best way to treat the Service men and women when they come back."[1]

There were two classes of ex-servicemen in particular for whom Bevin made a special plea. One was the disabled; the other was the young man who had left school at seventeen and a half with the expectation perhaps of going on to a university, only to find himself called up.

[1] Speech at the lunch given by the Multiple Shops Federation, 9 March 1945.

"These men have in numerous instances become leaders. I do not want to see them back in the queue at the Employment Exchange as labourers looking for odd jobs. We want a whole-hearted co-operation with the new Appointments Department which I have set up in the Ministry of Labour. I hope that you will look round and find the right jobs so that they will not lose that great faculty for leadership which the State has developed in them through the Services. Assuming that a man has the leadership, but not the knowledge, then if you cannot supply the knowledge ... our Training Department will help and try to give it."[1]

Any chance, however, of carrying out the change to civilian employment in orderly fashion, Bevin declared, would be wrecked if there was too great a rush to scrap controls. The Tory press was running a strong campaign to end all controls and restore the individual's freedom to choose his own job. It sounded fine, Bevin said, but the fact was that without control of labour, the essential industries would be unable to get the manpower they needed to restore the economy or meet the banked-up demand for goods of every sort. And if there had to be control of labour, he insisted, then there would also have to be control of other things as well, control of raw materials and prices, rationing and the distribution of industry.

This had nothing to do with socialism.

"I have repeatedly said that I shall not be a party to using wartime controls as an indirect means of achieving a political objective. I take the honest view that if an industry is to be nationalised, the right thing to do is to tell the public that. Let it be decided by the electorate and if they support you go on with it. But I do not want to use any kind of control developed in the war to ensure any political aim."[2]

Bevin looked on the question of controls as a purely practical issue. No one needed to be a socialist to recognise that, at a time when everything was in short supply, from food to houses, and industry was still in process of changing from war to peace production, controls were an essential protection which any Government, whatever its views on nationalisation, would have to keep, if the country was to avoid chaos. "If I wanted to be popular," he remarked, "I could go to the House of Commons tomorrow and say 'All the controls have gone.' But I would rather go out of office for many years to come than sacrifice this nation."

[1] Address to the Birmingham Chamber of Commerce, 26 January 1945.
[2] Address at the annual lunch of the British Plastics Federation, 18 April 1945.

His concern was not only with the immediate post-war period. If nationalisation had to be set on one side until after a general election, he continued to preach, as he had throughout the war, a revolution in attitude, "a mental revolution" which he believed to be necessary whether industry was in private or public ownership. "This country," he told his Headquarters Staff at the Ministry of Labour, "cannot survive unless a new approach is made to the proper utilisation of its productive capacity."[1]

What did Bevin mean by this? The answer is not to be found—it never was with Bevin—in a systematic exposition of his ideas, but in a series of remarks flung off in discussion of practical questions which illuminate an underlying social philosophy. This began not with the question of ownership but—in an older tradition—with the under-valuation of men. This, in Bevin's eyes, was the great sin of nineteenth-century industrialism: it had treated men as cheap and expendable, as labour not as human beings.

"In the nineteenth century we subordinated the man to the machine: we were so proud of our technical successes that we neglected our chief asset— the men who made all this possible, and the results were Durham, the Black Country and the valleys of South Wales."[2]

The consequences of this lop-sided development—technically advanced but socially backward—were seen in the First World War.

"We then found ourselves able to make the goods, but unable to organise and plan their production on a scale required by modern war, and we did not know how to handle labour without the whip of the sack or scarcity of jobs. We were forced into controls because of our lack of knowledge of organisation and large-scale planning: we were forced into personnel management under the Munition Workers' Committee and on both sides we had to improvise quickly to save our skins."[3]

These lessons were forgotten once the war was over and had to be painfully relearned in the Second World War when the nation was forced to recognise a truer scale of values, with manpower not as the least valued, but the most precious of its resources. From this had

[1] Address to the Headquarters Staff of the Ministry of Labour, Central Hall, Westminster, 6 March 1945.
[2] Address to the Institute of Industrial Administration, 12 May 1945.
[3] Ibid.

followed a new concern with welfare, health and better conditions of working, full employment and the recognition of the workpeople's right through their representatives to have a voice in matters that affected them closely.

All these reforms had been introduced, and accepted by the Cabinet, on the grounds of wartime necessity, but Bevin had never disguised his belief that they were long overdue and must form part of any far-sighted social policy in an industrialised country. He pleaded with both sides, employers and workers, not to let the gains slip through failure once again to appreciate their importance. In particular he urged them to maintain and extend the pattern of joint negotiation and consultation.

"Voluntary in its nature, democratic in its functioning, united by law and yet stronger than law because it rests on men's words, it stands stronger than anything else man has yet conceived. . . . I believe, however much the political battle may rage, however much parties may divide, this great machinery of industrial relations which has stood us in such good stead during the years of war, will see us back to an orderly industrial society, at the end when victory is won."[1]

4

If there was one subject on which Bevin continued to hammer as long as he was Minister of Labour it was the importance of adequate training. In Manchester and Birmingham he warned employers that they could not count on a big return of their former operatives. Contrary to their experience between the wars, they were going to find themselves faced with a shortage of labour; they would have to recruit and train their own skilled men.

He illustrated the price paid for the pre-war neglect of skill by the shortage of trained men he had found in 1940. At that time, he reminded an audience of Birmingham industrialists, he had been able to find no more than 14,000 trained designers and draughtsmen in the whole country. This was a major weakness in British industry. "Nearly everybody in the production trades has waited for the engineer to come and sell new machines to him, and it was the Germans and the Americans who turned up to do it." British industry

[1] Address to the Multiple Shops Federation, 9 March 1945.

ought to be training its own designing staff and telling the engineering firms what it wanted: in a world where Britain would have to live by its exports, good designers would be as necessary as they had been in war.[1]

It was not the lack of native ability but the lack of opportunity and encouragement. "I had a great experience over this on radar. We were the most advanced in the world in the laboratory, certainly in the Sciences, but the mechanics had not been trained. No one had sent out to industry telling them what the right line of development was. The result was that while we had all the knowledge, we could not deliver a thing."[2]

"Then one night I said I would make an appeal on the radio to everybody who had made radio a hobby. I had over 200,000 applications, and I arranged to get trained in the short period of a year something like 78,000 radio mechanics. What amazed me was the number of people from technical and secondary schools who had been forced by economic circumstances into jobs they did not like: there was this great reservoir of ability in the country. . . .
"I have been more than ever convinced during this war that the mass of people employed in the capacity of what was called labourers ought to be reduced to a minimum. It ought to be the duty of every industry in the country to impart knowledge as widely and as profusely as possible."[3]

Bevin recalled an occasion in the Metropolitan Club in New York when he found himself sitting next to an industrialist who was in need of tool-makers. "I said to him: 'Will you tell me what a tool-maker is worth to you in terms of capital?' He replied: 'I have never thought of the matter in that way,' and began to make some calculations in pencil. Finally he replied: 'I think the capital value of a good British-trained tool-maker is £15,000.' At 5 per cent that makes £750 a year calculated in terms of capital."[4]

Linked with the proper evaluation of men and their skill was another governing idea: expansion, an expansion not merely of production, but, even more important in Bevin's eyes, of consumption, an "expansion in the conception of life itself". One key to this was housing. Bevin was one of the first ministers to grasp the importance

[1] Speech at the Birmingham Chamber of Commerce, 26 January 1945.
[2] Speech at the Birmingham Rotary Club, 26 January 1945.
[3] Address to the N.J.I.C. for the Motor Vehicle Retail and Repair Trade, 18 January 1945.
[4] Ibid.

housing would acquire after the war. He was the driving force behind the plan to create and train an adequate labour force for a much expanded industry and devised the Class B scheme for releases from the Services largely in order to get the housing programme under way at the earliest possible moment. He told the Federation of Building Trade Employers that the job they had to do could not be treated as a return to normal business. "You have a mission, a national mission, a mission of vital importance to the welfare and social development of our people."[1]

If they wanted proof of this, he said, let them think of the several million marriages that had taken place during the war: only a very small percentage of these young married people had got homes at all. "Houses for the people" had got to take priority over anything else in domestic reconstruction; not simply more houses, but more spacious and better built ones.

"The better the house, the better the people . . . People said, take the men and women out of the slums, and they will still be dirty. But the next generation is not. The children who have grown up in the new environment will not be. New gardens and a new environment will make an enormous difference."[2]

He preached the same message when he went to Manchester in January to open an exhibition of new designs in furnishing fabrics and household textiles which he had challenged the cotton industry to organise. Revealing an unexpected side of his mind he talked about the importance of harmonious colour schemes and good lighting. "The attempts to develop good taste among children as part of their education have been extremely limited. Work in this sphere has been lacking in imagination."[3] He urged the organisers of the exhibition to send it on tour so that young people setting up house could enlarge their ideas of what could be done, and on VE day itself, 8 May, he went out of his way to open another exhibition, organised by the wallpaper industry, and to renew his plea for better homes more imaginatively decorated. When someone raised the cry about low rents as the first necessity, he retorted that the Labour Movement was wrong

[1] Address to the A.G.M. of the National Federation of Building Trade Employers, 24 January 1945.
[2] Ibid.
[3] Speech in Manchester, 5 January 1945.

to think the working classes could be housed on a nineteenth-century rent. "If I had the choice of building houses of 1,200 or 800 square feet, and there was a few shillings a week difference in the rent, I would rather have the bigger houses and fight for wages to enable them to pay the additional rent."[1]

He pursued the same theme of an expanding conception of life when he presided over the opening session of a British Association conference on the place of science in industry. The gap between the discovery and application of new inventions, he argued, was too wide. It was not only the gap between scientific research and the application of its results in industry, but also between the introduction of new processes in industry and their products becoming available to the mass of ordinary people. It would help a great deal to overcome the resistance of working men to innovation, he declared, if besides the disturbance of their habits and security of employment, they could see a chance of enjoying the benefits of innovation themselves in their daily life. In war the speed with which new discoveries had been developed for general use was amazing, but in peacetime the process was long drawn out.

"I want to get the benefits of the advances of science made available quickly to the people, so that they will have a vested interest in scientific discovery. The masses would then show a keen interest in the work of scientists, be receptive to new ideas and welcome scientific progress."[2]

Talking to the Birmingham Rotary Club,[3] he applied his argument on a still wider scale. He reminded his 1945 audience of the address he had given to the Bristol Rotary Club in 1919,[4] in which he had pointed to the consequences which would follow if they continued to revolutionise and increase production while leaving the standard of consumption over the greater part of the world stationary at a desperately low level. He had been ridiculed by the business men in his audience, but his forecast had been confirmed. Over-production coupled with under-consumption had led to economic crisis and persistent unemployment, while the poverty and misery in which the greater part of the human race lived continued to provide the most fertile breeding grounds for revolution and war.

[1] Speech at the Wallpaper Exhibition lunch, 8 May 1945.
[2] Royal Institute, London, 12 January 1945.
[3] 26 January 1945.
[4] See Vol. I, pp. 114–115.

"You cannot have peace while you perpetuate poverty. You cannot have confidence unless you take steps to prevent people being driven by hunger and necessity, or what is more vital, the great contrasts in life between the most advanced and the backward."

The richer countries could not limit their efforts to raising the standard of life of their own peoples: they would have to undertake, in their own interests, the task of expanding the conception of life in the under-developed countries. And this was a task, he added, in which Britain with the responsibility for so many millions in her Asian and African possessions had the opportunity to take the lead.

5

The final manpower debate of the war took place on 16 May 1945, a few days after VE Day. There could not have been a greater contrast with the manpower debates of 1940–1. This time there was hardly a note of criticism and one speaker after another congratulated Bevin on the way in which he had handled the preparations for de-mobilisation. He used the occasion to make two important announcements and to review his resettlement plans as a whole. He was able to give a starting date for releases from the Forces six weeks earlier than had been expected—18 June—and a forecast that three-quarters of a million men would be released by the end of the year. Perhaps most important of all, he was able to vindicate his promise to defend the interests of "the ordinary unknown man" against those with special claims: releases under Class B would not amount to more than 10 per cent of those under Class A, and would not begin until the Class A scheme had got well under way.

Bevin laid particular emphasis on the need for industrial training: schemes had already been worked out in thirty different industries, and discussions were going on with another thirty. "The one thing I have set myself," he declared, "is to have as few ex-soldiers as possible general labourers. This country must never again, for its own sheer defence, let the training of skilled personnel fall to the level which it did in 1939."[1] He still refused to give an overriding preference in employment to ex-servicemen, if only because of the difficulty of

---

[1] House of Commons, 16 May 1945, Hansard, Vol. 410, col. 2532.

distinguishing between those who had served on a fighting front and those who, although in uniform, had been exposed to no greater danger than dockers and other civilian workers carrying on their jobs under bombing. But the Government was not going to let an ex-serviceman or anyone else stand in a queue outside the Labour Exchange.

"This time the Government is accepting responsibility for employment on the authority of this House. . . . Last time it was just a case of demobilisation and the men fending for themselves without our taking very much re-sponsibility for them other than for unemployment pay. . . . The scheme that I have arranged is that, within three weeks of a man signing on, if there is no re-opening up of the factory or nothing doing, they have to call the man in and train him for something else or find other work for him."[1]

In fact, as Bevin warned the House, there was likely to be a shortage rather than a surplus of labour after the war, a prospect which few people in the country had yet recognised. To make sure that this shortage did not lead to a scramble for men and an un-balanced labour force, his last act as Minister of Labour was to sign a new Control of Engagement Order requiring all jobs to be filled through the Ministry's employment exchanges. His final appeal was to both sides of industry to continue the practice of consultation and co-operation which they had built up during the war.

"The old game of merely stating a case and one side saying 'No' in three weeks' time is not my conception of negotiation and adjustment. . . . I hope employers will not raise bogeys like 'managerial functions' which have handicapped us so much, because this attempt to draw the line as to when and what you discuss does more to harm a settlement than anything else, and the more responsibility you can draw into industrial relations between both sides, the better for this transition period."[2]

This was the last debate in which Bevin took part as Minister of Labour and National Service. There had been times in his first two years of office when it seemed impossible for anyone occupying this position to survive or at any rate to emerge with his reputation un-scathed. Bevin did both. He held his office without a break for five years (the only member of the War Cabinet to do so besides Churchill)

[1] Ibid., cols. 2536-7.
[2] Ibid., col. 2542.

and at the end he had a far greater reputation than at the beginning.

Out of a total population of no more than thirty-three million men and women between the age of fourteen and sixty-four he had drafted five million into the Forces, providing an Army of two and three-quarter million men as well as a first-class Air Force and Navy. At the same time he had actually raised, instead of reducing, the working population, from under twenty million before the war to over twenty-two million, providing an industrial output second only to America's. It was this double feat, a high degree of military as well as industrial mobilisation, which is so striking, when either by itself might have seemed a considerable achievement.

To build up and maintain this double mobilisation over a period of five years called for great powers as an organiser and was one of the biggest organising jobs ever undertaken by a British minister. But it called for other qualities as well. Every one of the issues with which Bevin was concerned—the call-up, conscientious objection, direction of labour, conscription of women, wages, industrial disputes, the order of release from the forces—was highly charged and could have produced a major row at any time. It took courage and determination for any minister, knowing this, to refuse to play safe and to put the demands on people up to the limit he believed they would stand—it also required great confidence in his own judgment.

Even more remarkable is the fact that all this was done on a basis of consent, without any apparatus of coercion, with no special courts and only a negligible number of prosecutions. This was the vindication of Bevin's policy of "voluntaryism", which he had so stubbornly defended against his critics, and it remains as a striking illustration of what can be achieved by democracy when it is combined with resolute leadership.

To be responsible for the mobilisation of the country's manpower might have seemed enough for most men. But he had done a great deal more than that as Minister of Labour.

In addition to mobilisation, he had borne the chief responsibility for the Government's labour and industrial policy and had discharged it in such a way as to avoid industrial trouble on the scale of the First World War or a runaway inflation. He had done this without abandoning the framework of collective negotiation, had supplemented it, where necessary, by the statutory regulation of wages and conditions, and had introduced a series of reforms which, taken together, produced a radical change in the conduct of industrial relations.

He had used the trade unions to help find the labour needed for the war factories and for emergencies, and in the process established a new relationship between the trade-union movement and the Government which was to have permanent consequences for both parties.

He had taken charge of demobilisation, developed this into a plan for resettlement as well and handled an operation which in 1918–19 had produced real trouble with such success that it went through with hardly a hitch in 1945.

Finally, while bearing a full share with the other members of the War Cabinet of the responsibility for the war and putting loyalty to the coalition before party interest, to the anger of not a few members of the Labour Party, he had shown himself more active than any other senior minister in pushing on with plans for the post-war period of reconstruction and securing an all-party commitment to a policy of full employment.

He had done all this as a man of sixty whose health was already showing signs of strain[1] and at the same time had found the energy to make frequent tours of the country, visit innumerable works of one sort or another and deliver over three hundred speeches.[2]

At no time had he gone out of his way to court popularity. From the nature of his job, which involved him in calling up or issuing orders to almost every man and woman in the country, he had to stand up to a great deal of criticism. He never attempted to placate his critics, to evade responsibility or to avoid decisions which he believed right. He made mistakes and he made enemies, but there was no doubt about the final verdict: the *Manchester Guardian* was not exaggerating the reputation he had won when it said that "he came out of the war second only to Churchill in courage and insight".[3]

It was the end of a great tenure of office.

---

[1] Lord Moran says of Bevin in the spring of 1945: "He was already a very sick man, suffering from alarming attacks of heart block in which he would lose consciousness. But full of guts, he was determined to do his job and would not give in." Moran, p. 248.

[2] The total for the five years May 1940—May 1945 is 340, not counting speeches in the House of Commons.

[3] *Manchester Guardian*, 24 May 1945.

CHAPTER 12

The 1945 Election

I

IN THE opening months of 1945 all parties were busy preparing for the election which everyone expected would follow the defeat of Germany. In January, Morrison had been appointed chairman of a Special Campaign Committee and drew up the Labour Party manifesto *Let Us Face the Future*; in February Ralph Assheton, as chairman of the Conservative Party, made a speech at Leeds which, with its warnings against nationalisation and bureaucracy, was taken as a foretaste of the Tories' election campaign. Other speeches by Churchill, Attlee and Morrison the following month all contained sharp attacks on their political opponents. To everyone's surprise, however, the sharpest attack of all came from Bevin.

Bevin's attitude to politics had changed a good deal in the past fifteen months, but he was still the least inclined of the Labour ministers to look at matters from a party point of view. Several things, however, combined in the spring of 1945 to make him determined to leave no one in any doubt where he stood.

The main lines of the Conservatives' election propaganda were already emerging: Churchill the Man who won the War, the National Leader who put Country before Party versus "The Socialists", the men who bore the responsibility for breaking up the National Coalition and who continued to put Party and Class before Country. The charge that Labour was responsible for the break-up of the coalition at the end of the war rankled with Bevin, if only because it blurred the difference between a genuine national coalition in wartime, and a Conservative Government disguised as a national coalition in peacetime. Nor was he in a mood, however much he admired Churchill, to see his reputation exploited for election purposes by the Party which had repudiated him in the 1930s.

Beaverbrook's press campaign against controls was part of the same pattern. Bevin not only distrusted Beaverbrook, but resented—and perhaps was jealous of—his influence with Churchill. And at the Conservative Party Conference in March Churchill seemed to identify himself with Beaverbrook's views when he declared:

"Controls under the pretext of war or its aftermath, which are in fact designed to favour the accomplishment of totalitarian systems, however innocently designed, whatever guise they take, whatever liveries they wear, whatever slogans they mouth, are a fraud which should be mercilessly exposed to the British public."[1]

Bevin was no less annoyed by other remarks of Churchill's in the same speech which bear an interesting resemblance to his later "Gestapo" broadcast during the election. The Prime Minister derided the Labour Party for adopting, "much to the disgust of some of their leaders", a programme of nationalisation, and went on to describe "these sweeping proposals" as destructive of "the whole of our existing system of society . . . , the creation and enforcement of another system . . . borrowed from foreign lands and alien minds." The *Manchester Guardian*, with some justice, declared Churchill's speech to be composed of "reach-me-downs from the tub-thumping twenties", a reversion to the Churchill of 1926 and the General Strike.[2]

Churchill's attempt to drive a wedge between the Labour Party leaders over nationalisation was accompanied by the promise to bring into a reconstructed Government after the break-up of the coalition "men of goodwill' from any or no party, a remark which was widely interpreted as a bid for Bevin's support.[3]

If this was Churchill's intention, he could hardly have pursued it more maladroitly than by sending to the T.U.C. a few days earlier "a final reply" refusing their request for an amendment of the 1927 Trades Disputes Act.[4] Churchill's statement, in his reply, that the overwhelming mass of Conservatives would never agree to such an amendment underlined the difference between the two parties on an

[1] Reported in *The Times*, 16 March 1945.
[2] *Manchester Guardian*, 16 March 1945.
[3] See the *Daily Mail* on 9 April, after Bevin's repudiation of such a suggestion: "Many reasonably shrewd political observers believed that, when the time arrived, he would join forces with the Prime Minister in a new nationally based government."
[4] The reply was received by the T.U.C. on 10 March. T.U.C. 77th Annual Report, 1945, pp. 206–7.

issue on which Bevin had always felt strongly and which he had himself raised with the Prime Minister on several occasions.[1]

Bevin appears to have said nothing of his intentions to his ministerial colleagues, whether Labour or Conservative, but by the time he went up to Leeds to address the Yorkshire Regional Council of the Labour Party on 7 April he had made up his mind to destroy for good any illusions entertained by the Tories—or the Left—that he was prepared to play the role of Ramsay MacDonald in 1931. In doing so, he delivered the most forthright attack on the Tories heard from any Labour minister since the coalition had been formed five years before.

He dealt first with the charge that Labour had recklessly decided to break up the coalition at the end of the war:

"The fact is," he retorted, "that Parliament was elected ten years ago on a lie—a self-confessed lie. It has run five years beyond its course, and the Conservative Party are afraid to face the electors solely on their own record before the war."

The Cabinet had been unanimous against a coupon election like that of 1918 and the only honest course was for each party to face the electors on its own programme. There were fundamental differences on economic and domestic issues which could not be settled by a coalition:

"We must put the issues to the electorate and let them decide. . . . That is honest and straightforward and now let me remind my good friend, the Prime Minister, that he is no stronger in the country now than Lloyd George was in 1918 and he led the Conservatives for only two and a half years. Then he went out of office—and never came back."

Bevin turned from Mr. Churchill's future to his own:

"There have been a lot of suggestions about my future. I do not know why. I do not wear 'Loyalty' on a band on my arm, but nevertheless I have been loyal to this Party for 40 years.

"The Tories are saying that the Labour Party has not got the men to govern. Believe me, if I were to join them, you would be surprised at the virtues which

---

[1] For example, on 21 March 1944, Bevin had written to Churchill suggesting an independent inquiry presided over by a judge into the question of whether the civil service unions should not now be allowed to affiliate to the T.U.C. for industrial purposes only. The prohibition of this by the 1927 Act was one of the provisions most resented by the trade-union movement. The Prime Minister did not accept Bevin's proposal.

would be attributed to me in tomorrow's press. A lot of the Tory propagandists seem to have forgotten that I am a member of the Labour Party. I have been a member through all its vicissitudes. I have witnessed its ups and downs, the treachery of some of its leaders in the past. .... As far as I am concerned I still abide by the Party decision[1]—whatever it may be. Those of us in the Trade Union Movement know how to accept majority decisions."

Bevin then went on to discuss the issues at the forthcoming election. The Tory Party was relying on short memories and hoping that people would forget its record during the twenty years between the wars. For the greater part, almost the whole of this time, the Tories had been responsible for the government of the country: they could blame the opposition for all sorts of mistakes, but the fact was that it had been they who possessed access to information, the power and the majority to get their policy accepted by Parliament. And their record was a miserable one. In foreign policy they had brought the country to the verge of disaster by their anxiety to do a deal with Hitler; in domestic policy they preferred to keep millions of unemployed drawing the dole rather than employ them on useful work.

The effect of their twenty years of misrule was seen in 1940 when the British were brought face-to-face with their danger and weakness. If Labour had hesitated at this juncture and national unity had not been established, Britain today might well be an occupied country.

"I have a profound admiration for the Prime Minister as a war leader—unfettered. I gave him my loyalty in that position: I never gave it to him as leader of the Conservative Party. It is not for me to belittle what any one man in the coalition Government has done—but I assure you this has not been a one-man Government. Notwithstanding all the limitations which a coalition brings, it has been a good team which has accepted full responsibility for bringing this country through its horrible difficulties and set-backs."

Bevin was prepared to defend the achievements of the coalition, particularly in his own field of industrial and social questions. But no one was going to thank them for that: the election would be fought over the unsettled questions, first of all, the question of public ownership.

"Now why do we ask for public ownership? We want industry to serve the public and not merely a few monopolies. The tendency of modern industry is to develop into monopolies and I regard private monopolies, responsible to nobody but themselves, as a danger to the State, a positive danger to the

[1] I.e. on the election and the course to be taken by the Party after the election.

community. . . . When a business becomes a monopoly or fails, owing to its structure, to provide a decent standard of life for its workpeople or to serve the public by providing goods or services at reasonable prices, then the Government ought to step in."

Bevin named three groups of industries which required such action by the State, "if we are to recover our position as a nation": the power and fuel industries, including coal, electricity, gas and oil; steel; and the transport industries, including railways and commercial road transport. He then went on to talk of the second major issue: housing. The need, he remarked, was on such a scale that he doubted whether it could be met by the local authorities. One measure he suggested was a national housing credit, on the same scale as the credits for war, which would make money available for house purchase and building at 2 or $2\frac{1}{2}$ per cent, instead of the $4\frac{1}{2}$ to 6 per cent charged by the building societies.

Finally, he turned to controls and asked the Conservatives,

"What do you mean by the announcement that you will take off controls? . . . I ask then: do you intend to take controls off food, the price control off clothing, furniture and other things? Do you intend to take control off raw materials? Are you prepared, as the price of taking off controls, to see an unbalanced production in the country and the loss of many export facilities? If you do mean this, come out into the open— do not try to put the onus and odium of maintaining certain restrictions on the Labour Party. . . . Is the Conservative Party prepared to risk rising prices, the halving of the value of all the savings certificates which represent deferred buying?
"I have . . . the feeling that this desire to get rid of controls is coupled with a desire to get rich quickly at the expense of the community. At the end of the last war we saw this sorry spectacle of profiteering immediately the war was over. . . . The Labour Party is as anxious as anybody to get rid of personal restrictions . . . but not at the expense of the nation."

Bevin's speech, promptly and wittily answered by Brendan Bracken, the Minister of Information, led many newspapers to conclude that the coalition was about to break up. Churchill, however, remained unruffled. At a Cabinet meeting on 9 April, which Low made the subject of one of his happiest cartoons ("Just a Big Happy Family", p. 379) he was reported to have read slowly through the *Evening Standard* account of Bracken's speech while the rest of the Cabinet, including Bevin and Bracken, waited in uneasy silence, then thrown the paper under the table and started the Cabinet's business

Bevin speaking on foreign policy at the Blackpool conference of the Labour Party, June 1945.

The 1945 election. *Above.* Bevin speaking at a meeting in his constituency of Central Wandsworth. *Below.* The three party leaders, Bevin, Attlee, Morrison, after the Labour victory.

without saying a word about the exchanges between ministers. At question time in the House the following day, the Prime Minister was at his most urbane and ended rumours of an immediate crisis by expressing the commonsense view that Socialists were free to advocate socialism and Conservatives conservatism, provided that no statement was made by ministers reflecting upon the actual policy pursued by the coalition Government.

The coalition in fact survived another six weeks and personal relations between Bevin and Churchill appear to have been undisturbed by the episode. One thing which no doubt impressed Churchill was the strong support Bevin gave in this same Leeds speech to removing questions of defence and foreign policy as far as possible from the party conflict and treating them on a bi-partisan footing. On 21 April, as Chancellor of Bristol University, Churchill conferred honorary degrees on Bevin and Alexander, spoke warmly of the Minister of Labour's work and took the opportunity for a heart-to-heart talk about the political situation.[1]

Nevertheless, Bevin had made it clear beyond any shadow of doubt that he would not allow himself to be separated from the Labour Party and that he was not a candidate for a non-party or "National Labour" seat in any Government Churchill might form at the break-up of the wartime coalition. This was a matter of consequence for the Labour Party as well as for Bevin personally. As the political correspondent of the *Manchester Guardian* wrote,

"The Tories have done quite a lot to encourage the notion that Labour was leaving the coalition Government against Mr. Bevin's will . . . and that it was quite on the cards he would throw over his party and remain at Mr. Churchill's side. . . . Now it is made plain what no one who knew him doubted —that he remains completely loyal to the Labour Party. . . .
"Had he stayed, the consequences might have been disastrous for Labour unity. Some other Labour ministers might have been tempted to do likewise, and 1931 would have been repeated. That would have sealed the hopes of Labour for another decade at least. Now Labour goes into the fight as a completely united party."[2]

[1] It was apparently on this occasion that Churchill said to Bevin, "After all, I won't be able to go on for more than a year or two," a remark which, Bevin commented later, "made one wonder why he wanted to put in a Tory Government for five years." (Bevin's speech at Shipley 8 June 1945.)
[2] *Manchester Guardian*, 9 April 1945. Lord Moran records Bevin as saying of his Leeds speech: "We have a political Rasputin (Beaverbrook), who very thoroughly and scientifically has been putting it about that I was going to be another Ramsay

2

Bevin had sometimes spoken as if he meant to retire from politics when the war was over, and his disagreements with the parliamentary party and his reluctance to talk about the political future had encouraged this impression. His Leeds speech, however, put an end to any such idea, and a new question was at once asked,[1] whether he was now a candidate to replace Attlee as Party Leader, and in the event of a Labour victory, as Prime Minister.

There had been a good deal of criticism of Attlee in the Party during the war on the grounds that he allowed himself to be completely overshadowed by Churchill, and failed to maintain the separate identity of Labour in the coalition. Laski, who became chairman of the National Executive in the election year, was one of those most eager to see him replaced, and Morrison was generally thought to be ready, when the time came, to stand against him. Bevin distrusted Morrison, and was on good terms with Attlee: but suppose he became a possible contestant himself, what would happen then?

Further material for such speculation was provided by the announcement that, while Attlee accompanied Eden to the founding

---

MacDonald. I was asked about this, even by a large trade union. I didn't want to do it. I hesitated for a long time but when I had made the speech my stock went right up." Lord Moran's conversation with Bevin took place after lunch at the country house of Reginald Purbrick, M.P., an Australian industrialist. Moran's description of the occasion is worth quoting as a snapshot picture of Bevin in the early summer of 1945: "When Bevin entered the dining-room of Purbrick's house he stopped at the door to take in the lovely Georgian silver laid out on an exquisite lace centrepiece. A great grin spread over his untidy features as he rubbed his hands together. 'I always like,' he said, 'to return to the atmosphere of the proletariat.'

"During lunch Bevin drank a great deal and became very talkative. Beaming on the company, he rattled on and soon began to talk about what he wanted for 'his people'. After the war seventeen million would get three weeks' holiday every year with pay. He had a plan with an architect to build a thousand flats at Hastings where working people could go for their holidays and get a bath and a bed. He was going to have circular glass shelters on the Front, so that they could sit by the sea even in winter. Someone blurted out: 'What's wrong with the working classes?' Bevin gave a great guffaw. 'Well, they aren't here,' he snorted." Moran, pp. 248–9.

[1] E.g. by the lobby correspondent of the Press Association immediately after the Leeds speech.

conference of the United Nations at San Francisco, Bevin would act as liaison between the Cabinet and the Parliamentary Labour Party. Did this mean, political correspondents asked,[1] that Bevin had taken the lead from Morrison as the future chief of the Party?

Such speculations were wide of the mark. The arrangement was an informal one, proposed to Bevin by Attlee in a pencilled note pushed across the table at a Cabinet meeting:

"I am telling the P.M. to look to you for guidance in my absence. It would be well for you to keep in touch with Herbert if there are any questions of elections coming up, as he is on the National Executive.

C.R.A."[2]

The unexpected result was to bring Bevin and Morrison into closer co-operation not closer rivalry, and Dalton reported to Attlee on his return that they "had been getting on surprisingly well together while he had been away."[3]

In any case, whatever others might think, Bevin did not regard himself as a candidate for the Party leadership or the Prime Minister-ship, certainly not while Attlee was still leader. Trevor Evans came nearer the truth when he wrote in the *Daily Express* that, if the Labour Party were successful at the polls, Bevin's ambition was to become Minister of Labour and Reconstruction, and if they were defeated, Director-General of the I.L.O.[4]

While Attlee and Eden were still in San Francisco, the war in Europe suddenly, and incredibly, was over. On 30 April 1945 Hitler shot himself: a week later General Jodl signed the instrument of Germany's unconditional surrender at Rheims. On 8 May the nation celebrated VE day, and the Cabinet, after being entertained at the Palace, was photographed, with Churchill, Bevin and Anderson, its three outstanding members, flanking the King. After this they appeared on the balcony of the Ministry of Health to be greeted by the huge crowds waiting in Whitehall. When Churchill wanted Bevin to come forward and share the applause with him he refused: "No, Winston: this is your day." And when he finally stepped through the

[1] E.g. in four of the leading provincial papers: *Yorkshire Post, Birmingham Post, Liverpool Post* and *Western Mail.*

[2] The original note survives among Bevin's papers.

[3] Dalton, *The Fateful Years,* p. 458.

[4] *Daily Express,* 10 April 1945.

windows with the other members of the Cabinet, he led the crowd in three cheers for the Prime Minister, and beat time for the inevitable chorus of "For He's a Jolly Good Fellow".

Bevin's admiration for Churchill was unaffected. However sharp their political disagreements once the coalition broke up, these never affected the pride which he felt—and to the scandal of the Left did not hesitate to express—at having served under Churchill in the most famous of British administrations.

Victory in Europe, however, was only the beginning of the formidable task which now confronted any Government that came into power. The day immediately after VE Day, 9 May, Bevin attended a meeting of the National Council of Labour and urged the need to continue conscription after the end of the war and, despite its traditions, to commit the Labour Party to this in advance of the election. The first question of all to be decided, however, was the date of the election.

Up to this point it had been universally assumed that this would follow the defeat of Germany: but when? The Conservatives favoured June; Labour a date in October. The Conservatives argued that to prolong the coalition, once an election had been decided on, was to paralyse government: the sooner the new Government could take office the better. Labour denounced the Conservatives' proposal as a trick to exploit the victory mood in favour of the wartime leader. They declared themselves in favour of avoiding a rushed election and waiting until the autumn when many more men would be home from overseas service and more complete registers would be available.

The Prime Minister himself, however, liked neither June nor October. The nearer he approached a return to party politics, the more reluctant he became to see the break-up of the wartime coalition. "I was deeply distressed," he wrote later, "at the prospect of sinking from a national to a party leader. With all the new and grave issues pressing upon me, I earnestly desired that the national comradeship and unity should be preserved till the Japanese war was ended."[1]

This was a new proposal not heard before and it was little liked by Churchill's Conservative colleagues. The strongest argument in favour of it was the international situation, the dangers of which, however, particularly the strained relations between the Western

[1] Churchill, Vol. VI, p. 512.

Allies and Russia, were little appreciated as yet outside the inner circle of government. It would mean postponing the election and prolonging the ten-year-old Parliament for another six or twelve months, a course which the party managers on both sides and most political observers regarded as impracticable as well as undesirable.

When Bevin and Morrison (in Attlee's absence) came to talk over matters with him on the 11th, Churchill admitted that he was under strong pressure from the Conservative Party "to take it quickly" and hold a June election. He certainly preferred June to October, if there had to be an election in 1945, but his real wish was to continue the coalition of all parties until Japan was defeated.

The two Labour leaders had consulted the National Executive before seeing the Prime Minister, and they pressed Labour's case for an October election as strongly as they could, but without success. Their reactions to Churchill's suggestion of prolonging the coalition differed considerably. Morrison opposed it vigorously throughout: Bevin (like Attlee) was impressed by Churchill's argument that, with the international situation full of uncertainty, it was foolish to break up abruptly a team which had weathered so many storms.

Nothing could be settled until Attlee returned from San Francisco on the 16th. Late that night Attlee had a long talk with Churchill. Both men agreed in wanting to keep the wartime Government until victory in the Pacific as well as in Europe was secured; both recognised that the political pressures in favour of a return to party government might be too strong for them.

Following his talk with Attlee, Churchill drew up a letter to the Labour leader, the final version of which appears in his memoirs under the date 18 May. Rejecting Labour's proposal of an October election, Churchill offered them a choice: either an extension of the coalition until the end of the Pacific war or an immediate election. Attlee was shown the draft of the letter and, after consulting Bevin, secured the insertion of a sentence which promised that, if the life of the coalition was extended, the Government would do its utmost to implement the proposals for social security and full employment already laid before Parliament. With this addition, Attlee believed that he had a better chance of persuading his Labour colleagues to accept the course which he and Bevin as well as Churchill thought to be most in the national interest.

It is important to be clear what Attlee was proposing. Whatever

Churchill may have hoped, Attlee is emphatic that neither he nor Bevin was prepared to go back on the decision taken the previous October, that Labour should fight the next election as an independent party.[1] The issue was not coalition or independence, but the date at which the change was to be made and the election fought: in practice a difference between five months off, i.e. October (which was the original Labour proposal), and six months or whatever length of time the Japanese war was expected to last. Rejecting Michael Foot's later claim that more than this was involved and that "the two most prominent of Labour's leaders had to be hauled out of [the coalition] by the scruff of their necks," Attlee wrote:

"Bevin and I disagreed with our colleagues, not about leaving the coalition, but about when the election should be held. The Japanese war was still on and we were advised it might last another six months. We were also against having a snap election in July which we thought was not fair to the Servicemen. We therefore hoped for a continuation of the National Government until October. There was never any question of either Bevin or myself being opposed to Labour fighting the election on its own or favouring a post-election coalition."[2]

When the National Executive met at Blackpool, immediately before the Party Conference, Attlee took up Churchill's offer and recommended that Labour should stay in the coalition and postpone the election until the Japanese as well as the Germans were defeated. Bevin supported him, although he disliked the suggestion of a referendum which Churchill made in his letter as a substitute for an election. Only three members of the Executive, however, were ready to accept this recommendation[3]: all the others, led by Morrison and with strong support from Nye Bevan and Shinwell, were opposed. The political mood of the Party, Morrison argued, was such that it would be impossible to persuade the delegates to the Conference to agree to a continuation of the coalition, a view which the Chief Whip, Whiteley, confirmed was equally true of the rank and file of the parliamentary party.[4]

[1] See the account in Francis Williams' *A Prime Minister Remembers*, c. 5. This agrees substantially with the account in Churchill's, Dalton's and Morrison's memoirs.
[2] Attlee's review of the first volume of Michael Foot's *Aneurin Bevan* in *The Observer*, 21 October 1962. For the opposite view, see c. 15 of Foot's book.
[3] Dalton, p. 458.
[4] Shinwell, *Conflict without Malice*, p. 170.

Attlee and Bevin made no attempt to press the matter and it was left to the Labour members of the War Cabinet to draft the answer to Churchill, renewing the offer to continue the coalition until October, but not beyond. The Party Conference meeting in private session the same afternoon overwhelmingly endorsed the Executive's view.[1] The exchange of letters between Attlee and Churchill settled the matter. On 23 May, while Attlee and Bevin were addressing the Labour Party Conference, Churchill drove to the Palace and submitted his resignation.

Five years to the month after it had been formed, the wartime coalition was at an end and the election fixed for the first week in July.

3

The Tory propagandists did not wait for Attlee's reply to start attacking Labour as the party which broke up the coalition, a charge to which the Labour propagandists at once retorted that Churchill's offer had been no more than a trick designed to manœuvre them into an unfavourable position.

Nobody resented the Tory line more than Bevin. He seems to have convinced himself that Churchill's proposal of a referendum instead of a general election, a proposal which he immediately took against, had been introduced solely in order to make impossible Labour's acceptance of Churchill's offer to continue the coalition. At his adoption meeting on 7 June, he declared that Labour ministers had been "thrown out by a device that made their position intolerable", adding, "I do not believe political life needs to be conducted in the gutter." Further remarks led Churchill to issue a statement repudiating "Mr. Bevin's inaccurate and offensive references".

Little of this would be worth recalling if it were not to underline the fact that, once Bevin had made up his mind that the split had to come, he threw himself into the party fight with as much vigour as Churchill on the other side and rapidly became one of the most controversial figures of the campaign.

This did not rule out private civilities behind the scenes. On 28 May

[1] Dalton, p. 459. His account agrees with that given by Francis Williams, *A Prime Minister Remembers* and by Morrison. The text of the letter is given by Francis Williams, pp. 64–7.

the wartime Cabinet assembled at No. 10 and Churchill, with no attempt to conceal his emotion, thanked them for all they had accomplished together—"The light of history will shine on all your helmets." It was at this meeting that Churchill invited Attlee to accompany him, as Leader of the Oppposition, to the Potsdam Conference. The same day he wrote to Bevin, "You know what it means to me not to have your aid in these terrible times. We must hope for reunion when Party Passions are less strong." The next day (29th) he wrote a second letter to offer him the Companionship of Honour:

My dear Bevin,

In view of your remarkable work at the Ministry of Labour, which involved considerations far removed from ordinary political or Party services, it would give me great pleasure if you would allow me to submit your name for a Companionship of Honour. Greenwood was proposed to me six months ago by Attlee and I agreed but advised waiting until hostilities with Germany ceased. I am glad to say that Attlee himself is willing to have his name put forward. The advantage of the C.H. is that it does not involve any title and, as far as I can see, having already borne it myself for a quarter of a century, it is no serious burden.

I cannot think of anyone in the country in any party who would not be gratified by your acceptance. I shall quite understand if you feel you would rather not.

Yours very sincerely,
Winston S. Churchill.

To this Bevin replied a day later:

Thank you for your letter of 29 May. I appreciate your kindly thought in suggesting that I allow my name to go forward for a Companionship of Honour.

I have given the matter careful consideration and whilst I repeat that I thank you for your suggestion, I prefer not to accept. The job I have undertaken, like thousands of others, during the war has been in the interests of the Nation and I do not desire special Honours.

Amongst the letters which Bevin received on the break-up of the coalition was a warm one from Andrew Duncan, the Minister of Supply,[1] and another from Attlee. After praising his patience in face

[1] Recalling their association on the Production Executive "over which you presided so perfectly," Duncan wrote: "I shall never forget the pleasure it always was to see you at work in Cabinet and in Committees when your imagination, resource and recognition of straight issues made your contributions so invaluable and your counsels such a strength" (28 May 1945).

'Just a Big Happy Family'
Evening Standard, 12 April 1945

379

'Applicants for the Caretaking Job'
Evening Standard, 25 May 1945

of opposition and the great services he had rendered to the country and the Labour Movement, Attlee added, "You have always been to me such a splendid loyal colleague," signing himself "Yours ever, Clem".

Bevin's reply has a terseness worthy of Attlee himself:

"The five years have been a great experience and worthwhile. We have faced many great problems together, and have overcome them. One thing it should have done is to remove the inferiority complex amongst our people.
"I want to talk to you soon as to what would happen if we are returned. I think we should get our minds clear on many subjects."[1]

Bevin's successor as Minister of Labour was R. A. Butler. The choice had been awaited with interest and produced one of Low's best cartoons of the period (see page opposite). Butler immediately made it clear that he would carry out Bevin's plans and policies, and Churchill's policy statement issued on 11 June took the unusual step of referring to Bevin by name: "The broad and properly considered lines of the demobilisation proposals which Mr. Bevin has elaborated with much wisdom," Churchill promised, "will be adhered to."[2] At the opening of the West End office of the Resettlement Advice Service, Bevin at Butler's express wish was given the position of honour.

All this, however, belonged to the past: what mattered now was the future, who was to govern Britain in the next five years of reconstruction. The Labour Party had got off to a good start: its conference was in session at Blackpool when the coalition broke up and the 1,100 delegates received their marching orders on the spot and returned home with their enthusiasm at its height. Bevin spoke towards the end of the conference, winding up the debate on foreign affairs which Attlee had opened, saying nothing about his work during the war and intervening in a field in which, as the debate on Greece at the last Labour Conference (December 1944) showed, his views by no means coincided with those of the Left.

According to Morgan Phillips, the Labour Party's National Secretary, it was largely by chance that Bevin was asked to speak on foreign affairs: he happened to be the senior minister most easily available on the day of the debate. Whatever the explanation, it proved to be an inspired choice. While avoiding any attempt to make

---

[1] Attlee to Bevin, 26 May; Bevin's reply, 31 May.
[2] *The Times*, 11 June 1945.

party capital out of foreign affairs, he delivered a speech which was greeted with one of the biggest standing ovations of the conference and at once started speculation whether Bevin was staking a claim for the Foreign Office. There is no evidence that Bevin entertained such an idea, but his views on foreign policy were his own and he developed them both at Blackpool and during the election campaign with a vigour which attracted a lot of attention.

Bevin started his speech with a warning that foreign affairs would face a Labour Government with its "most vexed and difficult problems". One of the great mistakes of the last peace settlement had been to leave Russia out of it and he quoted Beaconsfield's remark, "Britain and France joined together is an insurance for peace; Britain, France and Russia joined together is a security for peace." Nobody had any doubt this time that the security for peace would rest with Britain, the United States and Russia:

"But you cannot remove the prejudices, the economic differences and the effect of internal economy easily. . . . Whoever the statesman may be who has to weld together these different conceptions into a world organisation to prevent aggression has a difficult task. You will not do it, comrades, by slogans; you will not do it by saying that one country is all angels and the other all devils. You have got to do it by patience, trying to understand the other man's mind and point of view and bring them together for one common purpose."

People did not agree about the causes of war:

"Some say it is economic; some say it is traditional ambition; some say that some nations get it into their heads that the only way they can get prosperous is by domination. In my mind, it is a combination of all three."

It was necessary, Bevin argued, to pay attention in future to economic problems as much as to political. "Malnutrition accounts at least for one of the fundamental causes of war . . . and the real centre of malnutrition is among the peasantry of the world." The Hot Springs conference had provided plans for dealing with this, but it needed a powerful Labour Government in Britain to back it up. Britain was the biggest food importer in the world and in a better position than anyone else to see that a fair international price was fixed for foodstuffs:

"If the industrial worker in countries like Britain is to maintain a decent standard of life, then you must be just to the peasant because he cannot buy the goods of the industrial worker unless his own price is right. . . . No one has lower purchasing power. . . . It is no use to talk about finding markets unless the standard of life of the masses of the world is improved. . . . The receiving and the producing ends must be brought into harmony and no agreements entered into which prevent this or leave every state in the world to maintain their economy in their own way."

That meant not only bulk purchase and international commodity agreements, but efforts like those of Bretton Woods to reach agreement on monetary stability and like those of the I.L.O. to fix minimum standards and maximum hours in as many trades as possible.

"Bound up with this is the whole question of raw materials, and we must get international control of a very large area of raw material. I am not prepared to leave the whole of West Africa to the economic control of the United Africa Company where it is not possible to secure improvement unless we squeeze the company."

From the economic Bevin turned to the political side of foreign policy:

"Collective security," he reminded the conference, "involves commitments and I do beg Labour not to bury its head in the sand. It is no use talking about an international police force unless you supply policemen and decide the means by which you will supply them. . . . I do not know how the World Organisation will turn out, whether it will be effective . . . but we must introduce another National Service Act for a limited period until we know exactly what is going to happen."

It was no good trying to avoid conscription by talking of volunteering:

"So far as I can see, the 'voluntary' method means that the British Army was maintained in peacetime largely as a result of unemployment. . . . With the experience I have gained, I will not be a party to misleading the country or the conference. . . . There must be a universal scheme without the deferments which have had to operate during this war."

India was the first area to which Bevin referred, offering the suggestion that it would be best for Indians themselves to cut through all the legal difficulties and proceed by means of an unwritten constitution. Far from wanting to withhold it from India, Labour wanted her to take over more and more responsibility for her own government. The greatest obstacle was psychological, Indian suspicion of Britain's intentions. As a contribution to removing this, Labour would close the India Office and transfer its business to the Dominions Office, making the latter responsible for all the Overseas Commonwealth.

Europe was going to present no less of a problem than India. Whatever steps were taken to destroy Germany's war potential, Britain could not leave sixty million Germans derelict; they must grow their own food, if nothing else, since neither Britain nor the U.S.A. could manage to feed them. A peace conference was essential: "You cannot settle the problem of Europe by long-distance telephone calls and telegrams. Round the table we must get, but do not present us with *faits accomplis* when we get there." Bevin showed what he was thinking of when he went straight on to speak of Poland:

"People have stood on the rostrum this week and said that the Poles are fascists. Some are. I knew General Sikorski, no one will tell me he was a fascist. Neither is Mikolajczyk nor the Polish Socialist Party. Hurling epithets at one another will not do.
"In all the states of Europe, East or West, we are anxious to create a situation where there can be free and democratic elections, where they can choose their own Governments. If, as a Labour Party, we get through, we pledge ourselves never to use these small states as the instruments of foreign policy against the big states."

The difficulties with Russia sprang from a lack of confidence and fear on both sides. The first job of a Labour Government would be to try and remove the fear.... "And may I say that we have no bad past to live down. ... If we had been in power in 1939 we would have sent our Foreign Secretary to Moscow, and not just an official."

It was in connection with France, however, not Russia that Bevin used the phrase, "Left understands Left, but the Right does not," referring to the leftward trend of French politics at the time of the Popular Front and the lack of sympathy with which it met in Britain. As so often in the next few years, he went out of his way to speak of the losses and difficulties of the French and to affirm his faith in France's recovery of her old status in the world. As to Italy, they had deliberately avoided concluding a final treaty until the North was liberated;

but they would not make the same mistake they had made with Germany last time and go on treating Italy as if Mussolini were still in power.

His final and unexpected plea was not to belittle the work the League of Nations had done, but to carry it forward, "in order that the dreams of a Parliament of Men may eventually be realised".

"Great as are the struggles before us, Labour is ready to face them. . . . The people returning us to power may be assured that we shall not funk the issue in any respect."

4

Immediately after the Blackpool conference Bevin returned to London. Now that he had ceased to be a minister, he resumed office as General Secretary of the T.G.W.U., but in fact left Arthur Deakin to carry on until the election should settle what his future was to be.

The Wandsworth Conservative Association had adopted a public-school regular officer, Brigadier J. G. Smyth, decorated with the V.C. in the First World War, and Bevin, whose ministerial duties had left him less time than he would have liked to cultivate his constituency during the past five years, took nothing for granted in his campaign. At the 1937 by-election his predecessor, now Lord Nathan, had won the seat by no more than 485 votes after Conservative victories in 1931 and 1935. Central Wandsworth, therefore, ranked as one of the most marginal of Labour seats, with all the difficulties of an election fought in a London constituency, a strong candidate on the other side and Bevin himself without experience of winning a contested election. Fortunately, he was able to rely not only on the excellent organisation in the constituency,[1] but also on a band of helpers from Transport House where his former secretary, Miss Saunders, brought all her gifts as an organiser into play.

His election address was reinforced by a four-page newspaper, *Bevin and You*, headed by a quotation from *The Observer*: "The first British statesman to have been born a working man and remained one", and by a Labour Party pamphlet *Ernest Bevin's Work in Wartime*.

[1] His agent was Mr. Alan J. Herbert.

He spoke at a number of meetings in Wandsworth but he was one of the leaders of the Labour Party and had to spend much of his time out of his constituency speaking in other parts of the country. In the middle of June, he made a tour as far north as Derbyshire and back through the eastern counties, returning to London for a big Labour rally at the Albert Hall on 23 June.

The first round of the election was fought in the House of Commons which reassembled for a final short session at the end of May. For the first (and only) time in his parliamentary career "the Rt. Hon. Member for Wandsworth Central" took his place on the Opposition front bench and joined wholeheartedly in the barrage of questions directed at the Prime Minister and the caretaker Government. After the first day, however, he left the work of opposition in the more practised hands of Attlee, Morrison and Cripps, and made only one more intervention, in the debate on the Finance Bill.

Most of his speech was devoted to one of Bevin's favourite grievances, the system of motor-car taxation and its bad effects on British car design, particularly the proliferation of small cars, the failure to achieve the economies of standardisation and the failure to produce cars for export. The most interesting part, however, was his remarks on the Bretton Woods proposals. Pointing out that less than two million people were employed in Britain's export trades in comparison with fourteen million producing for the home market, he declared:

"I shall never be a party to any international agreement which, either as a result of the fluctuation of exchange, speculation or international action of that kind . . . would prevent me from insulating the home market from the violent repercussions which will break down the home price level when these wide fluctuations take place on the international price level.

"I will join with anyone," Bevin concluded, "in finding a national basis for an international price level, properly organised, provided it does not reflect itself in depressing the standard of life on the home market. As yet, neither the Chancellor nor Lord Keynes has ever been able to persuade me that there are sufficient safeguards in the Bretton Woods proposals to achieve that object."[1]

Neither Bretton Woods, however, nor any other part of foreign policy, economic or political, played much part in the 1945 election. The Conservatives, staking everything on Churchill's prestige as the national war leader, made a great deal of the danger to the nation of

[1] House of Commons, 4 June 1945, Hansard, Vol. 411, col. 581.

losing his experience in handling foreign affairs, but the *policy* to be pursued, whether economic or political, was not an issue, it was largely taken for granted.

The main battleground was on domestic questions, Labour attacking the Tories as the party responsible for all that had gone wrong—particularly mass unemployment—between the wars, a party which because of its close involvement with financial and property interests was incapable of planning in the public interest, and no more to be trusted now to provide adequate housing and social services or full employment than it had been in 1918. The Tories retorted by drawing a picture of Labour as inexperienced and irresponsible, prepared to sacrifice "sound" financial policies and the need for greater productivity to rash doctrinaire experiments with nationalisation and to strangle free enterprise with bureaucratic controls. On both sides there was a strong appeal to class fears and class resentment.

The two most controversial features of the campaign were Churchill's "Gestapo" broadcast in which he denounced Socialism as an alien principle "inseparably interwoven with totalitarianism", an attack on freedom as well as property, and the storm which blew up over Harold Laski's statement, as chairman of the Labour Party Executive, that if Attlee attended the Potsdam Conference it should be in the role only of an observer, without the Labour Party being committed to continue the foreign policy of "a Tory-dominated coalition".

Bevin was not involved in either of these controversies, but held strong views on both. He wrote to a friend that he thought Churchill was out of his senses to accuse men like Attlee and himself, with whom he had shared the government of the country until a few weeks before, of wanting to set up a Gestapo. As for Laski's intervention, this only confirmed Bevin's low opinion of him as a political meddler. His wrath, however, was concentrated on Beaverbrook on whom he and the other Labour leaders laid the responsibility for the Tory campaign to brand the Labour Party as totalitarian. At Belper on 12 June Bevin declared:

"I have no quarrel with the Prime Minister but I have had enough these last five years of Lord Beaverbrook. . . . I object to this country being ruled from Fleet Street, however big the circulation, instead of from Parliament."[1]

[1] *The Times,* 13 June 1945.

Attlee spoke even more sharply of "Lord Beaverbrook's record of political intrigue and instability" and of "the power of great wealth exercised by irresponsible men of no principle"; but the press preferred to dramatise the quarrel between the Labour Party and Beaverbrook in personal terms of Bevin v. the Beaver. On 16 June *The Economist* wrote:

"The only life that has been put into the election comes from the personal duel of Lord Beaverbrook and Mr. Bevin. Lord Beaverbrook is apparently the guiding spirit of the Conservative campaign. In the last few days he has been giving his audience the impression that if it were not for Mr. Bevin's labour controls, there would be unlimited housing, food and clothing by tomorrow morning. This of course is simply untrue—but it may not be any the less effective electioneering for that. Mr. Bevin on his side has also been hitting hard without any respect for persons—apart from Lord Beaverbrook and Mr. Lyttelton, the motor manufacturers and the steel industry have come under his lash. He has exaggerated and he has been less than perfectly fair; but he has also talked a lot of common sense. He is building up a big reputation for knowing what he wants and how to get it.[1]

Apart from the continuing control of labour for which other Conservatives besides Beaverbrook attacked him, Bevin's other principal controversy was with Duncan Sandys over housing. At Brentford Bevin declared that the Labour Party was planning to build four or five million houses. Sandys called this "an electioneering trick" and asked how many years such a programme would take. To this Bevin retorted that it was not enough to work on an emergency housing plan for two years, as Sandys and Willink were doing:

"I have urged over and over again that they should make up their *total* demand for housing whether I could supply the labour or not; then I would have a target to work to in order to fulfil the obligation in just the same way as I did in 1940–41 when I had to man the factories of this country for shell production."[2]

Bevin was one of the few national speakers who talked about foreign policy in more than conventional terms, developing the lines of thought he had set out at Blackpool. His views on domestic issues were those he had expressed in his speech at Leeds, including the need to nationalise power and fuel (especially the coal industry), inland

[1] *The Economist*, 16 June 1945.
[2] *Daily Herald*, 20 June 1945.

transport and the steel industry, the last of which he attacked for inefficiency and artificially high prices.

The 1945 election was the first in which broadcasting played a great role: Bevin was one of a team of ten speakers put up by Labour. Following the line adhered to by all the Labour speeches, he eschewed the fireworks and exaggerations of Churchill's broadcasts and made a serious attempt to discuss issues of policy. If this robbed individual broadcasts of much interest, collectively it gave Labour a decided advantage with an electorate which, most observers agreed, was in a serious mood and irritated by old-fashioned electioneering.

As a platform speaker the historians of the 1945 election record that Bevin was reported more prominently and at greater length than any other of the Labour leaders.

"The hero of the Opposition papers was undoubtedly Mr. Bevin. . . . The *Manchester Guardian* was interested in his foreign policy. The popular press[1] tended to concentrate more on his domestic policy. The vigour of his speeches appealed to these papers. . . . He was boosted by the opposition press as the man who could get things done, and in particular who could get houses built."[2]

Bevin spent most of the last week before polling in London, in his constituency. On the day before the election, Churchill spoke at Wandsworth on a last-minute tour of the London constituencies. "It is no good pretending," he confessed, "that I have not got a liking for Ernie Bevin, although I deprecate a habit he has got into of quoting private conversation, because I say an awful lot in private conversation." He added that he was convinced the Tories were going to win: "I feel it in my bones."[3]

5

The poll was taken on 5 July but the votes not counted until the 26th. This was to allow time for the overseas service votes to be brought in. Twenty-four constituencies in the north of England and Scotland where annual holidays were fixed took advantage of the delay to hold

---

[1] I.e. the *Daily Herald, Daily Mirror, News Chronicle* and *Reynolds News.*
[2] R. B. McCallum and Alison Readman, *The British General Election of 1945* (1947).
[3] *Manchester Guardian,* 5 July 1945.

the poll a week late and Bevin travelled north with Attlee and Morrison for a further week's electioneering.

Once that was over, an unnatural political silence descended on the country. Bevin expected a Labour victory, but he had no idea of the size the majority would be. In any case, it was impossible to make plans in advance. Attlee, on Churchill's invitation, left for Potsdam to take part in the conference with Stalin and Truman. Bevin took advantage of the lull to return to Blackpool for the biennial conference of his union, picking up the threads of many contacts and moving a resolution in favour of the public ownership of coal, steel, electricity and inland transport.

The count took place on 25 July and the first results were published during the morning of the 26th. Bevin started his day in Wandsworth and by 11.30 a.m. knew that he had been returned by 14,126 votes to Brigadier Smyth's 8,952, a Labour majority of over 5,000, compared with the 1937 by-election figure of 485 and a Conservative majority of over 4,000 in 1935. Soon afterwards, Bevin went over to join Attlee and Morrison at Transport House where it became clear by the afternoon that Labour had won a striking victory.

When the results were complete, the Labour Party had received just under 12 million votes, the Conservatives 8.66 million, the Liberals 2.23. In terms of seats, the Labour advantage was much greater: Labour returned to the House with 393 seats to the Conservatives' 189 and the Liberals' 12. If account is taken of various minor groups in alliance with the major parties, Labour could call on 399 votes against the Tories' 213, with 28 Liberals and Independents in between: an overall majority of 158.

The swing against the Conservatives was particularly marked in London and the other big English towns. Amongst the ministers defeated were Harold Macmillan, Brendan Bracken, Ernest Brown, and Leopold Amery (in Birmingham, where the swing from Tory to Labour was as high as 23 per cent). The size of the majority won by Labour was by no means unusual: in 1918, in 1924, in 1931 and in 1935 the Conservatives and their allies had commanded even larger majorities. But it was the first time Labour had come anywhere near these figures in the House of Commons and the *swing* from Conservative to Labour was on a scale to which only the elections of 1832 and 1906 provided a parallel. The war had stirred up a conservative nation into one of its rare bouts of radicalism. For the first time the

Labour Party was the beneficiary of the radical mood which in the past (in 1868 and 1906) had carried reforming Liberal Governments into power. Its misfortune was the fact, still concealed by the glow of victory, that its arrival in office with ambitious schemes for reform should coincide with the lowest point to which British national wealth and power had fallen since the Napoleonic wars. The history of the first Labour Government with a majority was to be dominated by this coincidence.

For the moment, however, nothing mattered beside the majority itself—that and the composition of the new Government.

Since the Blackpool conference, a determined effort had been made by a section of the Party to get rid of Attlee. Immediately after the conference, Laski, who was chairman of the National Executive, wrote to Attlee to tell him of the widespread feeling in the Party "that the continuance of your leadership of the Party is a grave handicap to our hopes of victory in the coming election". Laski called upon him

"regretfully, to draw the inference that your resignation of the leadership would now be a great service to the Party. Just as Mr. Churchill changed Auchinleck for Montgomery before El Alamein, so I suggest, you owe it to the Party to give it the chance to make a comparable change on the eve of this greatest of our battles."

It was to this letter of Laski's (dated 27 May 1945) that Attlee made the famous reply: "Dear Laski, Thank you for your letter, contents of which have been noted."[1]

Laski, however, was not to be put off by this rebuff and together with Ellen Wilkinson and Maurice Webb, chairman of the Labour Parliamentary Committee, continued to canvass actively for a free vote of the Parliamentary Party with Morrison as their candidate in place of Attlee. Morrison, in his memoirs,[2] represents himself as concerned solely with the democratic right of the Labour Party to choose its own leader. Whatever his motives, he gave Attlee formal notice that, "if I am elected to the new Parliament, I should accept nomination for the leadership of the Party."[3]

On the afternoon of 26 July, Morrison, Attlee, Bevin and Morgan

[1] Francis Williams, *A Prime Minister Remembers*, p. 7.
[2] *Autobiography*, pp. 245–6.
[3] Morrison to Attlee, 27 July 1945.

Phillips were together in the Party Secretary's room at Transport House when a message arrived from Churchill conceding victory, congratulating Attlee and informing him that he had already arranged to hand his resignation to the King that evening at 7 p.m. At the same time, Churchill added, he would recommend the King to send for Attlee and invite him to form a Government. Morrison stuck to his view that Attlee had no right to accept such an invitation until a meeting of the new Parliamentary Labour Party had been held to elect its leader. Morrison claimed that Cripps, who rang up in the course of the afternoon, supported his view.

Attlee, however, was unperturbed. He was sure of his constitutional position:

"If you're invited by the King to form a Government, you don't say you can't reply for forty-eight hours. You accept the commission, and you either bring it off successfully or you don't, and if you don't you go back and say you can't, and advise the King to send for someone else. It used to happen often in the nineteenth century."[1]

No less important, he was sure of Bevin's support. When Arthur Deakin, prompted by Laski, had suggested that Bevin himself might stand for the leadership, with Morrison as his deputy, Bevin turned on him: "How dare you come and talk to me like that?"[2] Disloyalty was the blackest crime in his calendar, and all his old suspicions and dislike of Morrison were aroused by the manœuvres to get rid of Attlee. He saw Attlee as the Campbell-Bannerman of the Labour Party, the man best fitted by temperament to hold a team of ministers together as C. B. had done after the Liberal victory of 1906. To Bevin who, as he told Dalton, wanted no more personal leadership like Churchill's— or MacDonald's—this was a strong recommendation. People had voted for the Labour Party in the belief that Attlee would be Prime Minister: to change leaders now, or even to hesitate, would create a highly damaging impression of a divided Party, ill-fitted to govern the country. So far as he himself was concerned, he would agree to serve under no other leader.

While Morrison was out of the room taking the telephone call from Cripps, Bevin asked Morgan Phillips: "If I stood against Clem, should

[1] Francis Williams, *A Prime Minister Remembers*, p. 4.
[2] Bevin's account to Dalton, in the latter's *The Fateful Years*, p. 467.

I win?" Morgan Phillips replied: "On a split vote, I think you would." Then Bevin turned to Attlee and said: "Clem, you go to the Palace straightaway."

This settled, Bevin started to ruminate about his own position. "I don't know what you have in mind for me, Clem. I was Minister of Labour during the war, and I'd do that again if you want me to, but I'd rather not. I could go as President of the Board of Trade, but the job I'd like would be Chancellor: finance and taxation have been a special interest of mine ever since the Macmillan Committee." When Attlee asked whom he would make Foreign Secretary, Bevin answered, "Hugh Dalton." Attlee made no reply.

On the way downstairs to join his family, however, he remarked to Morgan Phillips: "I thought from E. B.'s speech at Blackpool that he wanted the Foreign Office."[1] That evening at a Victory Rally in Central Hall, Morgan Phillips, still under the impression of this remark, said to Bevin: "You'll be going to Potsdam." But Bevin was sure that he was going to be Chancellor. "Don't you believe it," he told Morgan Phillips. "Flo and I have packed and we're off to Devon."

Meanwhile at 7.30, Mrs. Attlee drove her husband to the Palace in their family car. After the commission to form a new administration had been offered and accepted, the King asked Attlee whom he had in mind for the Foreign Office, a post that had to be filled at once so that the new Foreign Secretary could accompany the new Prime Minister to Potsdam. Attlee replied that he had been thinking of Dalton. "His Majesty," Sir Alan Lascelles noted, "begged him to think carefully about this, and suggested that Mr. Bevin would be a better choice."[2]

This incident was later blown up out of all proportion to its importance. The King was exercising his undoubted right to advise, and there is no evidence at all either that he "insisted", or that Attlee came to his decision under pressure from the King or anyone else. He had in fact been thinking for some time of Bevin as Foreign Secretary and Dalton at the Treasury. The new factor was his discovery that afternoon that Bevin would prefer to go to the Treasury and it was

[1] This account of the afternoon of 26 July is based on accounts given to the author by Morgan Phillips and Attlee. See also the accounts in Dalton, *The Fateful Years*, pp. 467–75: Francis Williams, *A Prime Minister Remembers*, c. 1.

[2] Sir John Wheeler-Bennett: *King George VI* (1958), p. 638, footnote.

under the influence of Bevin's remark that he gave his answer to the King.

Nor did he at once revert to his original plan. When Dalton arrived from Bishop Auckland the next morning (the 27th) he found Attlee inaccessible and went to see Bevin instead. Although full of wrath against Morrison's intrigues, Bevin was otherwise pleased, and told Dalton with a grin that he would leave Attlee to inform him what was in his mind for the two of them. Attlee meanwhile had refused a claim to the Foreign Office from Morrison, and with the help of William Whiteley had persuaded him to accept the office of Lord President of the Council together with the leadership of the House of Commons. When Dalton saw him just before lunch, Attlee told him that he would "almost certainly" be offering him the Foreign Office and putting Bevin at the Treasury, an arrangement as satisfactory to Dalton (who had been Henderson's Under-Secretary in 1929–31) as it was to Bevin.

In the course of the afternoon, however, Attlee changed his mind. No doubt he had heard other views in favour of Bevin besides those of the King, but the decisive factor was his belief that Bevin would stand up to difficulties with the Russians better than Dalton: "I thought affairs were going to be pretty difficult and a heavy tank was what was going to be required rather than a sniper."[1] A second consideration was that, if Morrison was going to play a leading part in home affairs, it would be better to keep him and Bevin as far apart as possible. "Ernie and Herbert didn't get on together. If you'd put them both on the home front, there might have been trouble; therefore it was better Ernie should operate mainly in foreign affairs."[2] Morrison's acceptance of his offer of the morning seems to have tipped the scale in favour of going back to his first idea and ignoring Dalton's and Bevin's preferences. At any rate, when Dalton saw him again at 4 o'clock on the 27th Attlee told him: "I'm reconsidering it. I think it had better be the Exchequer."[3]

When Attlee persuaded Bevin to go to the Foreign Office is uncertain; but the next morning, when the first group of Ministers arrived at the Palace to kiss hands and receive the seals of office, Bevin did not conceal his disappointment. In answer to Jowitt's

[1] Lord Attlee in conversation with the author.

[2] Francis Williams, p. 5, confirmed by Lord Attlee in conversation with the author.

[3] Dalton, p. 469.

congratulations, he pointed to Dalton and said: "I wanted the job he's got" and on the way back he asked Dalton: "Have you thought much about the income tax?" When Dalton admitted that he had not, Bevin went on: "Well, don't fix your mind till I get back from Potsdam. I've got some ideas I should like to talk to you about."[1]

From the Palace the Labour leaders[2] drove straight to Beaver Hall in the City where an enthusiastic crowd of close on 400 M.P.s greeted them. As soon as they were seated, William Whiteley, the Chief Whip, rose and said: "The Foreign Secretary", and for the first time Bevin stepped forward in his new role. In an unprepared speech which, Chuter Ede noted in his journal, was "finely phrased and magnificently delivered", Bevin moved a vote of confidence in the new Prime Minister. It was seconded by Greenwood and supported by George Isaacs as chairman of the T.U.C. When Attlee rose to reply, the meeting gave him a standing ovation which lasted for several minutes. The significance of all this was not lost on those who knew the efforts which had been made to oust Attlee, and the part Bevin had played in frustrating them. With a touch of irony, Attlee handed the meeting over to Morrison and followed Bevin out of the Hall.

By the afternoon the new Secretary of State for Foreign Affairs was making his first flight in the company of the Prime Minister to meet Stalin and Truman at Potsdam and to represent Britain for the first time at an international conference. The third and final phase of his career had begun.

---

[1] Ibid., p. 470.

[2] The other appointments announced on the 28th besides Bevin, Dalton and Morrison were: Cripps, President of the Board of Trade; Jowitt as Lord Chancellor; Greenwood as Lord Privy Seal.

Bibliography

I. UNPUBLISHED SOURCES

1. Bevin's Papers: copies of letters, memoranda, etc., which he kept from the war period. These are to be deposited in the Bevin Library of Churchill College, Cambridge.
2. Bevin's Speeches: texts of all his wartime speeches.
3. Lord Attlee's Papers: deposited in the Library of University College, Oxford.
4. Lord Chuter-Ede's Diary, deposited in the British Museum.

2. SPEECHES AND PROCEEDINGS

1. Parliamentary Debates, House of Commons. Hansard, 5th Series, Vols. 364–411.
2. *The War Speeches of Winston Churchill*, compiled by Charles Eade, 3 vols. (Cassell, 1951.)
3. *The Job to be Done*, a selection of Ernest Bevin's speeches in the first half of the war. (Heinemann, 1942.)
4. The Labour Party: Report of the Annual Conference: 39th, Bournemouth, May 1940; 40th, London, June 1941; 41st, London, May 1942; 42nd, London, June 1943; 43rd, London, December 1944; 44th, Blackpool, May 1945.
5. T.U.C.: Report of the Proceedings of the Annual Trades Union Congress: 72nd, Southport, October 1940; 73rd, Edinburgh, September 1941; 74th, Blackpool, September 1942; 75th, Southport, September 1943; 76th, Blackpool, October 1944.

3. OFFICIAL DOCUMENTS AND REPORTS

Note: I have not thought it necessary to list all the Statutory Regulations and Orders issued by the Ministry of Labour during the war: a complete list with dates is in the Ministry of Labour Report for 1939–45.

Ministry of Labour and National Service: Report for the years 1939–45. Cmd. 7225 (1947).
Ministry of Labour and National Service: Industrial Relations Handbook (H.M.S.O. 1944).
Price Stabilisation and Industrial Policy. Cmd. 6294 (July 1941).
Proposals for the Reform of the Foreign Service. Cmd. 6420 (January 1943).
Training for the Building Industry. Cmd. 6428 (1943).

Employment Policy. Cmd. 6527 (May 1944).

Social Insurance. Part I. Cmd. 6550 (September 1944).

Social Insurance. Part II. Workmen's Compensation. Proposals for an Industrial Injury Insurance Scheme. Cmd. 6551 (September 1944).

The reallocation of manpower between the Armed Forces and civilian employment during the interim period between the defeat of Germany and the defeat of Japan. Cmd. 6548 (1944).

Reallocation of manpower between civilian employment during any interim period between the defeat of Germany and the defeat of Japan. Cmd. 6568 (1944).

Reports of the Select Committee on National Expenditure, 1940-41.

Interim Report by the Committee on skilled men in the Services. Cmd. 6307 (1941). Second Report, Cmd. 6339 (February 1942).

4. PAMPHLETS

Report of the Special Conference of Trade Union Executives, 25 May 1940 (T.U.C. 1940).

ERNEST BEVIN: *The War and the Workers* (Labour Party 1940).

ERNEST BEVIN: *The Truth about the Means Test* (Labour Party 1940).

The Trade Unions and Wage Policy in Wartime (T.U.C. 1941).

The Old World and the New Society (Labour Party 1941).

ERNEST BEVIN: *Square Meals and Square Deals* (Labour Party 1943).

ERNEST BEVIN: *A Survey of the War Situation and Post-War Policies* (T.G.W.U. 1943).

Let us Face the Future (Labour Party 1945).

Ernest Bevin's Work in Wartime (Labour Party 1945).

5. THE PRESS

The Times, Manchester Guardian, News Chronicle, Daily Herald, Evening Standard, The Observer, The Economist, New Statesman and Nation, Tribune.

6. OFFICIAL HISTORIES

History of the Second World War: U.K. Civil Series, edited by W. K. Hancock.

W. K. HANCOCK and M. M. GOWING: *British War Economy* (1949).

M. M. POSTAN: *British War Production* (1952).

H. M. D. PARKER: *Manpower* (1957).

P. INMAN: *Labour in the Munitions Industries* (1957).

J. D. SCOTT and RICHARD HUGHES: *The Administration of War Production* (1955).

W. H. B. COURT: *Coal* (1951).

C. B. A. BEHRENS: *Merchant Shipping and the Demands of War* (1955).

C. M. KOHAN: *Works and Buildings* (1952).

S. M. FERGUSON and H. FITZGERALD: *Studies in the Social Services* (1954).

History of the Second World War: U.K. Military Series, edited by J. R. M. Butler.

J. R. M. BUTLER: *Grand Strategy*, Vol. II (1956).

JOHN EHRMAN: *Grand Strategy*, Vols. V and VI (1956).

SIR LLEWELLYN WOODWARD: *British Foreign Policy in the Second World War* (1962).

7. MEMOIRS

C. R. ATTLEE: *As It Happened* (Heinemann 1954).

THE EARL OF AVON: *The Eden Memoirs*, Vol. 2, *The Reckoning* (Cassell, 1965).

LORD BEVERIDGE: *Power and Influence* (Hodder and Stoughton 1953).

The Memoirs of Lord Chandos (Bodley Head 1962).

WINSTON S. CHURCHILL: *The Second World War*, Vols. II—VI (Cassell, 1949–1954).

HUGH DALTON: *The Fateful Years, Memoirs 1931–1945* (Muller, 1957).

P. J. GRIGG: *Prejudice and Judgment* (Cape, 1948).

LORD HALIFAX: *Fulness of Days* (Collins, 1957).

ARTHUR HORNER: *Incorrigible Rebel* (MacGibbon and Kee, 1960).

The Memoirs of Lord Ismay (Heinemann, 1960).

GEORGE MALLABY: *From My Level* (Hutchinson, 1965).

LORD MORAN: *Winston Churchill, The Struggle for Survival 1940–1965.* (Constable, 1966).

LORD MORRISON: *Herbert Morrison, an autobiography* (Odhams, 1960).

LORD REITH: *Into the Wind* (Hodder and Stoughton, 1949).

EMANUEL SHINWELL: *Conflict without Malice* (Odhams, 1955).

ERNEST THURTLE: *Time's Winged Chariot* (Chaterson, 1945).

JOHN G. WINANT: *Letter from Grosvenor Square* (Hodder and Stoughton, 1947).

EARL WINTERTON: *Orders of the Day* (Cassell, 1953).

The Memoirs of the Earl of Woolton (Cassell, 1959).

WOODROW WYATT: *Distinguished for Talent* (Hutchinson, 1958).

FRANCIS WILLIAMS: *A Prime Minister Remembers* (Heinemann, 1961).

SIR ARTHUR BRYANT: *The Turn of the Tide* (Collins, 1957).

SIR ARTHUR BRYANT: *Triumph in the West* (Collins, 1959).

8. BIOGRAPHIES

SIR JOHN W. WHEELER-BENNETT: *King George VI* (Macmillan, 1958).

SIR JOHN W. WHEELER-BENNETT: *John Anderson, Viscount Waverley* (Macmillan, 1962).

TOM DRIBERG: *Beaverbrook, A Study in Power and Frustration* (Weidenfeld and Nicolson, 1956).

Bibliography

JANET BEVERIDGE: *Beveridge and his Plan* (Hodder and Stoughton, 1954).

TREVOR EVANS: *Bevin* (Allen and Unwin, 1946).

FRANCIS WILLIAMS: *Ernest Bevin* (Hutchinson, 1952).

KENNETH YOUNG: *Churchill and Beaverbrook* (Eyre and Spottiswoode, 1966).

COLIN COOKE: *The Life of Richard Stafford Cripps* (Hodder and Stoughton, 1957).

KINGSLEY MARTIN: *Harold Laski* (Gollancz, 1953).

FRED BLACKBURN: *George Tomlinson* (Heinemann, 1954).

J. T. MURPHY: *Labour's Big Three* (Bodley Head, 1949).

9. POLITICAL AND ECONOMIC STUDIES

V. L. ALLEN: *Trade Unions and the Government* (Longmans, 1960).

V. L. ALLEN: *Trade Union Leadership*, based on a study of Arthur Deakin (Longmans, 1957).

LEOPOLD AMERY: *Thoughts on the Constitution*, 2nd edition (Oxford University Press, 1953).

SIR JOHN ANDERSON: *The Machinery of Government* (Oxford University Press, 1946).

SAMUEL H. BEER: *Modern British Politics* (Faber, 1965).

SIR GILBERT CAMPION and others: *British Government since 1918* (Allen and Unwin, 1950).

BYRUM E. CARTER: *The Office of Prime Minister* (Faber, 1956).

(ed.) D. N. CHESTER: *Lessons of the British War Economy* (Cambridge University Press for the National Institute of Economic and Social Research, 1951).

G. D. H. COLE: *A History of the Labour Party from 1914* (Routledge, Kegan Paul, 1948).

MARGARET COLE: *The Story of Fabian Socialism* (Heinemann, 1961).

C. A. R. CROSLAND: *The Future of Socialism* (Cape, 1956).

HERBERT A. DEANE: *The Political Ideas of Harold J. Laski* (New York, Columbia University Press, 1955).

JOHN EHRMAN: *Cabinet Government and War* (Cambridge University Press, 1958).

(ed.) ALLAN FLANDERS and H. A. CLEGG: *The System of Industrial Relations in Great Britain* (Blackwell, Oxford, 1961).

W. L. GUTTSMAN: *The British Political Elite* (MacGibbon and Kee, 1963).

SIR GODFREY INCE: *The Ministry of Labour and National Service* (Allen and Unwin, 1960).

SIR IVOR JENNINGS: *Cabinet Government*, 3rd edition (Cambridge University Press, 1958).

K. G. J. C. KNOWLES: *Strikes* (Blackwell, Oxford, 1952).

D. GEORGE KOUSOULAS: *Revolution and Defeat, The Story of the Greek Communist Party* (Oxford University Press, 1965).

D. F. MACDONALD: *The State and the Trade Unions* (Macmillan, 1960).

W. J. M. MACKENZIE and J. W. GROVE: *Central Administration in Britain* (Longmans, 1957).

R. T. McKENZIE: *British Political Parties* (Heinemann, 1955).

J. P. MACKINTOSH: *The British Cabinet* (Stevens, 1962).

HERBERT MORRISON: *Government and Parliament* (Oxford University Press, 2nd edn., 1959).

R. H. S. CROSSMAN: *New Fabian Essays* (ed.) (Turnstile Press, 1952).

R. PAGE ARNOT: *The Miners in Crisis and War* (Allen and Unwin, 1961).

HENRY PELLING: *A Short History of the Labour Party* (Macmillan, 1961).

SIDNEY POLLARD: *The Development of the British Economy 1914–1950* (Arnold, 1962).

B. C. ROBERTS: *National Wages Policy in War and Peace* (Allen and Unwin, 1958).

J. B. SEYMOUR: *The Whitley Councils Scheme* (P. S. King and Son, 1932).

ANDREW SHONFIELD: *Modern Capitalism* (Oxford University Press for the R.I.I.A., 1965).

ERIC L. WIGHAM: *Trade Unions* (Home University Library, O.U.P., 1956).

F. M. G. WILLSON (ed. D. N. CHESTER): *The Organisation of British Central Government 1914–1956* (Allen and Unwin, 1957).

(ed.) G. D. N. WORSWICK and P. H. ADY: *The British Economy 1945–50* (Oxford University Press, 1952).

Index

Date Due

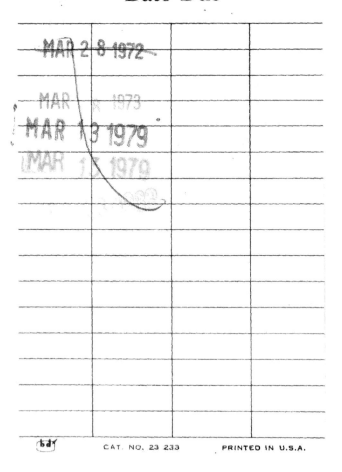